TRANSITION METAL COMPLEXES AS DRUGS AND CHEMOTHERAPEUTIC AGENTS

CATALYSIS BY METAL COMPLEXES

VOLUME 11

TRANSITION METAL COMPLEXES AS DRUGS AND CHEMOTHERAPEUTIC AGENTS

by

NICHOLAS FARRELL

The University of Vermont, Burlington, Vermont, U.S.A.

KLUWER ACADEMIC PUBLISHERS

DORDRECHT / BOSTON / LONDON

Library of Congress Cataloging-in-Publication Data

CIP

Farrell, Nicholas.
Transition metal complexes as drugs and chemotherapeutic agents / by Nicholas Farrell.
p. cm. -- (Catalysis by metal complexes)
Includes index.
ISBN 9027728283
1. Transition metal complexes--Therapeutic use--Testing.
2. Transition metal complexes--Physiological effect. I. Title.
II. Series.
[DNLM: 1. Metals--pharmacokinetics. 2. Metals--therapeutic use.
QU 30 F245t]
RM666.T74F37 1989
615'.26--dc19
DNLM/DLC
for Library of Congress 88-13633

ISBN 90-277-2828-3

Published by Kluwer Academic Publishers,
P.O. Box 17, 3300 AA Dordrecht, The Netherlands.

Kluwer Academic Publishers incorporates the publishing programmes
of D. Reidel, Martinus Nijhoff, Dr W. Junk and MTP Press.

Sold and distributed in the U.S.A. and Canada
by Kluwer Academic Publishers,
101 Philip Drive, Norwell, MA 02061, U.S.A.

In all other countries, sold and distributed
by Kluwer Academic Publishers Group,
P.O. Box 322, 3300 AH Dordrecht, The Netherlands.

Printed in The Netherlands

CONTENTS

PREFACE

When this book was first conceived as a project the expanding interest in the clinical use of platinum and gold complexes made a survey of the relevant biological properties of metal complexes timely and appropriate. This timeliness has not diminished during the gestation and final publication of the manuscript. The introduction contains an explanation of the layout and approach to the book, which I wrote as an overall survey of the wide variety of biological properties of metal complexes. Hopefully, the reader will see the parallels in mechanisms and behavior, even in different organisms.

The writing was considerably helped by the enthusiasm and confidence (totally unearned on my part) in the project of Professor Brian James and I owe him my special thanks. I also owe a great debt of gratitude to my colleagues, and especially to Eucler Paniago, of the Universidade Federal de Minas Gerais, for their comprehension and for the initial leave of absence which allowed me to begin the project. To those who read some or all of the manuscript and made suggestions, Bernhard Lippert, Kirsten Skov, and Tom Tritton, as well as the editor's reviewer I am also grateful. As usual, the final responsibility for errors or otherwise rests with the author.

NICHOLAS FARRELL

INTRODUCTION

Inorganic and coordination complexes have a long history of use as chemotherapeutic agents. The successful application of inorganic complexes as drugs involves the recognition of their bioinorganic modes of action coupled with the traditional pharmacokinetic parameters of uptake, distribution and excretion. Further, the rational approaches to chemotherapy and particularly the notion of selective toxicity [1] must be placed in an inorganic context. The concept of selective toxicity, and its scientific basis, is of particular use in describing the actions of drugs on an invading organism, be it viral, bacterial, parasitic or ultimately malignant tumours caused by the cancerous growth of the hosts' cells. The achievement of some form of selectivity is critical to the successful use of any agent as a drug or modifier of biological response.

An overall scheme for metal complexes in bioinorganic chemistry can start by consideration of the roles of endogenous and exogenous metals in a flow diagram such as:

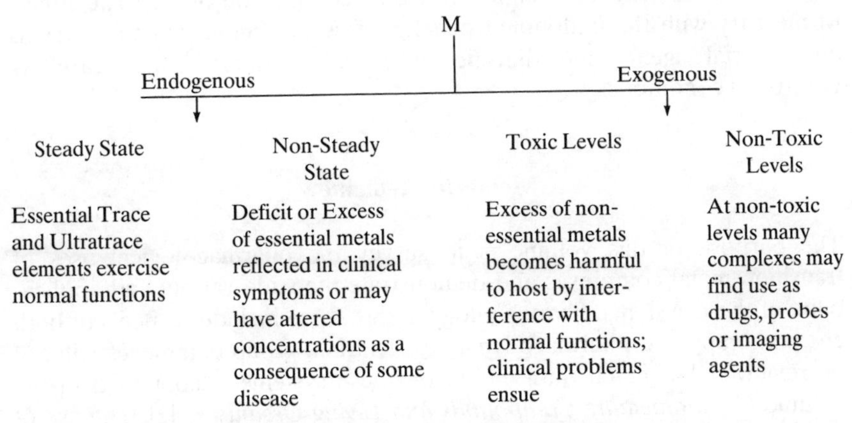

In a steady state the trace metals function in their normal roles but an imbalance in concentration, either by excess or deficiency, can cause disease or be a reflection of disease. The deficiency, whether caused by diet or genetic factors reflected in altered biochemistry of the individual, can often be treated by administration of the metal in an appropriate pharmaceutical form. Where excess trace metal is present, such as copper in Wilson's disease, specific chelating agents may be used to reduce the concentration to the physiologically desirable level.

The appreciation of altered trace metal content in certain diseases, such as enhanced serum copper concentrations in certain cancers and in rheumatoid arthritis, is an aspect of increasing interest in determining whether manipulation of this balance can have a therapeutic effect. The possibility exists that these elevated levels are physiological responses to disease states. A further aspect of trace metal concentration relevant to the concept of selective toxicity is the fact that, as all organisms require trace metals for growth (Fe) and cell division (Zn), selective depletion of these essential elements may be more destructive to the invader (bacterium, parasite, virus). Specific agents may be designed to achieve this goal.

Chelating agents are also necessary to counteract the effects of introduced exogenous metal, where such a concentration presents health hazards. Again, the agent must be specific for the metal in question and not deplete essential elements simultaneously. While we normally consider exogenous metals as harmful, the careful consideration of toxic doses and allowable concentrations can allow for a beneficial role of complexes as drugs or imaging agents. The considerations of safe use in these cases are, of course, identical to those of any organic substance of natural or synthetic origin. The dichotomy for the exogenous metals of malevolent *versus* beneficial effect is well illustrated by comparison of the toxic effects of mercury with the undoubted benefits of certain mercury complexes as antibacterial agents and diuretics when administered under carefully controlled conditions.

Metals in Medicine

The purpose of this volume is to survey the pharmacological uses of transition metal complexes and demonstrate the wide variety and applicability of these systems. Pharmacology in this sense includes aspects of both chemotherapy and pharmacodynamics [1]. Main group complexes will not be reviewed but much information on these systems is contained in the volume *Organometallic Compounds and Living Organisms* [2]. The use of platinum salts in cancer, gold salts in rheumatoid arthritis, silver salts as topical antibacterials, and a simple iron complex in heart disease is witness to this diversity.

The organization of the book is as follows: Chapter 1 outlines the modes of action of metal complexes on DNA, a primary target for any drug. Chapters 2—4 deal with aspects of the antitumour effects of platinum complexes and much emphasis is placed on this compound, partly due to my own interests but also because, as a clinically used drug, a daunting amount of information on its chemistry, biochemistry and pharmacology has been collected. Hopefully, this knowledge can be placed in the overall context of metal complexes as drugs. Chapter 5 reviews the principal aspects of the platinum-pyrimidine blues, an excellent example where interest in biological effects of platinum complexes led to exciting new coordination chemistry. Chapter 6 deals with antitumour effects of other transition metal complexes where the diversity of mechanisms of action is hopefully apparent. In Chapter 7 the aspects of metal-catalysed oxygen toxicity (see below) are summarized and Chapter 8 reviews the use of metal complexes as radiosensitizers, which in some cases also involves aspects of oxygen radicals. Chapters 9 and 10 review activity of metal complexes as antibacterial, antiviral and antiparasitic agents. Chapter 11 reviews those complexes used to combat arthritis and, finally, Chapter 12 contains information on miscellaneous effects such as mercurial diuretics, neurotoxic effects and vasodilation by nitroprusside, the very disparate effects of which are nonetheless united by mechanisms involving enzyme inhibition.

The outline written in this manner tends to emphasize differing properties of different structural classes of complexes, although in many cases the underlying mechanisms may be formally the same. In this way, for example, the possibilities of metal sequestration leading to antibacterial or antiviral effects by diverse chelating agents are treated in the appropriate section, rather than collecting all examples of metal sequestration together. For this reason, the individual chapters are independent and may be read in any order; the references are listed at the end of each chapter, despite a small amount of overlap. An Appendix gives the structures of the nucleic acid precursors and analogues and a Glossary explains uncommon or unusual terms and definitions.

A further use of metals in medicine is their use as imaging agents. The uses of technetium radioisotopes for tumour imaging and paramagnetic metal chelates as NMR imaging agents are of increasing interest. While the absorption, metabolism and excretion of these complexes are governed in the same manner as any other type of drug they do not fall within the scope of this book because we do not expect or want biochemical reactivity in these complexes; their utility lies in their specific tissue localization.

Chemotherapeutic Agents and Drugs

The term chemotherapy, as defined by Paul Ehrlich (the undisputed 'father' of chemotherapy) means the use of drugs to injure an invading organism without damage to the host. Historical accounts of the development of chemotherapy are to be found in Albert's *Selective Toxicity* [1] and in *The Molecular Basis of Antibiotic Action* [3]. The utility of any substance as a chemotherapeutic agent will clearly depend on how its action favors the host over the invading organism or 'uneconomic species'. By selective attack the drug can augment the host's natural defense, which will be responsible in the last resort for elimination of unwanted species.

Chemotherapy uses toxicity in the service of man and requires an essentially irreversible process (cell death). Normally a different organism is attacked — virus, bacterium, parasite — but in cancer the host's own malignant cells are attacked. The achievement of selectivity in this latter case is clearly of great difficulty. The requirement for an irreversible process distinguishes chemotherapeutic agents from drugs which modify biological responses or alter biochemistry reversibly in the one organism, without the necessity of cell kill. Drugs which act, for instance, on the central nervous system or as muscle relaxants are not strictly chemotherapeutic. In the case of metal complexes we can readily distinguish between the chemotherapeutic complexes of cisplatin, silver sulfadiazene, and antibacterial mercury complexes from such drugs as the gold antiarthritic salts and mercury diuretics.

Mechanisms of Toxicity

Selective toxicity may be achieved through differences in accumulation, or comparative biochemistry, or through comparative cytology [1]. In the latter case the example of penicillin acting on the bacterial cell wall is a classic example. Behind each clinical use or biological effect of metal complexes or chelating agents is a bioinorganic mechanism, which will vary depending on the use and structure of the complex. Metal complexes may be treated similarly to organic complexes in their selective accumulation and comparative biochemistry. For example, the requirement of essential metals for cell growth means that alteration of intrinsic metal concentration may preferentially affect the growth of the invading organism.

Metal Accumulation and Activation

Accumulation of metals may take many forms and aspects of chelation may affect both uptake and activity. Thus, while we consider chelating agents as substances which eliminate metals, in some cases a metal chelate is more lipophilic and more easily absorbed than the chelating ligand (Chapter 9.2). The metal chelate may dissociate once inside the cell with either the ligand being cytotoxic (as in isoniazid, Chapter 9.2) or the metal being cytotoxic (Ag, Chapter 9.1.2). Chelating agents may also be used to selectively decrease the concentration of essential trace elements; inhibition of viral replication by zinc chelation is an example (Chapter 10.2.1). As corollary to this, the free chelating agent may bind metals intracellularly with subsequent activation of the metal. Examples here are the bleomycins (Chapter 7), thiosemicarbazones (Chapter 6.2.2) and the antibacterial effect of oxine (Chapter 9.2).

The activation of metal complexes in this manner is noteworthy because the examples above all involve production of active oxygen, independent of the final cellular target. The toxic side effects from reduction products of oxygen, arising from normal metabolism of oxygen, have been the subject of a number of volumes [4, 5]. The intrinsic role of metals in the activation and deactivation of oxygen is paralleled in examples of complexes with both toxic and beneficial properties. Natural defences against these reduction products include catalases, peroxidases and the superoxide dismutases. The role of superoxide dismutase in disease has been extensively discussed [6, 7]. Decreased concentrations of superoxide dismutase in some tumours raise the possibility of selective use of superoxide, or further reduction products, to achieve cytotoxicity (See Chapter 6.2.3). Indeed, this concept in general is a reflection of the natural defence against bacteria by leukocyte production of superoxide. Alternatively, increased defence against superoxide production, by the enzyme itself or simple complexes, can have beneficial effects as in radiation protection (Chapter 8.4) and arthritis relief (Chapter 11.2).

The administration of robust complexes such as those of Pt or Au will be of use when their reactivity with the target is pharmacokinetically appropriate — the window of reactivity must be recognised and will depend on chemical factors such as hydrolysis and ligand exchange as well as uptake and distribution. Uptake of metal complexes may be best for neutral species, since no transport process need be activated but it is well to remember that this is not an absolute requirement. The charged amine complexes, close analogues of cisplatin, enter cells (Chapter 3.1) and indeed lipophilicity may enhance this uptake; an example is the ruthenium (II) chelate of (3,4,7,8-)tetramethyl-1,10-phenanthroline (Chapter 6.1.3).

Biochemical Targets for Metal Complexes

In terms of comparative biochemistry, many chemotherapeutic agents act by inhibition of synthesis of DNA, a natural target due to its predominant role in cellular replication. The modes of metal complex binding to DNA are summarized in Chapter 1. An important point here is that the simple fact of binding does not indicate how cytotoxic a given complex may be. Weak binding will be more easily reversed in either a chemical or biochemical manner and will require in any case higher and thus more acutely toxic doses. Metal complexes may act by direct binding (Pt, Chapter 4; Ag, Chapter 9.1.2) or redox chemistry producing active oxygen which can then cleave DNA upon binding of the complex (bleomycin, Chapter 7).

Selective inhibition of DNA synthesis over RNA and protein synthesis, as examined by standard assays, does not automatically imply binding to DNA. An example is rhodium acetate which may exert its influence on DNA synthesis by reaction and inhibition of precursor enzymes (Chapter 6.1.2). Inhibition of the synthesis of essential precursors is also well demonstrated by thiosemicarbazone–iron complexes, which may be formed *in vivo* (Chapter 6.2.2).

The number of proteins capable of metal interaction is enormous but actions of gold in arthritis (Chapter 11), and mercury in diuretics (Chapter 12.1), are two examples which may be traced to enzyme inhibition. The antibacterial effect of mercury is also probably due to enzyme inhibition (Chapter 9.1.1), contrasting with the inhibition of DNA synthesis by silver (Chapter 9.1.2).

Summary

Examination of the wide variety of biologically active transition metal agents and their diverse structures may help to show further possibilities for the design of specific, selectively acting metal agents. Thus, the various demonstrated effects are summarized for each disease type and it is hoped that the continuing similarity (and indeed formal coincidence) in apparently widely disparate structures will be appreciated as a reflection of the same basic phenomena. Only by placing chemotherapeutic effects in this bioinorganic context can we see the overall nature of these effects. The continuing appreciation of both the bioinorganic nature of cell synthesis and manners in which this may be interfered with to produce a chemotherapeutic effect should also lead to even wider study of these aspects as well as to the development of more sophisticated metal complexes based on the knowledge accumulated from the presently used clinical agents.

References

1. A. Albert: *Selective Toxicity*, 6th Ed., Chapman and Hall, London (1979).

2. J. S. Thayer: *Organometallic Compounds and Living Organisms*, Academic Press, London (1984).
3. *The Molecular Basis of Antibiotic Action*, 2nd Ed., Eds. E. F. Gale, E. Cundliffe, P. E. Reynolds, M. H. Richmond, and M. J. Waring, Wiley, New York (1981).
4. *The Biology and Chemistry of Active Oxygen*, Eds. J. V. Bannister and W. H. Bannister, Elsevier, New York (1984).
5. *Oxygen Free Radicals and Tissue Damage*, Ciba Foundation Symposium 65, Excerpta Medica, Amsterdam (1979).
6. *Superoxide and Superoxide Dismutases*, Eds. A. M. Michelson, J. M. McCord, and I. Fridovich, Academic Press, London (1977).
7. *Superoxide Dismutase*, vols. I and II, Ed. L. W. Oberley, CRC Press, Boca Raton (1982).

CHAPTER 1

INTERACTION OF METAL COMPLEXES WITH DNA

A basic aim of chemotherapy is clearly to obtain a selective effect between a normal host cell and the invading cell, be it bacterial, protozoal, or cancerous. This selectivity may be achieved by differences in distribution, comparative biochemistry, or comparative cytology [1], the damage induced eventually rendering the attacked cell more susceptible to the host's defense mechanisms.

The predominant role of DNA in cellular replication and transmission of genetic information makes the nucleic acids a primary target for drug action, and many drugs are considered to act fundamentally at this level [2]. The inhibition of synthesis of DNA may be direct, involving binding and/or subsequent damage such as strand breakage to the target, or indirect, implying interference with synthesis of critical precursors necessary for DNA synthesis [3]:

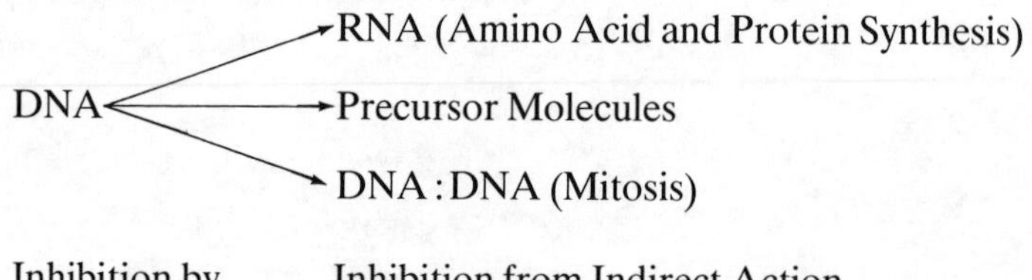

The studies reviewed in Chapter 6 show that metal complexes act by both mechanisms. For indirect action the interference with either cellular biochemistry or small molecule precursors of the nucleic acids is required. In the case of direct action the binding of the molecule (drug) to the polymer is a requisite.

Furthermore, for cancer, the reciprocity between anticancer drugs and potential carcinogens, and their role in DNA modification, have been long recognized [4]. This is obviously no less true for metal complexes in general and observations have been made on this point [5, 6]. Indeed, the juxtaposition of two volumes of *Metal Ions in Biological Systems* on carcinogenicity [7] and antitumour properties [8] of metal ions and complexes serves to underline this interrelationship. For these reasons, the subject of drug–DNA interactions is of considerable importance.

In the case of metals (strictly speaking metal–aqua complexes) and other metal complexes, the subject of their interaction with DNA is relevant for a number of reasons, besides those discussed here, and these are summarized with some examples in Table 1.I. These systems range from *in vivo* aspects such as the endogenous presence in the nucleus of metal ions, the role of zinc in RNA polymerase, the mutagenesis and eventual carcinogenesis of metal ions, the diverse uses of metal ions as probes of polynucleotide structure, as well as the cytotoxic effects of metal complexes which form the subject of this monograph. Many of the above subjects have been reviewed in two useful volumes [9, 10]; structural aspects of binding of metal ions with DNA have been reviewed in these volumes [11–13] and separately [14]. The *Metal Ions In Biological*

TABLE 1.I
Relationships of metals and metal complexes with polynucleotides

Effect	Examples	Ref.
Concentration in nucleus	Cu, Zn	17
Role in replication, transcription and translation	Zn/DNA, RNA polymerases	18–20
Mutagenesis	Cr(VI), Ni(II)	21
Drug action by DNA modification	*cis*-$[PtCl_2(NH_3)_2]$	[a]
	Bleomycin	[b]
Paramagnetic ion substitution	Cr(III), Co(III)–ATP complexes	22
Fluorescent probes	Lanthanides	23
Heavy metal labelling	t-RNA Structure	24, 25
Base sequencing	$K[PtCl_3(DMSO)]$	26
Selective precipitation and separation	Hg^{2+}	27
Spectroscopic and structural probes of conformation	$[Co(NH_3)_6]^{3+}$ $[M(1,10\text{-phen})_3]^{n+}$ [Fe-EDTA-Intercalator]	[c]

[a] See text, Chapter 4.
[b] See text, Chapter 7.
[c] See text, this chapter.

Systems series also has various volumes of particular relevance [7, 8, 15, 16].

Common to all the effects summarized is the need to understand the nature of the physical interactions between metal complexes and polynucleotides. With respect to the identification of DNA as a plausible target for metal-based drugs, the interpretation of the interaction of species such as cisplatin (Chapter 4), and indeed bleomycin (Chapter 7), will hopefully allow the application of such findings to drug development and further, if only slightly, the goal of 'rational' drug synthesis.

1.1. DNA and Conformation

The familiar structure of DNA is the right-handed helix, with the phosphates on the outside and the bases paired by hydrogen bonding in the interior. This fundamental arrangement allows for the existence of various conformations which are interconvertible and whose relative stability is affected by factors such as primary sequence, humidity, the nature of the co-crystallizing cation, and also chemical reagents such as alcohols [28].

The dimensions for various forms of DNA and RNA have been extensively tabulated [28], and a brief summary is given in Table 1.II. A small

TABLE 1.II
Parameters for polynucleotide helixes[a]

Helix	Pitch (Å)	Base Pairs per Turn	Translation (Å) per residue	Rotation (°) per residue	Base Tilt (°)
A-DNA	28.15	11	2.55	32.7	20
B-DNA (Na)	34.6	10	3.46	36	0
B-DNA (Li)	33.7	10	3.37	36	2
C-DNA	31.0	9.3	3.32	39	6
Z-DNA	45.0	12	3.52 (G—C) 4.13 (C—G)	−51.3 (G—C) −8.5 (C—G)	7 7
RNA	29—30	10 (11)	2.9 (2.64)		

[a] Adapted from References 12 and 28. The base displacement in Å of A-DNA is 5, while that of C-DNA is 1.5.

number of parameters can be used to describe the conformations. The rise per residue (or axial translation) gives the distance along the helix axis between adjacent base pairs. The total number of base pairs per turn describes the dimensions of the helix, as does the helical pitch, which is the distance for one complete turn of the helix. The angle made by the plane perpendicular to the helix axis and the planar base pair is the base tilt. The base displacement refers to the distance from the helix axis to the center of a base pair. Differences in these parameters are reflected by changes in the width and depth of the helical grooves.

The most common B-form contains the bases perpendicular to the axis with ten residues per helix turn and zero displacement. The major variant is the A-form, and one important difference is in the sugar residue. Four of the five sugar (ribose or deoxyribose) ring atoms are coplanar, with the fifth atom being significantly displaced, either on the same side (*endo*), or the opposite side (*exo*), as the 5′ carbon. In B-form DNA the out of phase atom is the C2′, while in the A-form the C3′ is displaced:

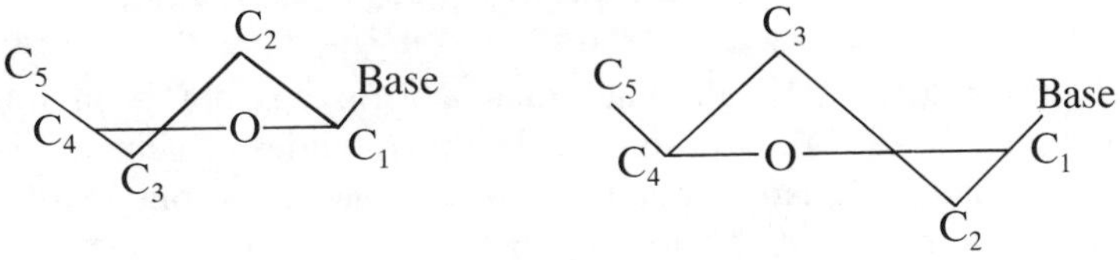

C_2 (*endo*) C_3 (*endo*)

The consequences of this apparently minor puckering alteration are that there are now 11 residues per turn in the A-form, the center of the base pair is greatly displaced by nearly 5 Å and the bases are tilted at an angle of 20° from the perpendicular (Figure 1.1). Because the 2′-OH group in RNA requires a C3′ *endo* pucker, RNA double helices also assume forms similar to that of A-DNA. In C-DNA the base configuration is similar to that of the B-form with a nonintegral number of residues per turn (9 1/3), a displacement of about 1.5 Å from the helix axis, and a slight base tilt of 6°.

In recent years, Z-form DNA has been a focus of attention. This fascinating form was first observed in crystal structures of cytosine–guanine copolymers $(dCpG)_n$, $n = 3$ [29] and $n = 2$ [30]. These copolymers are self-complementary and thus form double helices in solution, and a surprising result was obtained in that the helix was left-handed with a zigzag backbone, hence the 'Z' nomenclature. The solution results indicate that the copolymer undergoes structural changes under high salt or high alcohol conditions and this may result from a transition

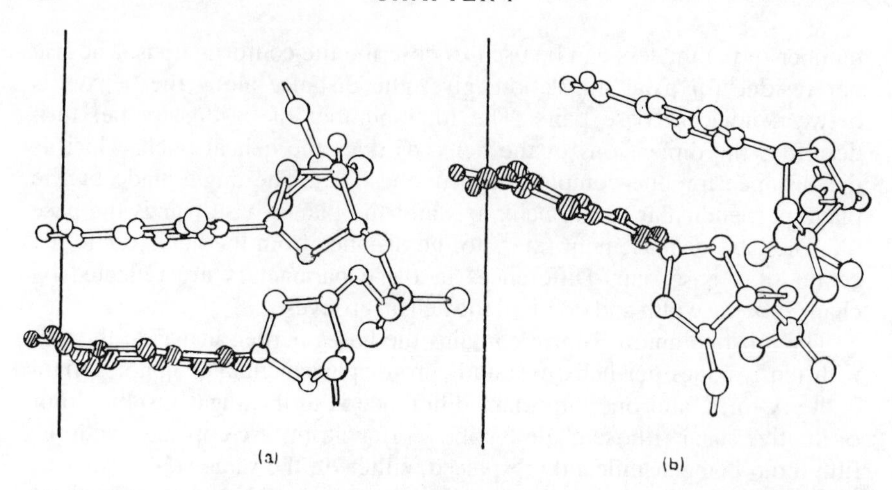

Fig. 1.1. Idealized structures of B- and A-DNA dinucleotides showing the difference in base tilt. In B-DNA the bases are horizontal and perpendicular to helix axis and in A-DNA the bases are inclined to 20°. From Reference 28.

to the Z-form [31], although such a form has not been found in the presence of the A—T base pair. The possibilities of genetic regulation by conformational changes due to local sequence is just one reason why these copolymers are of such interest [32].

The Z-form shown schematically in Figure 1.2. is characterized by a number of unique features. The asymmetric repeating unit is a dinucleotide, rather than the mononucleotide of B-DNA; deoxyguanosine is in the *syn* rather than the *anti* conformation with a C3′ *endo* pucker. The stacking of the cytosine bases results in the situation that the guanine residues can no longer stack but instead interact with the O1′ of an adjacent sugar.

Other forms of DNA such as the compacted Φ form have also been described [33, 34]. This form occurs in association with positively charged proteins or polypeptides such as histones. A pertinent point here is that in nuclei of eukaryotic cells, DNA does not occur unassociated as in bacteria. Instead, it is present in the highly condensed form of chromatin; this essentially consists of periodic aggregates of histone proteins that package the DNA helix into particles called nucleosomes, which are considered to be the primary level of structural organization within the chromatin fibrils [35].

Finally, in many cells, DNA is isolated in the closed circular or superhelical form. Sources of this DNA include bacterial, tumour and animal viruses, bacteria and sperm [24]. Because the closed circular form,

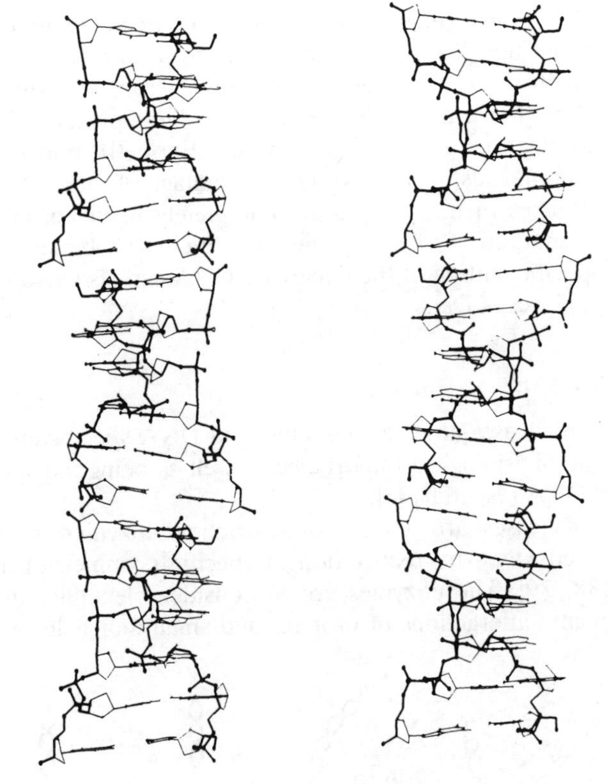

Fig. 1.2. Schematic representation of Z-DNA. From Reference 29.

Form I, may undergo various transformations that are readily assayed, this DNA is of particular utility as a probe of binding of small molecules. The basic feature is that, upon cyclization of a helical duplex, the total number of turns of one strand about the other remains a topological constant; therefore, changes in the number of Watson—Crick turns in the closed form will be uniquely related to the number of supercoil turns. This is expressed [36, 37] as

$$t = a - B$$

where t is the superhelix winding number, i.e. the number of revolutions made by the duplex around the superhelix axis, which is given by the difference between a, the topological winding number, the number of helical turns, and B, the duplex winding number or the number of Watson—Crick turns in the molecule in solution.

The possible transformations of Form I are depicted in Figure 1.3. Alkaline denaturation leads to an unwinding to a more relaxed, but still closed, circular form, Form I′ or I_0; a single cleavage, or a small number of these, gives a more expanded, nicked circular molecule, Form II. A double strand cleavage gives a linear species, Form III. Further cleavage or denaturation may also occur. Another advantage of these DNAs is that the base sequences in many cases are completely resolved, allowing for detailed examination of the sequence specificity of binding by small molecules. Specific cutting of the linearized DNA can also give short (200 base pair: bp) pieces of DNA.

1.1.1. NUCLEASES AND DNA

The bacterial nucleases are enzymes which cut DNA specifically, and they are of intrinsic physiological importance as well as being indispensable in recombinant DNA research [38].

The endonucleases are a class of restriction enzymes which cleave doubly stranded DNA by recognition of specific (symmetric) nucleotide sequences [38, 39]. The enzymes are of considerable utility in studying sequence-specific interactions of proteins and small molecules with DNA.

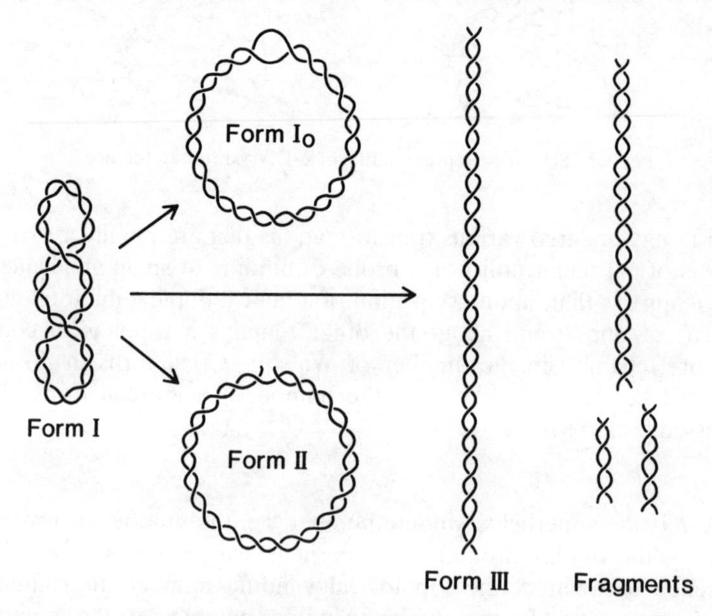

Fig. 1.3. The various forms of closed circular DNA and their conversion to linear forms. The closed circular form can be relaxed to closed circles with no turns (Form I_0) and a nicked circular DNA (Form II). Linearization gives Form III which may be specifically cut.

The binding of small molecules at or near the cutting site may inhibit the restriction, thus providing information on binding sites and specificity of the small molecule interaction. A further use is to study the binding of small molecules and drugs to the cut fragments after enzyme treatment since, as stated, in many cases the nucleotide sequence is known (from what are now standard techniques, such as those of Maxam-Gilbert or Sanger [40, 41]) and small sequences (100—200 bp) may be more appropriate for extrapolation of results to localised regions of mammalian DNA.

A directory of restriction enzymes is available [42] and some of the most commonly used are outlined below, with their recognition sequence. The sequence may be four or six base pairs; those binding at four-base sites cutting a particular DNA many more times than a six-base enzyme. Indeed, the remarkable specificity of these enzymes is emphasised by the fact that the Eco R1 recognition sequence is not present in T7 DNA (40 000 bp) and single, unique sites for Eco R1 and Bam H1 are found in SV40 DNA (5226 bp). The nomenclature is usually an abbreviation of the microorganism source (i.e. EcoR1 from *E. Coli* RY13) [42]:

Enzyme	Restriction sequence
Bam H1	5′ – G GATCC –
	– C CTAGG – 5′
Bgl I	5′ – GCCNNNN NGGC –
	– CGGNNNN NCCG – 5′
Eco R1	5′ – G AATTC –
	– C TTAAG – 5′
Hin dIII	5′ – A AGCTT –
	– T TCGAA – 5′
Hpa II	5′ – C CGG –
	– G GCC – 5′
Hae	5′ – GG CC –
	– CC GG – 5′

For metal complexes care should be taken that no inhibition of free enzyme occurs, which could affect the interpretation of the results.

The exonucleases cut DNA single strands sequentially from either the 3′ or 5′ terminus and may be specific for either singly or doubly stranded DNA. A commonly used probe is exonuclease III [43] which cuts DNA exonucleolytically from the 3′ end one base pair at a time. Given a DNA fragment of known sequence, it will be appreciated that a combination of the above techniques can give the exact binding site of any molecule bound to the polynucleotide. Further reading on this fascinating subject may be found in References 38 and 39.

1.2. Experimental Methods for the Study of Conformational Changes

Many of the physical properties of polynucleotides are changed upon alteration of their structure; this allows for the application of a wide range of techniques in such studies. Hydrodynamic properties such as viscosity, sedimentation coefficients, and buoyant density can all be readily monitored. Light and low-angle X-ray scattering are also well understood. Changes in spectroscopic properties are also conveniently studied. These have all been summarized in *Physical Chemistry of Nucleic Acids* [28], and the major effects which are discussed in this book are briefly reviewed below.

1.2.1. ULTRAVIOLET SPECTROSCOPY

One of the simplest measurements is the plot of ultraviolet absorbance of a system versus temperature. The UV absorbance of both double and single stranded polynucleotides increases with temperature; for a double helix this hyperchromicity is attributed to disruption of the ordered state and elimination of the stacking interactions of the base pair chromophores.

A useful and important parameter, readily obtained from these measurements, is the helix-to-coil transition temperature, T_m, the temperature at which half of the helical structure is lost or altered. The transition, also called melting or denaturation, is highly cooperative and is also pH dependent. Indeed, the transition occurs within 0.1 pH unit instead of several units as might be expected from a consideration of the titration values of the individual bases. Below the transition temperature separated complementary strands automatically reassociate (renaturation).

In general, interactions which tend to stabilize the double helix, e.g. neutralization of the charged phosphate backbone, lead to an increase in T_m, whereas destabilizing interactions such as base-binding and disruption of hydrogen bonding lead to a decrease in T_m. There is, however, no clear cut distinction between these modes of binding, especially for metals, and the T_m gives information on overall stability changes.

1.2.2. CIRCULAR DICHROISM AND OPTICAL ROTATORY DISPERSION

These techniques are especially useful for qualitative information on conformational changes; the interpretation with respect to defined changes is difficult and may be complicated [28].

The shape of the CD curve of DNA is defined as conservative with approximately equal positive and negative areas centered at 260 nm. The spectrum is different to that of the constituent nucleotides and also different from that of RNA. Factors such as tilting of the base pairs or unwinding of the helix may induce alterations in the optical rotation, thus allowing observation of a conformational change that would be undetected by the relatively nonspecific UV spectrum.

1.2.3. NUCLEAR MAGNETIC RESONANCE

High resolution NMR is becoming increasingly useful for studying conformational changes and small molecule interactions [44, 45]. Stacking interactions may be studied and the fact that the hydrogen bonding protons in double helical DNA and tRNA are slowly exchanging and lie downfield from the base protons allows for study of the interaction with small molecules [46]. The helix–coil transition and intercalation have been monitored in this manner. More recently, the observation of ^{31}P resonances and their shifts upon binding of small molecules has become extremely useful [47].

1.2.4. HYDRODYNAMIC AND LIGHT SCATTERING PROPERTIES

The alteration of the secondary structure of DNA upon binding of small molecules will be reflected in changes in the hydrodynamic and light-scattering properties of the polymer [28]. The sedimentation coefficients and intrinsic viscosity will be altered. In the case of intercalation the viscosity increases and sedimentation coefficient decreases; the increase in length of the adapted polymer induces the viscosity change. Denatured DNA will also exhibit an increase in buoyant density. Rotational motion may be studied by measurements of birefringence, dichroism and fluorescence depolarization. The scattering of electromagnetic radiation by complexed DNA will differ from that of the free polymer and quantitative measurements of change in shape may be obtained by measuring the mass per unit length.

1.3. Binding of Metal Complexes to DNA

Figure 1.4 shows the basic sites of interaction and possible modes of binding to DNA, as exemplified by common drugs and also metal complexes. The interactions may be summarized as:

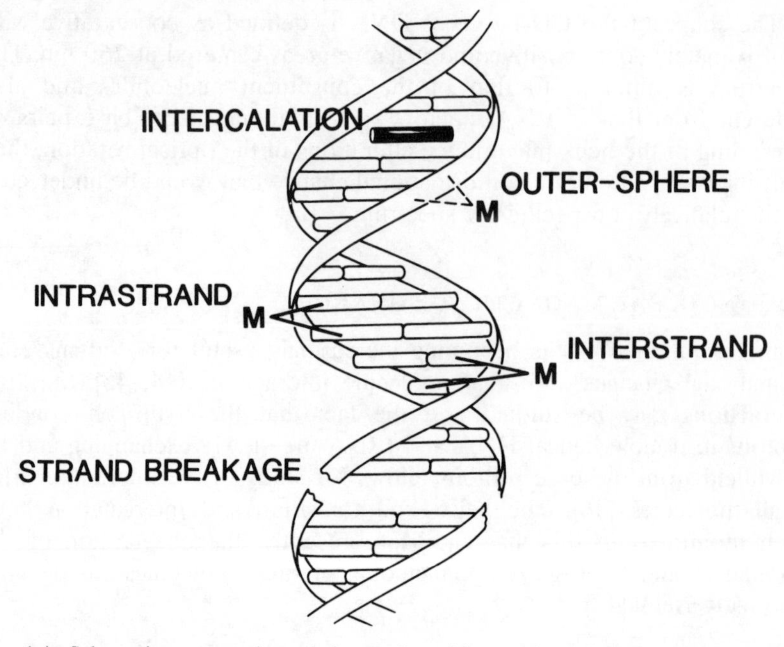

Fig. 1.4. Schematic presentation of the modes of interaction of metal complexes with DNA.

(1) Intercalation
(2) Covalent and noncovalent outer-sphere binding
(3) Covalent inner-sphere binding
(4) Strand breakage.

The ability of metal complexes to form more than one bond to the polynucleotide, especially in Form III, results in various forms of cross link between the complementary strands of the helix. The coordination number, geometry, and types of ligand all dictate a remarkable range of conformational changes and interactions with DNA, including base pair specificity.

1.3.1. INTERCALATION

The concept of intercalation was originally proposed by Lerman [48] to explain the binding of aromatic planar molecules such as aminoacridines, exemplified by proflavine (1a), and later phenanthridines, such as homidium bromide (ethidium 2b):

(1) a $R = H$, Proflavine
b $R = CH_3$, Acridine Orange

(2) a $R = CH_3$, Methidium
b $R = C_2H_5$, Ethidium

The basic, classical interaction involves the insertion of the planar molecule between two neighboring base pairs of DNA to which it is held by Van der Waals forces [49]. A further weak interaction of an electrostatic type between the phosphate anions and charged groups of the molecule may also be present. A nonclassical intercalation model involving bending of the helix has also been proposed [50]. The models are outlined schematically:

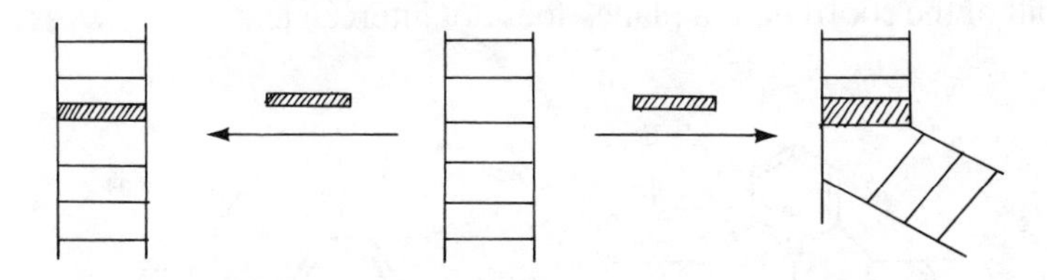

The classical view of intercalation requires that the helix be extended and locally unwound by the binding reaction. The hydrodynamic properties are affected with an increase in viscosity, resulting from an extension and stiffening of the double helix, as well as a decrease in sedimentation coefficient. The transition melting temperature, T_m, is also increased and this can be understood by considering that the intercalator–base pair stacking interaction stabilizes the helical over the unwound form.

The effect of binding of intercalators on closed circular DNA is highly informative. The intercalation model again requires that the helix be unwound, and in closed circular DNA this implies a change in the number of superhelical turns. After this initial unwinding, formation of turns in the opposite sense is observed. Knowledge of the unwinding angle of a particular intercalator allows the number of superhelical turns in any DNA to be measured. Conversely, given the number of tertiary turns in a closed circle, measurement of the number of molecules required to unwind these turns furnishes the unwinding angle for that particular intercalator. An unwinding angle of 26° for ethidium bromide (See structure above) is generally accepted as the standard [4, 51].

Many intercalators have biological activity and, in fact, the remarkable preponderance of planar aromatic cations with activity was noted even before the intercalation concept was fully formulated. The DNA binding and molecular pharmacology of these species have been reviewed in comprehensive articles [4, 52].

1.3.1.1. *Platinum—Terpyridine Complexes*

The intercalation of platinum complexes containing bidentate or tridentate aromatic amines was first proposed for binding to tRNA and then to DNA [53]. The structures of the relevant complexes of palladium and platinum are shown in Figure 1.5 and a requisite is the presence of planar ligands such as bipyridine, terpyridine, or *o*-phenanthroline, which are coplanar with the coordination plane formed by the four ligand donor atoms and the metal ion. The similar pyridine derivative, $[Pt(en)py_2]^{2+}$, where the pyridine ligands are forced by nonbonded steric constraints to be out of the coordination plane, does not intercalate.

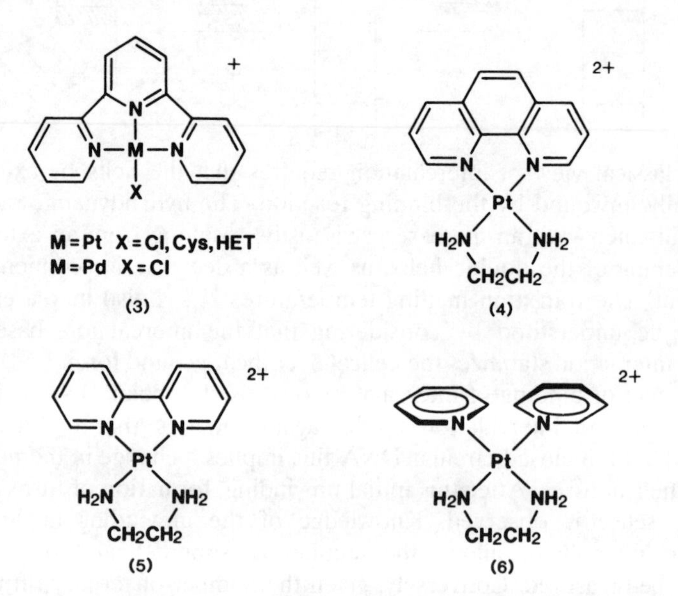

Fig. 1.5. Structures of platinum complexes of planar aromatic ligands studied as intercalators.

The intercalation, proposed from variations in physical properties such as T_m, viscosity, competitive experiments with ethidium, and studies on closed circular DNA [53—55] was confirmed by an X-ray crystal structure of the adduct of $[Pt(HET)(terpy)]^+$ and the dinucleotide dCpG [56], Figure 1.6. The major features elucidated include an unwinding angle of 23°, confirming an earlier proposal [55], the value being close to the 'standard angle' of 26° for ethidium; the sugar puckerings are similar also to those observed in other crystalline intercalator: dinucleotide complexes [57]. Several stabilizing intermolecular interactions were also noted: the platinum atom lies directly above and below the guanine oxygen O(6) with the terpyridine overlapping the base pairs; the side chain of the intercalating molecule is in the double groove of the helix. Hydrogen bonds exist between cytosine O(2) and the hydroxyl group of the ethanesulfide ligand and between the C5′ (OH) of deoxycytidine and the phosphate oxygens of an adjoining molecule.

A good example of the utility of metal complexes as probes of polynucleotide structure is to be found in this study. The fact that no intercalating molecules ever bind at a saturation ratio higher than one for every two base pairs led to a 'neighbor exclusion' hypothesis, whereby upon binding of an intercalator the neighboring base pair is excluded from binding; saturation is therefore achieved at a ratio of 1 : 2 intercalator/base pair. This model gives a repeating distance of 10.2 Å for the proximate

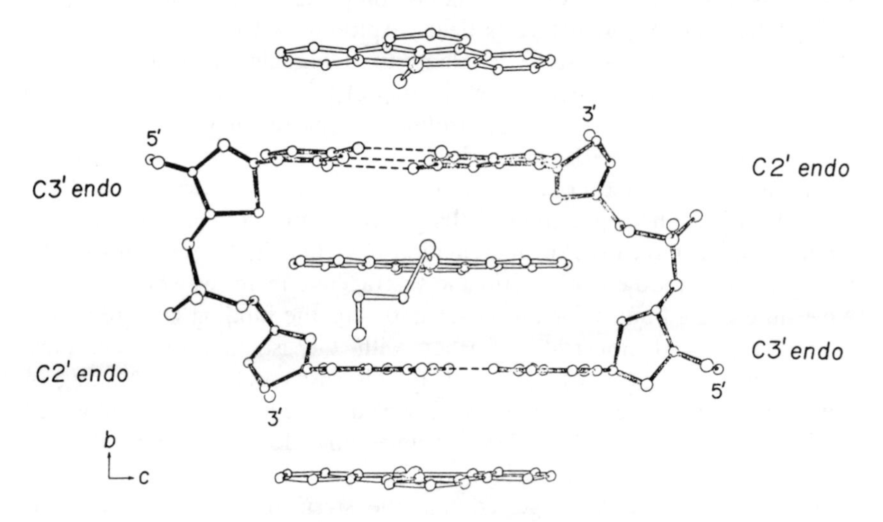

Fig. 1.6. Structure of the complex of the dinucleotide d(CpG) with the intercalator [Pt(terpy)(HET)]. From Reference 56.

intercalators. The presence of the electron dense heavy metal atom enhances that periodicity and allows for observation of a layer line (not as clear with other species because of the similar electron density of the intercalator and a base pair) that lends strong support to the hypothesis [58].

The X-ray crystal structure of a [Pt(terpy)Cl]—AMP complex has also been briefly reported [59]. In this case, the complex consists of a hydrogen bonded AMP base pair, that mimics the rare Hoogsteen model [60] with NH_2—N_1 and NH_2—N_7 hydrogen bonds (see Appendix 1) sandwiched between two platinum—terpyridine molecules. The nucleotide is not locked into a helix in this structure and so the sequence can be described as I—I—BP—I—I—BP, emphasizing the propensity of platinum complexes for stacking interactions [61].

1.3.1.2. *Metal—(1,10-Phenanthroline) Complexes*

The minimal overlap for an intercalating ligand such as an acridine and its relation to biological activity (DNA binding and antibacterial activity) have been thoroughly studied by Albert [1, 62]. For the platinum—terpyridine complexes, as shown above, the intercalator plane is coplanar with the coordination plane. The chelating ligands 1,10-phenanthroline (phen) and 2,2′-bipyridine (bipy) are also capable of sufficient overlap with base pairs, and use of their metal complexes as intercalating agents and probes of DNA structure is being explored in a series of interesting studies [14, 63]. In this case, coplanarity is not always possible as the geometry varies from octahedral $[M(phen)_3]$ to tetrahedral or square planar $[M(phen)_2]$. The phenanthroline complexes have been studied extensively for their biological activity [64].

Studies on the zinc complexes, $[ZnCl_2(phen)]$, $[Zn(phen)_2]^{2+}$, and the octahedral $[Zn(phen)_3]^{2+}$, showed that all the complexes unwound closed circular DNA, indicative of intercalation [65]. Octahedral chelates of the type $[M(L—L)_3]$ are enantiomeric and the racemic mixtures were shown to have different affinities for the DNA and thus the binding is stereoselective. This was demonstrated further with the isolated optically pure enantiomers of $[Ru(phen)_3]$ [66, 67]. In this case, preferential binding of the Δ-form over the Λ-isomer is observed, as evidenced by enhanced luminescence and in differences in unwinding closed circular DNA at equal concentrations. An explanation for these differences is shown in Figure 1.7, where it is suggested that the steric hindrance of the non-intercalating ligands of the Λ-isomer with the phosphate backbone results in diminished binding in comparison to the favored Δ-isomer.

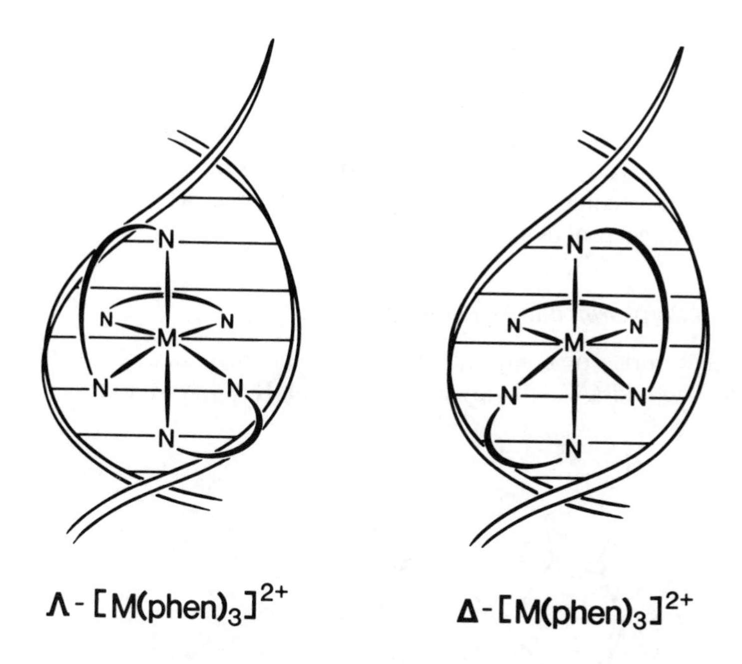

Fig. 1.7. The interactions of optical isomers of complexes of phenanthroline, $[M(phen)_3]^{2+}$, showing the differences in steric effects upon binding to DNA. From Reference 66.

The equilibrium binding constant for the racemic Ru chelate is, however, low in comparison to ethidium or $[Pt(phen)(en)]^{2+}$, (6.2×10^3, 3×10^5, and 5×10^4 M^{-1}, respectively) [66] and $[Ru(bipy)_3]^{2+}$ does not appear to bind in the same way [68]. The diminished affinity for DNA was explained by decreased overlap for the octahedral Ru complex in comparison to the square planar Pt species. A four base-pair model is also indicated, rather than the two base-pair (neighbor exclusion) model of the Pt-terpy system [66]. Introduction of phenyl groups in the phenanthroline ligand in the 4,7-positions produces even more sterically rigid structures [69] and the Ru—chelate interaction with DNA is now totally stereospecific, the Δ-isomer binding specifically to B-DNA [70]. Interestingly, both isomers bind to left handed Z-DNA [70]. The free 4,7-diphenylphenanthroline (DIP), however, does not bind to DNA [71]. Extension of this work has led to a chiral probe for A-DNA, using the chelate $[Ru(TMP)_3]^{2+}$ (TMP = 3,4,7,8-tetramethylphenanthroline) [72]. The enantiomeric preference is altered if the $[M(DIP)_3]^{n+}$ complexes (M = Ru, n = 2; M = Co, n = 3), are reacted with DNA which has been previously associated with a restriction endonuclease, confirming the conformational change of the DNA upon binding of the protein [73].

Chiral discrimination has also been observed in *cis*-[$RuCl_2(phen)_2$], where both intercalation and covalent binding are possible, although the binding is weak with a maximum binding ratio of 1/11 base pairs [74]. This series of complexes is of great interest, then, as spectroscopic and conformational probes of helical structure.

1.3.1.3. *Porphyrins and Metalloporphyrins*

A further series of complexes that has been suggested to intercalate is based on tetrakis(*N*-methylpyridyl)porphyrin (R = 4-*N*-Mepy):

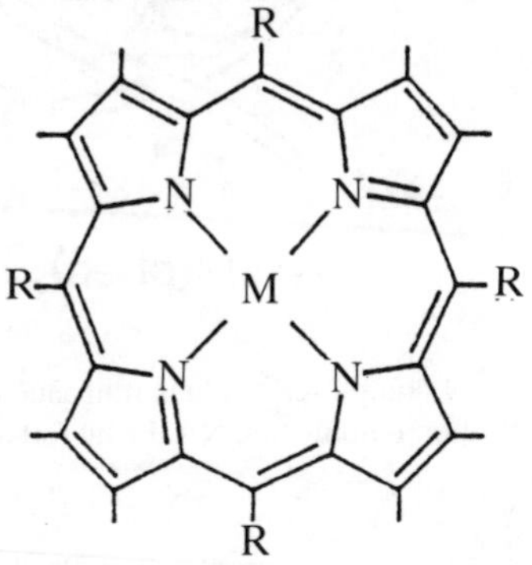

(7)

The intercalation of the free porphyrin with calf thymus DNA is judged to occur on the basis of studies on the effects on the UV spectrum, viscosity changes and interaction with closed circular DNA [75]. Extension to metal complexes [76] gave the interesting result that whereas the copper complex interacted with poly(dG—dC) and could be analyzed by the neighbor exclusion model of intercalation, complexes with Ni(II), Co(III), Zn(II), Fe(III), and Mn(III) did not react; however, all these reacted with poly(dA—dT) and also with DNA in a nonintercalative, presumably electrostatic manner. The presence of axial ligands for Zn, Co, Fe, and Mn is considered to inhibit the intercalation sterically. The copper complex, with no axial ligands, behaves like the free porphyrin and gives a complicated spectrum with poly(dA—dT) [77]. Studies with free nucleosides and nucleotides have also been reported and Coulombic and stacking interactions, but no covalent bond formation, were observed [78]. In all cases, the DNA interactions may be interpreted as a superposition of the homopolymers, and so the conclusion is that the nature, as well as the specificity, of the reaction may be dictated by minor structural changes, i.e. by axial ligands.

The metalloporphyrin system has now been examined by a wide variety of techniques and in fact presents a good example of how varying techniques can give information on modes of DNA binding. These methods include most of those outlined, especially NMR, CD, fluorescence and resonance Raman as well as physical studies such as viscosity and melting (see [162]). Study of an extensive series of metalloporphyrins recognised three groups with distinct binding characteristics [163] – Group I porphyrins give changes indicative of intercalation with DNA of greater than 40% GC base composition, Group II complexes appear to be outside binders but also induce DNA aggregation, while Group III porphyrins induce changes indicative of outside binding at both GC and AT sites. The great utility of NMR is demonstrated by its use in the study of sequence specificity in intercalation and the observation of selective binding to 5′-d(CpG)-3′ sequences [164].

With respect to the details of intercalation of this porphyrin and the 4,7-diphenylphenanthroline complexes discussed earlier, some rotation of the pyridine and phenyl rings into the plane of the phenanthroline and porphyrin rings, respectively, must occur for maximum overlap and intercalation to occur. The 2-*N*-methylpyridyl porphyrin, in agreement with the expected larger steric restraint, does not react with DNA. Even full planarity of one 4-*N*-methylpyridyl group, however, will barely give the minimal surface area for good overlap.

1.3.1.4. *Intercalators as Drugs*

Intercalators as a class contain many compounds with *in vitro* and *in vivo* antitumour activity [2, 46] but, although the binding is strong, the DNA–intercalator reaction is also kinetically labile and fixation through covalent bond formation with a transition metal is attractive. The use of known intercalators to tether metal complexes is demonstrated here by the acridine orange–platinum(ethylenediamine) complex, shown in Figure 1.8, where the acridine moiety increases and also modulates the DNA affinity of the platinum–amine [79]. Local structural changes due to intercalation alter the preferred Pt-binding sites (See also Section 1.3.3.2) [80]. Direct binding to the amines of proflavine has also been observed [81]. These species are expected to have interesting antitumour activities, consisting as they do of an active intercalator and a platinum moiety (See Section 3.5).

The intercalator may also act as a DNA-recognition ligand for the

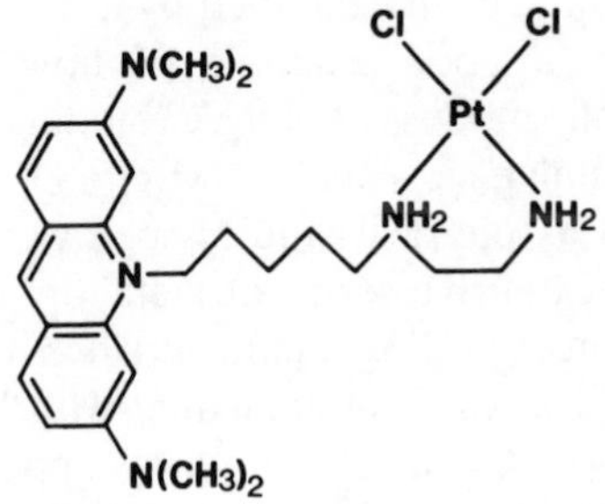

Fig. 1.8. Structures of Pt complexes linked to intercalators. An acridine orange linked to a platinum—ethylenediamine complex and the proposed structure of a complex with Pt directly bound to proflavine.

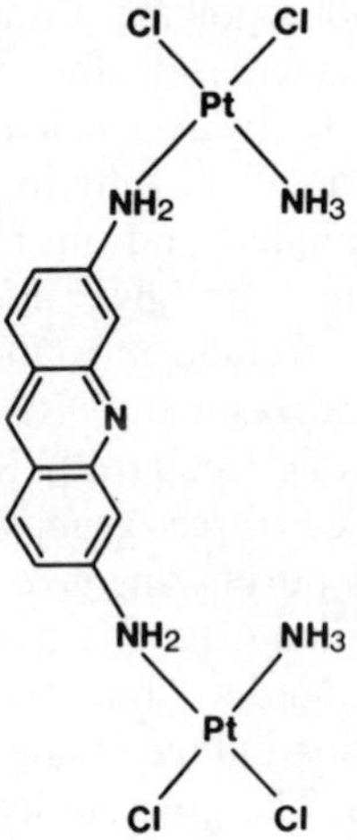

metal and activate hitherto 'inactive' moieties. The combination of intercalation and bond formation, given the right geometry, would also be a distinctly different lesion, in terms of biological damage, for instance, from those of platinum—amines. A 'synergistic' combination of DNA recognition (intercalator) and DNA fixation (metal) could substantially broaden the types of active metal species. A relevant point to consider is whether some minimal binding affinity must be achieved before cytotoxic effects are observed. The mere fact of binding to DNA, even with subsequent conformational change, does not imply a type of damage which will lead to cell kill, and this may be due to ease of repair or ease of reversibility of the interaction, apart from pharmacokinetic considerations.

1.3.1.5. *Redox Activity of Metallointercalators*

Metal complexes also offer the possibility of more than one oxidation state and offer the potential of redox chemistry associated with binding. Potentially damaging radicals can then be produced in the vicinity of the proposed target. Intercalators such as ethidium have also been used to tether iron species, which can produce radicals in the vicinity of DNA, reminiscent of bleomycin (Chapter 7). The reduction of Co(III) phenanthroline complexes with subsequent radical production also results in

DNA cleavage and these aspects are covered in Section 1.3.4. With respect to electron transfer, DNA-mediated photoelectron transfer between $[Ru(phen)_3]^{2+}$ and $[Co(phen)_3]^{3+}$ has been observed [82].

1.3.2. OUTER SPHERE BINDING

The negatively charged backbone of the DNA helix may interact with a variety of positively charged molecules such that, either through coulombic interactions or phosphate–oxygen binding, the overall charge is reduced with subsequent effects on stability and conformation. A study of the effects of divalent metal ions on the T_m of DNA divided the ions into three categories. The order of perturbation is Mg < Ni < Zn < Cd < Cu, with Mg being the most stabilizing and Cu the most destabilizing [83]. Thus, whereas magnesium binds primarily to phosphate — a stabilizing reaction — copper binds at the other extreme with a strong affinity for base binding.

Similarly to Mg, both Co(II) and Ni(II) exhibit increasing T_m with increasing metal ion concentration, while Mn(II) and Zn(II) give a maximal T_m at intermediate concentrations, although T_m is always positive. Analysis of the data, therefore, indicates an increasing affinity for base binding with increasing metal ion concentration. The initial phosphate binding, leading to a positive ΔT_m, followed by base binding and a negative ΔT_m, is very clear in the case of Cd. The binding of Au(III), administered as $AuCl_4^-$, has been interpreted in a similar manner with phosphate binding at low gold concentration, and base affinity increasing with increasing concentration [84]. The interaction of $[AuCl_3(py)]$ has been interpreted, however, as base binding [85] (see below). It is clear that while relative phosphate/base affinities vary for the different metal ions, what we are seeing is a gradation of properties from one extreme to the other, and no clear cut distinction (only phosphate or only base) should be made in most cases.

Study of the CD spectrum of DNA in the presence of Mn(II), Zn(II), and Co(II) indicated that the metal ions induced a B → C conformational change. The CD spectra are altered at very low concentrations, and the transformation is considered to occur via reduction of electrostatic repulsive interactions as the cations bind along the phosphate backbone and in the helical grooves [86].

1.3.2.1. *Metal–Amine Complexes*

A further set of complexes that has been shown to interact with DNA by outer sphere binding is of the inert metal–amine type, and these complexes are of interest not only because of the comparison with aqua complexes discussed above but also because of their similarity with cisplatin.

The influence of cobalt(III)—amine cations on native and denatured calf thymus DNA was first reported in 1972—3 [87, 88]. Subsequent studies on tRNA [89] and DNA [90] confirmed that such complex cations are very effective in stabilizing secondary and tertiary structures of nucleic acids. Karpel *et al.* have extended the studies on cobalt to a series of platinum metal derivatives and the results are summarized in Table 1.III [91].

The most studied system is that of $[Co(NH_3)_6]^{3+}$ and its ethylenediamine analogue. The inert nature of these species indicates reaction by an outer sphere mechanism and the main interaction is considered to be Coulombic attraction. Hydrogen bonding, either directly to phosphate or via water molecules, can also contribute to the overall stabilization. In this respect, it is of relevance that the amine ligand is sufficiently acidic to be deprotonated in the presence of base [92]. There is therefore some utility in considering the NH bonds in M—NH_3 as being slightly acidic as in NH_4^+ (albeit weakly), rather than covalent as in NH_3. The resolved enantiomers of the $[Co(en)_3]^{3+}$ cation showed no apparent differences in their binding [88]. An electrostatic component has been observed in the initial binding of $[Ru(NH_3)_5(H_2O)]^{2+}$ to DNA [93]. These interactions may be treated by standard theory [94].

In a study of the effect of cations on the B → Z transition of poly(dG—dC) · poly(dG—dC), the $[Co(NH_3)_6]^{3+}$ complex cation was found to be remarkably effective in inducing this change [95, 96]. At 0.02 mM in cobalt, the transition is complete and the complex is more

TABLE 1.III
Effect of metal-amine complexes on stability of polynucleotides.

Complex	dT_m/dlog[complex] (°C)	
	poly I—C	Calf Thymus DNA
$[Co(NH_3)_6]^{3+}$	23	12
$[Co(en)_3]^{3+}$	21	20
$[Pt(en)(en{-}H)]^{3+}$	10	9
$[Ir(NH_3)_5Cl]^{2+}$	17	—
$[Pt(NH_3)_4]^{2+}$	8	—

Adapted from Ref. 91.

potent than spermine or spermidine, which tend to aggregate, i.e. precipitate out, DNA at higher concentrations.

The B → Z transition is Na^+-dependent and the endpoint corresponds to 2 positive charges per phosphodiester group. The ability to induce this transition is remarkable in view of the fact that normal methods of induction involve high concentrations of salt (2.5M Na^+) or ethanol (60% vol/vol). The complex is now used routinely to induce the Z-conformation and as a probe for DNA structure.

The structure of $[Co(NH_3)_6]^{3+}$ bound to yeast phenylalanine tRNA has been resolved to answer questions as to the exact nature of binding [97]. There is selective binding to purine—purine sequences, especially those of guanine, and hydrogen bonding from the ammine ligands to N_7 and O_6 atoms of the guanine as well as to neighboring phosphates in three regions of the RNA. This is not a favorable interaction for adenine, due to the presence of the exocyclic NH_2, rather than C=O at position 6. No direct metal—nucleotide bonds are formed, in contrast to $[Co(H_2O)_6]^{2+}$, which binds to the N_7 of residue G15 [98]. Finally in this section, if one considers that the $[Co(H_2O)_6]^{2+}$ cation has been shown to induce the B → C transition, then the B → Z transition of $[Co(NH_3)_6]^{3+}$ presents a remarkable change in specificity upon ligand substitution, a specificity altered by binding mode.

1.3.3. INNER SPHERE (COVALENT) BINDING

A number of aqua metal ions bind covalently to the bases of DNA. Early examples of this mode of binding were mercury and silver and, from the very diverse literature, aspects of the binding of these metals have been reviewed in more detail [12—14]. In the case of mercury the destabilization normally apparent upon base binding is not present; in fact, the DNA is stabilized [99]. This unusual effect is explained by the fact that mercury substitutes the hydrogen bonds by binding primarily to the thymine base and thus forms very strong cross links, stabilizing the helix. In accord with this idea, Hg(II) stabilizes polyd(AT), where it can form T—Hg—T cross links, but destabilizes poly(dA) · poly(dT), where this cannot occur. The structure of the bis(1-methylthyminato)mercury(II) complex strengthens this interpretation, the Hg atom binding to N_3 of the pyrimidine in a linear mode N_3—Hg—N_3 [100]. The great preference of mercury(II) for AT-rich regions has been utilised in the separation of DNA molecules of differing base content [27].

Similar interactions occur with silver(I), the metal cross links again stabilizing the helix [101, 102]. The binding of silver occurs primarily at

G—C rich regions of DNA, in contrast to mercury, and separation founded on this basis can also be achieved [103]. Again a crystal structure of 1-(methylcytosine)silver(I) nitrate may serve as model for the DNA interaction [104]. This complex is also of interest because of its dimeric nature. The proposed inhibition of bacterial DNA synthesis by the Ag^+ ion (Chapter 9) is relevant here, although no attempts appear to have been made to correlate cytotoxicity with modes of binding.

Cross linking with Zn(II) and Cu(II) has also been proposed to explain their DNA binding [105, 106]. These base binding examples also underline the point that the increase in T_m which is normally associated with outer sphere binding (neutralization of phosphates) may, for metals, have its origin in a totally different form of binding.

1.3.3.1. *Binding of Platinum—Amine Complexes*

The above examples relate to simple metal ions without a coordination environment complicated by ligands other than water. As for outer sphere binding, ligand substitution alters the affinity and specificity of the binding interaction. One of the best defined series is that of the mixed platinum—chloroamines, from $PtCl_4^{2-}$ to $[Pt(NH_3)_4]^{2+}$, (Figure 3.1, page 68). Because of the relevance to the antitumour action of some of these complexes (Chapters 2 and 3), and the fact that differential antitumour activity is considered to be related to differential DNA binding, the physical aspects of the binding will be discussed here, as a very clear example of how closely related structures may differ in their DNA binding, and to recognize and summarize the various distinct modes of binding. More detailed examination of the Pt—DNA interactions and the interpretation of those lesions essential to cytotoxicity will be discussed in Chapter 4. These studies now include physical and structural studies on small oligonucleotides and analysis of Pt—DNA adducts from degradation of the polymer into small fragments.

All the platinum complexes bind to DNA, the tetraammine very weakly in an outer-sphere manner [91], and tetrachloroplatinate (II) in a destructive, nonspecific manner [107]. Clearly substitution by amines gives complexes with highly specific interactions with DNA. There is general agreement on the physical changes of DNA upon binding of either *cis*- or *trans*-$[PtCl_2(NH_3)_2]$. Some of these data have been mentioned in the principal reviews on platinum—nucleic acid interactions [108, 109]. The major conclusions are summarized here. For the *cis*-isomer, the UV spectrum shows a hyperchromic shift [110], indicative of interactions with the bases. There is a decrease in viscosity, with the major change occurring at small (<10% binding ratios) [111]. There is an increase in buoyant density and one of the earliest demonstrations of preferential binding was in this study, where a larger increase in buoyant density occurred for DNA of high G—C content, e.g. *M. lysodeikticus* (%G—C = 72) [112]. Further studies showed that the buoyant density of poly(dG) · poly(dC) increased almost 5 times over that of poly(dG · dC), outlining the importance of GpG sequences [113]. The preferential binding to guanine sequences was used to separate DNAs of varying base composition [113]. An increase in sedimentation coefficient is observed with [$PtCl_2$(en)] [114]. The interpretation of these physical results is that a localized denaturation occurs upon binding of relatively small amounts of platinum.

One of the clearest demonstrations of the differences in binding of *cis* and *trans* isomers comes from the CD spectra, Figure 1.9 [115—116]. While a slight increase in ellipticity occurs for the *cis*-isomer at low binding ratios, followed by a decrease at higher ratio, the binding of the *trans*-complex produces a linear decrease with increased binding. The

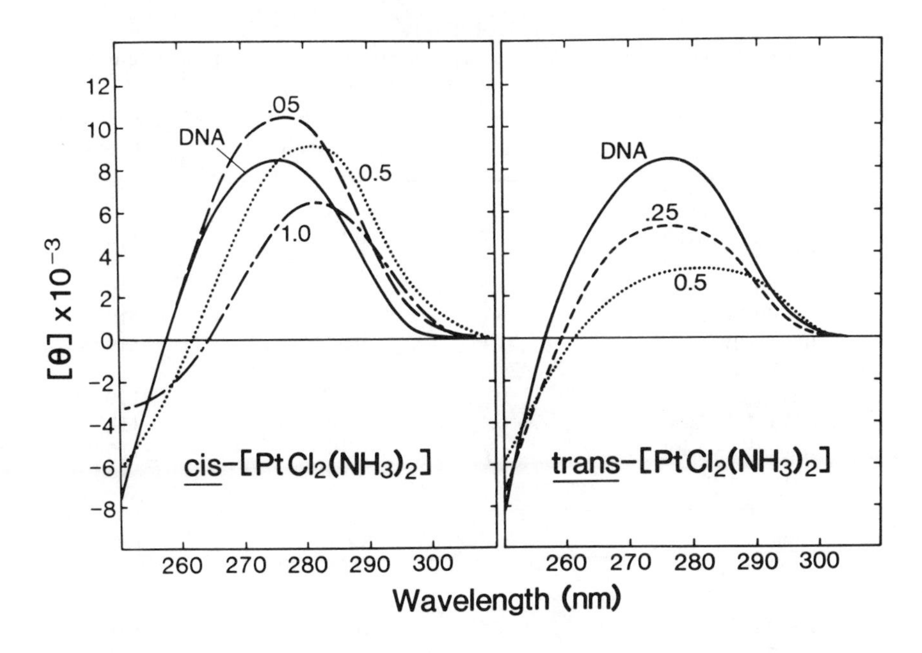

Fig. 1.9. Circular dichroism spectra of the interaction of the isomers of [$PtCl_2(NH_3)_2$] with Calf Thymus DNA. The numbers represent r_b, the Pt/P ratio. Note the initial increase for the *cis*-isomer. From Reference 115.

changes appear to be dependent on DNA source (i.e. base content) with the largest changes occurring for the G—C rich *M. lysodeikticus* DNA. The results have been interpreted in terms of a B → C transition with unwinding of the helix, decrease of the pitch and a decrease in groove size [118]. The CD spectral changes are matched by those of poly(G)—poly(C) indicating that the G—C rich regions are responsible for the changes in the DNA. Indeed, in a study of model complexes with simple polynucleotides the spectral changes were again only matched by 5'-GMP [118].

These differences are paralleled in the melting point behvaior of the DNA—amine complexes, which are summarized in Table 1.IV. Although the experimental conditions vary considerably the *cis*-isomers can be seen to produce a stabilization ($+\Delta T_m$) at low binding, followed by destabilization at higher binding, while the *trans*-isomers again show a consistent increase in T_m with increasing binding. The increase in T_m at low binding of the *cis*-isomer again varies with the base content of DNA, being highest for the G—C rich polymer [115]. Cooling leads to renaturation for both isomers, indicative of cross links which thereby keep the two strands in

TABLE 1.IV
Effect of amines and platinum—amine complexes on melting temperature, (T_m), of calf-thymus DNA.

Pt Complexes	ΔT_m (°C)	r (Pt/base pair)	Ref.
trans-[$PtCl_2(NH_3)_2$]	2.8	0.05	119
trans-[$PtCl_2(NH_3)_2$]	3.9	0.10	120
cis-[$PtCl_2(NH_3)_2$]	1.1	0.05	119
cis-[$PtCl_2(NH_3)_2$]	−4.0	0.10	120
cis-[$PtCl_2(NH_3)_2$]	−6.1	0.25	119
[$PtCl_2$(en)]	1.1	0.05	119
[$PtCl_2$(en)]	−27.0	0.25	119
[$PtCl_2$(R,S-dach)]	−9.6	0.10	119
[$PtCl_2$(R,R-dach)]	−6.4	0.10	119
[$PtCl_2$(S,S-dach)]	−6.8	0.10	119
[$PtCl(NH_3)_3$]Cl	−2.2	0.10	119

Amines $H_2N(CH_2)_nNH_2$	ΔT_m (°C)	Conc. (M × 10^2)	Ref.
$n = 2$	3.5	0.1	121
$n = 2$	2.0	4.0	122
$n = 3$	4.2	4.0	122
$n = 5$	6.8	0.1	121
$n = 5$	5.3	4.0	121

close proximity for recombination. Slight differences in the geometric, but not optical, isomers of 1,2-dach complexes are observed [112]. For [Pt(dien)Cl]Cl, only a small increase in T_m is noted (ΔT_m = 2 °C for r_b = 0.05, salmon sperm DNA) [116].

An interesting feature here is that free amines themselves also cause stabilization of the DNA helix, this being dependent on chain length. These results are also included in Table 1.IV because this is one of the few examples where there are comprehensive and comparative biological data for the free ligand as compared to the complex. There is some discrepancy in these studies on optimal chain length for stabilization but the basic electrostatic nature of the interaction is clear, and involves the phosphate oxygens. The relevance of this contribution to overall platinum complex binding requires clarification. In this respect it is noteworthy that the D and L isomers of 1,2-diaminopropane show different effects on the T_m of the poly(I)—poly(C) helix, the less sterically hindered amine stabilizing the helix to a greater extent [123]. It would be of interest to see if the different isomers of 1,2-diaminocyclohexane (dach), as a free ligand, behaved similarly (See Chapter 3.4.1).

The luminescent properties of platinum-modified DNA have also been studied, the fluorescence being quenched and phosphorescence enhanced [111]. Raman difference spectroscopy has shown that some loss of B conformation occurs on binding [124]. Competition studies with polynucleotides, using this technique confirmed the kinetic preference for guanine binding [125]. Changes in the ^{31}P NMR of platinated DNA show distinctive shifts, different from those obtained with intercalators [126].

In general, although there are clear differences in binding of the *cis*- and *trans*-isomers of [$PtCl_2(NH_3)_2$], no major changes in terms of conformation are observed and may reflect the localized nature of the changes involved. The transformation of $\psi(-)$ to $\psi(+)$ DNA by *cis*-[$PtCl_2(NH_3)_2$] has been reported [11]. The complex [Pt(dien)Cl]Cl tends to behave similarly to the *trans*-isomer but it is noteworthy that the dien complex, which is only capable of monodentate binding, facilitates the B → Z transformation of DNA in the presence of ethanol [127, 128]. Ethanol is one of the reagents commonly used to induce this change and the concentration needed is abruptly reduced upon binding of the Pt complex, although the fact that it is still required must be emphasized. In fact, the dien complex also binds very rapidly to DNA.

1.3.3.2. *Platinum—Amine Complexes and Intercalators*

Further insight into the nature of DNA binding to platinum complexes can be found by concurrent binding studies with intercalators. Equilibrium dialysis experiments showed that [$PtCl_2$(en)] binding to *E. Coli* DNA, poly d(G—C) or poly d(A—T) inhibits subsequent binding of 9-aminoacridine [129]. The inhibition was most extensive (complete at Pt/P ratios of 0.15) for the G—C polymer, and an interesting difference was found for poly(dA)—poly(dT) where only very high binding ratios inhibited intercalation, in contrast to the ready inhibition with polyd(A—T). Similar inhibition was found for ethidium bromide [130, 131]. For the series of complexes in Figure 3.1, different binding ratios were needed for equivalent inhibition [131]. The $[Pt(NH_3)_4]^{2+}$ cation does not inhibit and the monodentate binding of [Pt(dien)Cl]Cl inhibits only slightly. In general, these results confirm covalent binding and sequence specificity for platinum complexes.

Interesting results are obtained when the order of binding is reversed and the intercalator is bound before Pt binding. It must be expected that, since intercalation opens up the helix, the base sites would be more accessible to platination. This is confirmed and the kinetics of binding are unchanged in the presence of intercalator but the platinum is more easily removed [130]. The area of binding is, however, altered. Studies with plasmid pBR322 DNA showed that the shortening of the duplex induced by *cis*-$[PtCl_2(NH_3)_2]$ is not as effective in the presence of bound intercalator, although the unwinding is unaffected. The stiffening of the helix upon intercalation and the separation of the bases probably limit the binding modes of the platinum complex. The change in binding sequence site was also demonstrated by studies utilizing exonuclease III studies on a 165-base pair excision fragment of pBR322 DNA [132]. The results are, very briefly, that increased binding to an oligonucleotide guanine sequence, a d(G_6—C—G) site, is observed in the presence of intercalator in preference to smaller sequences bound in absence of intercalator. The switching of the nuclease-sensitive site implies sensitivity to local structure of the DNA, since the lack of binding to such a G-rich sequence is quite surprising. Once the helix is opened up, or the conformation altered by the intercalator, the site is platinated, as expected. Note the similarity to the above results on the Pt complex tethered to acridine orange. These results may offer a rationale for synergy of drugs, but more 'macroscopic' pharmacokinetic parameters will perhaps dictate this more than DNA binding. Once again, these results raise the possibility of enhancing the DNA affinity of hitherto unreactive platinum complexes and this may also be useful for study of the role of the amine ligand both in DNA binding and antitumour activity.

A further recent intriguing aspect is the possibility of ternary complex formation between Pt—DNA complexes and intercalators upon addition of intercalator to the Pt—DNA, followed by basification [165, 166].

1.3.3.3. *Restriction Enzymes and Platinum Complexes*

The use of closed, circular DNA and the potential of the restriction enzymes as probes are very well demonstrated by studies on Pt complexes, and have been exploited to give detailed information on the binding sites of these complexes, confirming the importance of binding to G-rich regions. Covalent binding of both *cis*- and *trans*-$[PtCl_2(NH_3)_2]$ results in alteration of the degree of coiling and electrophoretic mobility, indicating shortening of the DNA [133, 134], and this was confirmed by electron microscopy [133—135].

An early study on lambda bacteriophage DNA and Bam H1 restriction enzyme showed that platinum binding inhibited cleavage at the recognition sequence [136]. This was confirmed on pBR322 DNA [137]. The inhibition occurs at low Pt binding and is reversed by cyanide treatment, which removes platinum [111, 112, 137]. However, at high binding ratios there is some platinum that is not removable by cyanide treatment. Assuming random binding, and allowing that the cutting site is G—G, it has been calculated that a platinum moiety must be within 3 base pairs of the restriction site. Since the cutting site in this case is also expected to be a site for preferential binding to platinum, the physical presence of the complex may be sufficient for inhibition, independent of the mode of binding, and this is indicated by the fact that the *trans*-isomer, and in fact to some extent [Pt(dien)Cl]Cl, also inhibit the enzyme; the *trans*-isomer inhibits to about the same extent as the *cis*-isomer (at equal concentrations) [138].

The importance of sequence was shown by results on pSM1 and SV40 DNA [138, 139]. A survey of SV40 DNA with various restriction endonucleases gave an order of preference of Bgl 1 $\gg$ Bam H1 > Hpa II >, Kpn I > Eco R1. Here the platinated DNA was 'hypersensitive' to Bgl 1, which also has (dG) sequences on either side of the cleavage palindrome (completely GGCCG/AGG/CGGCC), in contrast to the other enzymes. Thus a clear importance is found, not just for cleavage site, but for sequence in the vicinity of the site. Other evidence indicating the differences is that more single stranded DNA is produced by cisplatin. Using the fact that the Pst 1 restriction enzyme (cutting sequence CTGCA/G, see above) has four cleavage sites on pSM 1 DNA, Lippard and co-workers [139] showed that one of these, the D—B junction which is bordered by a G-rich region, in this case a $(dG)_4$ sequence, was selectively inhibited and thus, at low binding ratios, more platinum will be concentrated there. The *trans*-isomer, in this case, does not produce this effect at similar binding levels.

By use of the exonuclease III, binding of platinum at guanine was also confirmed because stop sites caused by Pt binding were in guanine-rich areas [132]. A further experiment, which also demonstrates the use of excision fragments, showed that oligo(dG) sites were preferential stopping sites [141].

These accumulated results on Pt complexes demonstrate the uses of the techniques outlined in the first sections of this chapter in the study of M-DNA binding.

1.3.4. STRAND BREAKAGE

Strand breakage involves covalent bond cleavage of the helix backbone and may be either singly stranded or doubly stranded. In principle, either cleavage of the phosphate ester—sugar bond or a C—C break in the sugar skeleton will induce strand breaks [142]. The breakage may be monitored by changes in physical properties as outlined earlier. Radiation products, especially the hydroxyl radical, are particularly effective in this respect and much of the work on strand breaks centers on this aspect (See Chapter 8). The production of strand breaks is implicated not only in radiation damage but also in the mechanism of action of bleomycin, which requires Fe as cofactor (Chapter 7). Chemical induction of strand breaks by metal complexes can be initiated by electron transfer reactions centered on the metal ion and producing reactive oxygen radicals. Indeed, one of the earliest reports of DNA degradation by Fenton's reagent was in the early 1950s [143]. A number of metal systems which act in this way are now well defined. Besides the Cu and Fe-based systems discussed below the production of strand breaks by Ru(III) in presence of reducing agent (Chapter 6.1.3.1) and the light-induced cleavage by Co(III)-bleomycin (Chapter 7.3.2) are relevant.

1.3.4.1. *Copper—(1,10-phenanthroline) Complexes*

Perhaps the simplest coordination system known to produce DNA cleavage is the cuprous-1,10-phenanthroline system which was shown to cleave DNA in an oxygen dependent reaction [144]. The co-requirement for activity is a reducing agent, such as a thiol, and molecular oxygen. Subsequent work showed that this cleavage is remarkably specific for the 1,10-phenanthroline system, other Cu chelators such as neocupreine and 2,2′-bipyridine being ineffective [145—147]. The neocupreine (2,9-dimethyl-1,10-phenanthroline) complex is more stable in the Cu(I) state relative to Cu(II) [148], and actually inhibits the reaction.

Mechanistic work showed that catalase and some hydroxyl radical scavengers inhibit the reaction [145, 146]. Conversely a free radical generating system such as the xanthine oxidase—hypoxanthine system can substitute for the reducing agent and hydrogen peroxide can substitute for oxygen. The function of O_2, which serves as a precursor for H_2O_2, and the presence of the cuprous oxidation state, strongly implicate hydroxyl radical production as a mechanistic pathway. The further necessity for double helical DNA and the fact that intercalators such as ethidium bromide inhibit the reaction indicate that the Cu—phen complex is also bound by intercalation to the helix. Thus, radicals produced in the vicinity should be particularly damaging. The kinetics of oxidation of the Cu(I) state by both O_2 and H_2O_2 have been examined [149], and the results correlated with the kinetic data for the DNA degradation.

This system has also been shown to be dependent on the secondary structure of DNA, the A, B, and Z forms reacting at different rates [150]. The likely explanation is that the faster reacting B DNA forms a more stable complex with the catalyst. This artificial DNase activity has also been compared with cleavage by micrococcal nuclease, and shown to recognize the same sites but not all those cleaved by DNase 1, again implying some local conformational preferences [151]. Chromatin structure has also been probed [152].

1.3.4.2. *Cobalt—(1,10-phenanthroline) Complexes*

A further demonstration of metal catalyzed redox chemistry on DNA is that of the light induced cleavage of DNA by Co(III)—phenanthroline complexes [153]. The likelihood is that reduction of Co(III) to Co(II) initiates the oxygen activation leading to cleavage.

1.3.4.3. *Iron—EDTA—(DNA-binder) Complexes*

A number of molecules capable of cleaving DNA by metal catalyzed oxygen reduction have been prepared by Dervan and co-workers [154] by tethering an Fe(III)—EDTA complex to specific binders such as intercalators and small antibiotics. The structure of the parent complex is shown in Figure 1.10, and the other complexes are shown schematically.

One of the first systems described was Complex (10) which contains the intercalator methidium (MPE—Fe). Indeed, this complex and the tethered complex of Pt to acridine orange are formally analogous as they contain DNA—binder (intercalator) linked to a metal function (Pt—purine binding; Fe—radical production). In the presence of micromolar concentrations of Fe, dithiothreitol (DTT) as reducing agent, and in the presence

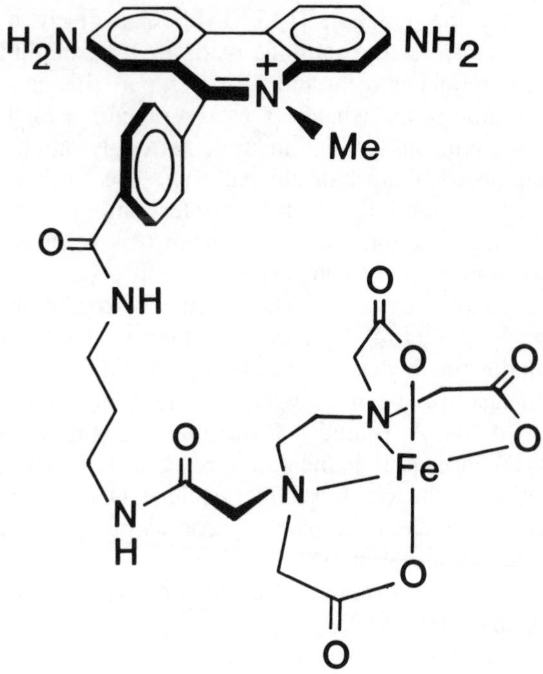

Fig. 1.10. Structure of an iron—EDTA complex linked to an intercalator. From Reference 154.

of oxygen, highly efficient cleavage of pBR322 plasmid DNA was observed [155]. The complex is more efficient by two orders of magnitude than the simple Fe—EDTA complex and approaches the efficiency of bleomycin. Both SOD and catalase inhibit the cleavage, implying that activation of oxygen produces local damage.

In contrast with bleomycin this cleavage is non-sequence-specific and this has been utilized (see below). In attempts to make this reaction more sequence-specific, the Fe(III)—EDTA molecule was tethered to distamycin, which is known to bind to double helical DNA with a strong affinity for AT-rich regions, Complex (11), Figure 1.11 [156]. In this case, use of the same chemical system (complex, DTT as reducing agent, and O_2), results in cleavage next to a four-base-pair A + T sequence, although the efficiency was not as high [156]. Study with restriction fragments from pBR322 also showed sequence-specific cleavage. The essential feature of the distamycin tether is the presence of three *N*-methylpyrrole carboxamides, and this chemical feature has been exploited to lengthen the

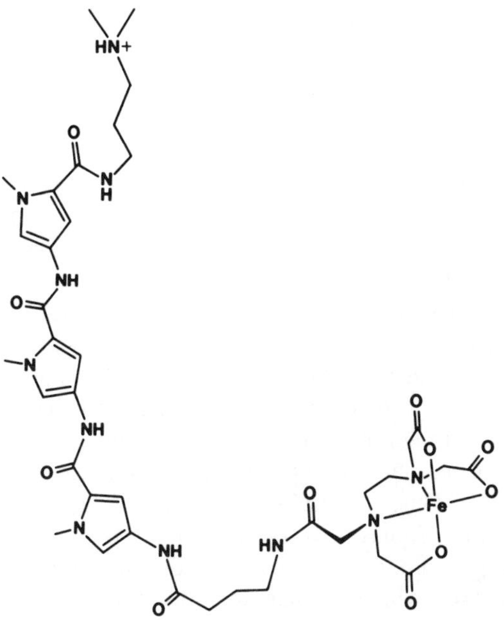

Fig. 1.11. Structure of an iron—EDTA complex linked to distamycin—[distamycin—Fe(EDTA)]. The Fe moiety may also be attached to the amine terminus. From Reference 154.

recognition chain by making a penta-*N*-methylpyrrole carboxamide linkage, which also cleaves 3—5 base pairs adjacent to a six-base pair AT site [157]. Extension to the bis(Fe—EDTA—distamycin) allows cleavage at an eight-base pair sequence [158]. Extensions to even larger spacers, including intercalating ligands, have been reviewed [154].

These innovative molecules can be of great use as probes of DNA structure and as artificial or synthetic 'restriction endonucleases' by cutting at very specific sites. The cleavage of chromatin and the mapping of antibiotic sites are just some of the uses explored for these complexes [154, 159—161]. Thus, the preferential binding of the antibiotic netropsin to restriction fragments was shown to be at AT-rich regions.

The antitumour activity of antibiotics, which may act by strand breakage through radical production from a metal-centered redox reaction, will be summarized in Chapter 7. The examples summarized here show the feasibility of design and use of a DNA-binding ligand (ranging from very simple to a molecular weight of 600) to deliver a metal to DNA, where production of damaging radicals can be brought about by metal-centered redox chemistry.

1.4. Summary

Metal complexes interact with DNA in a number of distinct manners and the binding causes distinct conformational changes on DNA, dependent on the exact mode of binding. Intercalation can be achieved by complexes with large planar ligands and representative series are those of 1,10-phenanthroline and metalloporphyrin complexes. Outer sphere binding is dominated by electrostatic interactions between the ligands of the complex and suitable groups such as phosphate on the DNA backbone. Exocyclic groups on the purine and pyrimidine bases can also be involved through hydrogen bonding to suitable ligand atoms. Inner sphere binding implies covalent bond formation of the metal complex to either phosphate or the nucleic acid bases. Where more than one metal—DNA bond is formed cross links result, which may be *inter*- or *intra*-strand.

Strand breakage, either single strand or double strand, may be effected by metal complexes which are capable of undergoing redox reactions. Reduction of the metal complex, subsequent reaction of the reduced species to produce active oxygen moieties has been demonstrated to produce strand breaks. Photolytic reduction of the complex can also result in strand cleavage.

The role of DNA in the transmission of genetic information makes it a primary target of drug action. The many metal binding modes on DNA vary widely in their ability to affect DNA synthesis but the variety of binding modes presents considerable scope for design of more efficient agents.

References

1. A. Albert: *Selective Toxicity*, Chapman and Hall, London, 6th Ed. (1985).
2. E. F. Gale, E. Cunliffe, P. E. Reynolds, M. H. Richmond, and M. J. Waring: *The Molecular Basis of Antibiotic Action*, Wiley, New York, 2nd Ed. (1981).
3. M. J. Cline and C. M. Haskell: *Cancer Chemotherapy*, W. B. Saunders, Philadelphia, 3rd Ed. (1980).
4. M. J. Waring: *Ann. Rev. Biochem.* **50**, 159 (1981).
5. M. J. Clarke: *Inorg. Chem.* **19**, 1103 (1980).
6. D. R. Williams: *Chem. Rev.* **72**, 203 (1972).
7. *Carcinogenicity and Metal Ions* (Metal Ions in Biological Systems v. 10, Ed. H. Sigel), Marcel Dekker, New York (1979).
8. *Metal Complexes as Anticancer Agents* (Metal Ions in Biological Systems v. 11, Ed. H. Sigel), Marcel Dekker, New York (1980).

9. *Metal Ions in Genetic Information Transfer* (Adv. Inorg. Biochem. v. 3, Eds. G. L. Eichhorn and L. G. Marzilli), Elsevier, Amsterdam (1981).
10. *Nucleic Acid—Metal Ion Interactions*, Ed. T. G. Spiro, Wiley (1980).
11. G. L. Eichhorn in Reference 7, p. 1.
12. J. K. Barton and S. J. Lippard: *Heavy Metal Interactions with Nucleic Acids* (Reference 8, pp. 31—113).
13. L. G. Marzilli, T. J. Kistenmacher, and G. L. Eichhorn: *Structural Principles of Metal Ion—Nucleotide and Metal Ion—Nucleic Acid Interactions* (Reference 8, pp. 179—250).
14. J. K. Barton: *Comments Inorg. Chem.* **3**, 321 (1985).
15. *Nucleotides and Derivatives: Their Ligating Ambivalency* (Metal Ions in Biological Systems v. 8, Ed. H. Sigel), Marcel Dekker, New York (1979).
16. *Inorganic Drugs in Deficiency and Disease* (Metal Ions in Biological Systems v. 14, Ed. H. Sigel), Marcel Dekker, New York (1982).
17. S. E. Bryan: in Reference 7, p. 87.
18. A. S. Mildvan and L. A. Loeb: *ibid.* p. 103.
19. L. A. Loeb and A. S. Mildvan: *ibid.* p. 126.
20. G. L. Eichhorn: *Complexes of Polynucleotides and Nucleic Acids* (Inorganic Biochemistry v. 2, Ed. G. L. Eichhorn), pp. 1210—1244. Elsevier (1973).
21. L. A. Loeb and R. A. Zakour: *Metals and Genetics Miscoding* (Reference 8, pp. 115—144).
22. W. W. Cleland and A. S. Mildvan: *Adv. Inorg. Biochem.* **1**, 163 (1979).
23. F. S. Richardson: *Chem. Rev.* **82**, 541 (1982).
24. T. R. Jack: *Heavy Metal Labelling of Nucleotides and Polynucleotides for Electron Microscopy Studies* (Metal Ions in Biol. Systems v. 8), pp. 159—182 Marcel-Dekker (1979).
25. M. M. Teeter, G. J. Quigley, and A. Rich: *Metal Ions and Transfer RNA* (Reference 8, pp. 145—177).
26. R. F. Whiting and F. R. Ottensmeyer: *Biochim. Biophys. Acta* **474**, 334 (1977).
27. U.S. Nandi, J. C. Wang, and N. Davidson: *Biochemistry* **4**, 1687 (1965).
28. V. A. Bloomfield, D. M. Crothers, and I. Tinoco, Jr.: *Physical Chemistry of Nucleic Acids* (Harper and Row), pp. 104—150 (1974).
29. A. H.-J. Wang, G. J. Quigley, F. J. Kolpak, J. L. Crawford, J. H. van Boom, G. van der Marel, and A. Rich: *Nature* (*London*) **282**, 680 (1979).
30. H. R. Drew, R. E. Dickerson, and K. Itakura: *J. Mol. Biol.* **125**, 535 (1978).
31. F. M. Pohl and T. M. Jovin: *J. Mol. Biol.* **67**, 375 (1972).
32. R. E. Dickerson: *Scientific American* **94**, 249 (1983).
33. S. Arnott, R. Chandrasekaran, D. W. L. Hukins, P. J. C. Smith, and L. Watts: *J. Mol. Biol.* **88**, 523 (1974).
34. C. F. Jordan, L. S. Lerman, and J. H. Venable Jr.: *Nature New Biol.* **236**, 67 (1972).
35. H. Busch: *The Cell Nucleus CHROMATIN, Part B* (v. 5, Ed. H. Busch), p. xxi. Academic Press (1978).
36. J. Vinograd, J. Lebowitz, and R. Watson: *J. Mol. Biol.* **33**, 173 (1968).
37. D. Glaubiger and J. E. Hearst: *Biopolymers* **5**, 691 (1967).
38. J. D. Watson, J. Tooze, and D. T. Kurtz: *Recombinant DNA: A Short Course*, Scientific American.
39. A. Kornberg: *DNA Replication*, W. H. Freeman and Co. (1980).
40. A. M. Maxam and W. Gilbert: *Proc. Natl. Acad. Sci. USA* **74**, 560 (1977).
41. F. Sanger and A. R. Coulson: *J. Mol. Biol.* **94**, 444 (1975).
42. Methods Enzymol. v. 65 (Eds. L. Grossman and K. Moldave).

43. S. G. Rogers and B. Weiss: *Methods Enzymol.* **65**, 201 (1980).
44. D. J. Patel: *Acc. Chem. Res.* **12**, 118 (1979).
45. G. Govil and R. V. Hosur: *Conformation of Biological Molecules. New results from NMR* (NMR Basic Principles and Progress v. 20, Eds. P. Diehl, E. Fluck, and R. Kosfeld), Springer-Verlag (1982).
46. J. Feigon, W. A. Denny, W. Leupin, and D. R. Kearns: *J. Med. Chem.* **27**, 450 (1984).
47. R. L. Jones and W. D. Wilson: *J. Am. Chem. Soc.* **102**, 7776 (1980).
48. L. S. Lerman: *J. Mol. Biol.* **3**, 18 (1961).
49. L. S. Lerman: *J. Cell. Comp. Physiol.* **64**, 1 (1964).
50. E. J. Gabbay, R. Scofield, and C. S. Baxter: *J. Am. Chem. Soc.* **95**, 7850 (1973).
51. W. R. Bauer: *Ann. Rev. Biophys. Bioeng.* **7**, 287 (1978).
52. W. D. Wilson and R. L. Jones: *Adv. Pharmacol. and Chemother.* **18**, 177 (1979).
53. K. W. Jennette, S. J. Lippard, G. A. Vassiliades, and W. R. Bauer: *Proc. Natl. Acad. Sci. USA* **71**, 3839 (1974).
54. M. Howe-Grant, K. Wu, W. R. Bauer, and S. J. Lippard: *Biochemistry* **15**, 4339 (1976).
55. S. J. Lippard, P. J. Bond, K. C. Wu, and W. R. Bauer: *Science* **194**, 726 (1976).
56. A. H.-J. Wang, J. Nathans, G. van der Marel, J. H. van Boom, and A. Rich: *Nature (London)* **276**, 471 (1978).
57. H. M. Sobell, B. S. Reddy, K. K. Bhandry, S. C. Jain, T. D. Sakore, and T. P. Seshadri: *Cold Spring Harbor Symposia on Quantitative Biology*, p. 87 Cold Spring Harbor Lab. New York (1977).
58. P. J. Bond, R. Langridge, K. W. Jennette, and S. J. Lippard: *Proc. Natl. Acad. Sci. USA* **72**, 4825 (1975).
59. Y.-S. Wong and S. J. Lippard: *J. Chem. Soc. Chem. Comm.* 824 (1977).
60. D. Prusiner and A. L. Sundaralingam: *Acta Cryst.* **B32**, 161 (1976).
61. G. C. Allen and N. S. Hush: *Prog. Inorg. Chem.* **8**, 357 (1900).
62. A. Albert: *The Acridines, Their Preparation, Properties, and Uses*, 2nd Ed. Arnold (London) (1966).
63. J. K. Barton: *Science* **233**, 727 (1986).
64. A. Shulman, G. M. Laycock, and T. R. Bradley: *Chem.-Biol. Inter.* **16**, 89 (1977) and references therein.
65. J. K. Barton, J. J. Dannenberg, and A. L. Raphael: *J. Am. Chem. Soc.* **104**, 4967 (1982).
66. J. K. Barton, A. T. Danishefsky, and J. M. Goldberg: *J. Am. Chem. Soc.* **106**, 2172 (1984).
67. C. V. Kumar, J. K. Barton, and N. J. Turro: *J. Am. Chem. Soc.* **107**, 5518 (1985).
68. J. M. Kelly, A. B. Tossi, D. J. McConnell, and C. OhUigin: *Nuc. Acids Res.* **13**, 6017 (1985).
69. J. K. Barton: *J. Biomol. Struct. Dyn.* **1**, 621 (1983).
70. J. K. Barton, L. A. Basile, A. Danishefsky, and A. Alexandrescu: *Proc. Natl. Acad. Sci. USA* **81**, 1961 (1984).
71. E. J. Gabbay, R. E. Scofield, and C. S. Baxter: *J. Am. Chem. Soc.* **95**, 7850 (1973).
72. H.-Y. Mei and J. K. Barton: *J. Am. Chem. Soc.* **108**, 7414 (1986).
73. J. K. Barton and S. R. Paranawithana: *Biochemistry* **25**, 2205 (1986).
74. J. K. Barton and E. Lolis: *J. Am. Chem. Soc.* **107**, 708 (1985).
75. R. J. Fiel and B. R. Munson: *Nuc. Acid Res.* **8**, 2835 (1980).
76. R. F. Pasternack, E. J. Gibbs, and J. J. Villafranca: *Biochemistry* **22**, 2406 (1983).

77. R. F. Pasternack, E. J. Gibbs, and J. J. Villafranca: *Biochemistry* **22**, 5409 (1983).
78. R. F. Pasternack, E. J. Gibbs, A. Gaudemer, A. Antebi, S. Bassner, L. de Poy, D. H. Turner, A. Williams, F. Laplace, M. H. Lansard, C. Merienne, and M. Perree-Fauvet: *J. Am. Chem. Soc.* **107**, 8179 (1985).
79. B. E. Bowler, L. S. Hollis, and S. J. Lippard: *J. Am. Chem. Soc.* **106**, 6102 (1984).
80. B. E. Bowler and S. J. Lippard: *Biochemistry* **25**, 3031 (1986).
81. N. Farrell, M. P. Hacker, and J. J. McCormack: *Proc. Am. Assn. Cancer Res.* **27**, 1143 (1986).
82. J. K. Barton, C. V. Kumar, and N. J. Turro: *J. Am. Chem. Soc.* **108**, 6391 (1986).
83. G. L. Eichhorn and Y. A. Shin: *J. Am. Chem. Soc.* **90**, 7323 (1968).
84. C. K. S. Pillai and U. S. Nandi: *Biopolymers* **12**, 1431 (1973).
85. C. K. Mirabelli, C.-M. Sung, J. P. Zimmerman, D. T. Hill, S. Mong, and S. T. Crooke: *Biochem. Pharmacol.* **35**, 1427 (1986).
86. S. Hanlon, A. Chan, and S. Berman: *Biochim. Biophys. Acta* **519**, 526 (1978).
87. F. Ascoli, M. Branca, C. Mancini, and B. Pispisa: *J. Chem. Soc.* (*Faraday Trans*) **168**, 1213 (1972).
88. F. Ascoli, M. Branca, C. Mancini, and B. Pispisa: *Biopolymers* **12**, 2431 (1973).
89. R. L. Karpel, N. S. Miller, A. M. Lesk, and J. R. Fresco: *J. Mol. Biol.* **97**, 519 (1975).
90. J. Widom and R. L. Baldwin: *J. Mol. Biol.* **144**, 431 (1980).
91. R. L. Karpel, A. H. Bertelson, and J. R. Fresco: *Biochemistry* **19**, 504 (1980).
92. F. Basolo and R. H. Pearson: *Reaction Mechanisms in Inorganic Chemistry*, 2nd ed., Wiley, New York (1967). C. H. Langford and V. S. Sastri: 'Mechanism and Steric Course of Octahedral Substitution' in Reaction Mechanisms in Inorganic Chemistry (Ed. M. L. Tobe v. 9) pp. 203–268 (See p. 223) Butterworths, London (1972).
93. M. J. Clarke, B. Jensen, K. A. Marx, and R. Kruger: *Inorg. Chim. Acta* **124**, 13 (1986).
94. G. S. Manning: *Acc. Chem. Res.* **12**, 443 (1979).
95. M. Behe and G. Felsenfeld: *Proc. Natl. Acad. Sci. USA* **78**, 1619 (1981).
96. H. H. Chen, E. Charney, and D. C. Rau: *Nuc. Acid Res.* **10**, 3561 (1982).
97. B. E. Hingerty, R. S. Brown, and A. Klug: *Biochim. Biophys. Acta* **697**, 78 (1982).
98. A. Jack, J. E. Ladner, D. Rhodes, R. S. Brown, and A. Klug: *J. Mol. Biol.* **111**, 315 (1977).
99. M. W. Lieberman, D. J. Harvan, D. E. Amacher, and J. B. Patterson: *Biochim. Biophys. Acta* **425**, 265 (1976).
100. L. D. Kosturko, C. Folzer, and R. F. Stewart: *Biochemistry* **13**, 3949 (1974).
101. R. Jensen and N. Davidson: *Biopolymers* **4**, 17 (1966).
102. T. Yamane and N. Davidson: *Biochim. Biophys. Acta* **55**, 609 (1962).
103. N. Davidson, J. Widholm, U. S. Nandi, R. Jensen, B. M. Oliveira, and J. C. Wang: *Proc. Natl. Acad. U.S.A.* **53**, 111 (1965).
104. L. G. Marzilli, T. J. Kistenmacher, and M. Rossi: *J. Am. Chem. Soc.* **99**, 2797 (1977).
105. C. Zimmer, G. Luck, H. Fritzsche, and H. Triebel: *Biopolymers* **10**, 441 (1971).
106. Y. A. Shin and G. L. Eichhorn: *Biochemistry* **7**, 1026 (1968).
107. B. P. Ulanov, L. F. Malaysheva, and Yu. Sh. Moshkovskii: *Biofizika* **12**, 326 (1967).
108. A. T. M.. Marcelis and J. Reedijk: *Recl. Trav. Chim. Pays-Bas* **102**, 121 (1983).
109. A. L. Pinto and S. J. Lippard: *Biochim. Biophys. Acta* **780**, 167 (1985).
110. P. Horacek and J. Drobnik: *Biochim. Biophys. Acta* **254**, 341 (1971).
111. L. L. Munchhausen and R. O. Rahn: *Biochim. Biophys. Acta* **414**, 242 (1975).

112. P. J. Stone, A. D. Kelman, and F. M. Sinex: *Nature* **251**, 736 (1974).
113. P. J. Stone, A. D. Kelman, and F. M. Sinex: *J. Mol. Biol.* **104**, 793 (1976).
114. K. V. Shooter and R. K. Merrifield: *Biochim. Biophys. Acta* **287**, 16.
115. R. C. Srivastava, J. Froehlich, and G. L. Eichhorn: *Biochimie* **60**, 879 (1978).
116. J.-P. Macquet and J.-L. Butour: *Eur. J. Biochem.* **83**, 375 (1978).
117. A. M. Tamburro, L. Celotti, D. Furlan, and V. Guantieri: *Chem.-Biol. Inter.* **16**, 1 (1977).
118. L. G. Marzilli and P. Chalipoyl: *J. Amer. Chem. Soc.* **102**, 873 (1980).
119. H. C. Harder: *Chem.-Biol. Interact.* **10**, 27 (1975).
120. K. Inyaki and Y. Kidani: *Inorg. Chim. Acta* **46**, 35 (1980).
121. H. R.Mahler and B. D. Mehrotra: *Biochim. Biophys. Acta* **68**, 211 (1963).
122. E. J. Gabbay: *Biochemistry* **5**, 3036 (1966).
123. E. J. Gabbay: *J. Am. Chem. Soc.* **90**, 5257 (1968).
124. G. Y. H. Chu, S. Mansy, R. E. Duncan, and R. S. Tobias: *J. Am. Chem. Soc.* **100**, 593 (1978).
125. S. Mansy, G. Y. H. Chu, R. E. Duncan, and R. S. Tobias: *J. Am. Chem. Soc.* **100**, 607 (1978).
126. W. D. Wilson, B. L. Heyl, R. Reddy, and L. G. Marzilli: *Inorg. Chem.* **21**, 2527 (1982).
127. B. Malfoy, B. Hartmann, and M. Leng: *Nuc. Acids Res.* **9**, 5659 (1981).
128. H. M. Ushay, R. M. Santella, J. P. Caradonna, D. Grunberger, and S. J. Lippard: *Nuc. Acids Res.* **10**, 3573 (1982).
129. I. A. G. Roos and M. C. Arnold: *J. Clin. Hematol. Oncol.* **7**(1) 374 (1978).
130. M. E. Howe-Grant, K. C. Wu, W. R. Bauer, and S. J. Lippard: *Biochemistry* **15**, 4339 (1976).
131. J.-L. Butour and J.-P. Macquet: *Eur. J. Biochem.* **78**, 455 (1977).
132. T. D. Tullius and S. J. Lippard: *Proc. Natl. Acad. Sci. U.S.A.* **79**, 3489 (1982).
133. G. L. Cohen, W. R. Bauer, J. K. Barton, and S. J. Lippard: *Science* **203**, 1014 (1979).
134. S. Mong, Y. Daskal, A. W. Prestayko, and S. T. Crooke: *Cancer Res.* **41**, 4020 (1981).
135. J.-P. Macquet and J.-L. Butour: *Biochimie* **60**, 901 (1978).
136. A. D. Kelman and M. Buchbinder: *Biochimie* **60**, 893 (1978).
137. H. M. Ushay, T. D. Tullius, and S. J. Lippard: *Biochemistry* **20**, 3744 (1981).
138. W. M. Scovell, L. R. Kroos, and V. J. Capponi: *Model for the Interaction of cis-Diamminedichloroplatinum(II) with Simian Virus 40 DNA* (Platinum, Gold and Other Metal Chemotherapeutic Agents, ACS Symposium Series 209, Ed. S. J. Lippard), pp. 101—121. A.C.S. (1983).
139. G. L. Cohen, J. A. Ledner, W. R. Bauer, H. M. Ushay, C. Caravana, and S. J. Lippard: *J. Am. Chem. Soc.* **102**, 2488 (1980).
140. B. Royer-Pokora, L. K. Gordon, and W. A. Haseltine: *Nuc. Acids Res.* **9**, 4595 (1981).
141. T. D. Tullius and S.J. Lippard: *J. Am. Chem. Soc.* **103**, 4620 (1981).
142. J. Hutterman, W. Kohnlein, R. Teoule, and A. J. Bertinchamps: *Effects of Ionizing Radiation on DNA*, pp. 204—251. Springer-Verlag (1978).
143. G. Scholes and J. Weiss: *Nature* **53**, 567 (1953).
144. D. S. Sigman, D. R. Graham, V. D'Aurora, and A. M. Stern: *J. Biol. Chem.* **254**, 12269 (1979).
145. D. R. Graham, L. E. Marshall, K. A. Reich, and D. S. Sigman: *J. Amer. Chem. Soc.* **102**, 5421 (1980).

146. B. G. Que, K. M. Downey, and A. G. So: *Biochemistry* **19**, 5987 (1980).
147. L. E. Marshall, D. R. Graham, K. A. Reich, and D. S. Sigman: *Biochemistry* **20**, 244 (1981).
148. B. R. James and R. J. P. Williams: *J. Chem. Soc.*, 2007 (1961).
149. S. Goldstein and G. Czapski: *Inorg. Chem.* **24**, 1087 (1985).
150. L. E. Pope and D. S. Sigman: *Proc. Natl. Acad. Sci. U.S.A.* **81**, 3 (1984).
151. B. Jessee, G. Cargiulo, F. Razvi, and A. Worcel: *Nuc. Acids Res.* **10**, 6873 (1982).
152. I. L. Cartwright and S. C. R. Elgin: *Nuc. Acids Res.* **10**, 5835 (1982).
153. J. K. Barton and A. L. Raphael: *J. Am. Chem. Soc.* **106**, 2466 (1984).
154. P. B. Dervan: *Science* **232**, 464 (1986).
155. R. P. Hertzberg and P. B. Dervan: *J. Am. Chem. Soc.* **104**, 313 (1982).
156. P. G. Schulz, J. S. Taylor, and P. B. Dervan: *J. Am. Chem. Soc.* **104**, 7863 (1982).
157. P. G. Schulz and P.B. Dervan: *Proc. Natl. Acad. Sci. U.S.A.* **80**, 6834 (1983).
158. P. G. Schulz and P. B. Dervan: *J. Am. Chem. Soc.* **105**, 7748 (1983).
159. M. W. Van Dyke and P. B. Dervan: *Nuc. Acids Res.* **11**, 5555 (1983).
160. M. M. Van Dyke and P. B. Dervan: *Biochemistry* **22**, 2373 (1983).
161. M. W. Van Dyke, R. P. Herzberg, and P. B. Dervan: *Proc. Natl. Acad. Sci.* **79**, 5470 (1982).
162. J. A. Strickland, D. L. Banville, W. D. Wilson, and L. G. Marzilli: *Inorg. Chem.* **26**, 3398 (1987).
163. D. L. Banville, L. G. Marzilli, J. A. Strickland, and W. D. Wilson: *Biopolymers* **25**, 1837 (1986).
164. L. G. Marzilli, D. L. Banville, G. Zon, and W. D. Wilson: *J. Am. Chem. Soc.* **108**, 4188 (1986).
165. J.-M. Malinge, A. Schwartz, and M. Leng: *Nuc. Acids Res.* **15**, 1779 (1987).
166. J.-M. Malinge and M. Leng: *Proc. Natl. Acad. Sci. U.S.A.* **83**, 6317 (1986).

CHAPTER 2

PLATINUM–AMINE COMPLEXES AS ANTICANCER AGENTS

The discovery of the antitumour activity of cisplatin, *cis*-$[PtCl_2(NH_3)_2]$, and its subsequent clinical development are now well documented [1, 2]. The present clinical utility of this complex and its analogues make them of great interest both from the chemical and pharmacological viewpoint.

2.1. The Discovery of Cisplatin as an Antitumour Agent

In 1965, Rosenberg and co-workers published their observations on the induction of filamentous growth in bacterial cells by platinum—amine complexes [3]. The production of filaments, long sausage-like rods, is caused by inhibition of cell division but not cell growth. This effect was observed upon application of an electric field but was shown to be due not to the electric current but rather to the electrolysis products of the platinum electrodes used in the experiment. Subsequent work identified the compounds shown in Figure 2.1, and it was noted that only the *cis*-forms were active, the *trans*-isomers failing to produce the effect. Neutral species were also found to be necessary to induce filamentation; corresponding charged species are strongly bacteriocidal [4].

A further bacterial effect of the neutral species observed shortly afterwards was that of induction of lysis in lysogenic bacteria [5]. Certain bacterial viruses (bacteriophages), upon invasion of a host, can develop in one of two ways: they can multiply and lyse the infected cell (lytic pathway) or their DNA can be integrated with that of the infected cell (lysogenic pathway). In the latter case the viral DNA, which retains the ability to multiply and lyse, is called a prophage, and the host cell is a lysogenic bacterium. A variety of agents that interfere with DNA replication in the host, including chemicals, X-rays and UV light, may induce the prophage to undergo lytic development. The platinum complexes were particularly effective in this respect, producing observable effects at less than 0.1 ppm.

Concurrent with this latter work, Rosenberg then made the intuitive step of testing the complexes on a tumour system, with the argument that

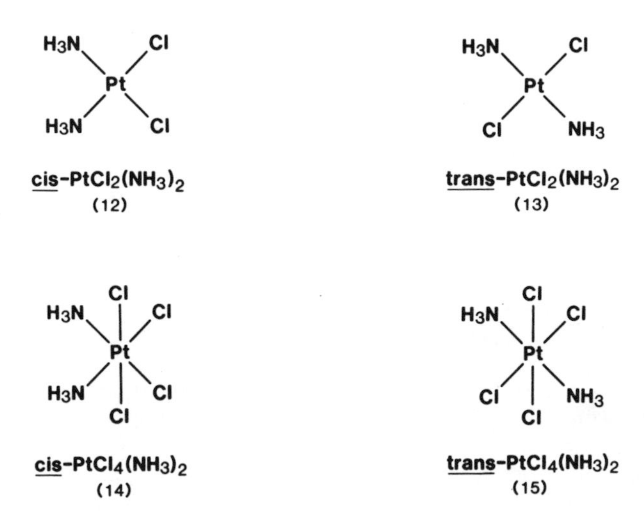

Fig. 2.1. Structures of original platinum complexes studied for antibacterial and antitumour activity.

the complexes might also inhibit cell division in rapidly growing tumour cells. Malignant cancers are of two major types: (a) solid tumours and (b) disseminated tumours, as exemplified by the leukemias and lymphomas of the blood and lymphatic systems. Activity was found in animal tumour systems representative of both these types: solid Sarcoma 180 and the leukemia L1210. The stereospecificity was again confirmed [6] and the *cis*-isomer is active against a wide variety of animal tumour systems, summarized in Table 2.I.

From these results the following conclusions may be made about the activity of cisplatin:

1. Exhibits a wide spectrum of antitumour activity against drug-resistant as well as drug-sensitive tumours.
2. Shows activity against slow-growing as well as rapidly-growing tumours.
3. Shows no strain or species specificity.
4. Exhibits activity against viral-induced, chemical-induced, and transplantable tumours.
5. Both solid and disseminated tumours are affected.

In relation to other established and clinically-used drugs, such as adriamycin and 5-fluorouracil, cisplatin shows at least equivalent activity in the NCI tumour panel [7], Table 2.II, and indeed this favorable comparison stresses the potential for metal-based drugs.

TABLE 2.I

Antitumour activity of cisplatin in representative animal systems.[a]

Tumour	Mean %T/C	Best Results	Rating[b]
Advanced Sarcoma-180 (solid)	—	100% cures	++
B-16 melanocarcinoma (solid)	180	8/10 cures	++
Primary Lewis Lung (solid)	155	100% inhibition	+
L1210 leukemia (disseminated)	178	%ILS = 379 4/10 cures	++
P388 leukemia (disseminated)	223	%ILS = 533 6/10 cures	++
ADJ/PC6	—	100% cures	
Ehrlich ascites (disseminated)	—	%ILS = 300	+
DMBA-induced mammary carcinoma (solid, chemical-induced)	—	77% total regressions	
Rous sarcoma (solid, viral-induced)	—	65% cures	

[a] Adapted from Refs. 2 and 7.
[b] ++ = significant activity; + = minimal activity (Ref. 7).

TABLE 2.II

Comparison of cisplatin with other drugs in NCI murine tumour panel.[a]

Compound	L1210 leukemia	B16 melanoma	$CD8F_1$ mammary	Lewis lung	Colon 38
Cisplatin	++	++	++	+	+
Adriamycin	++	++	++	?	—
Methotrexate	++	—	—	—	—
5-Fluorouracil	++	+	++	?	++
Bis(chloroethylnitrosourea)	++	+	++	+	+

[a] Adapted from Ref. 7 ++ = confirmed significant activity; + = minimal activity; — = no activity; ? = erratic.

2.2. Clinical Properties of Cisplatin

The first platinum drugs entered human clinical trials in 1971—1972. The trials culminated in 1978 in the United States with approval for use of cisplatin in the treatment of testicular and ovarian cancers, and later to bladder cancer. Using the definition of chemotherapeutic sensitivity as in Table 2.III [8], a summary of the present clinical utility is given in Table 2.IV.

The early results have been reviewed up to 1981 [9], and more recent results on clinical utility and toxic manifestations have also been well summarized [10]. The complex, although active in many tumours, is regularly curative in only one, testicular. The results, however, have been dramatic and, for instance, of approximately 300 patients in one long-term study 70% are considered as being probably cured [11]. Ovarian cancer is sensitive, and other positive changes may be classed as responsive and resistant.

A further feature of treatment with cisplatin is the marked synergy shown in combination with a wide variety of other chemotherapeutic agents such as 5-fluorouracil, cytarabine (ara-C) and bleomycin, which, on a practical level, allows for much more flexibility in the design of drug regimens. There is little doubt that, despite the toxic side effects common in one way or another to all antitumour agents and perhaps a reflection on our general 'naiveté' in our understanding of the control and treatment of cancer rather than the drawbacks of specific drugs, the platinum complex has been a very useful addition to the array of available chemotherapeutic agents.

A major obstacle to more widespread use of cisplatin is the persistence

TABLE 2.III
Classification of chemotherapeutic sensitivity.[a]

Chemotherapy Sensitivity	Cell Kill ($\log_{10}$)	Response Rate	Complete Remissions	Duration Response	Cures (%)
Unresponsive	0—2	<15%	None	—	—
Resistant	2—3	15—30%	None	Weeks	—
Responsive	3—4	30—60%	~5%	Months	Rare
Sensitive	4—8	50—80%	~50%	Months	5—20%
Curable	8—12	~100%	~100%	Years	>75%

[a] Adapted from Ref. 8.

TABLE 2.IV
Summary of clinical utility of cisplatin.[a]

Sensitivity	Type of Cancer
Curable	Testicular
Sensitive	Ovarian
Responsive	Head and Neck
Responsive	Bladder
Resistant	Cervix, Prostate Esophageal
Activity shown	Various e.g.: Non-small-cell lung Osteogenic sarcoma Hodgkins lymphoma
Limited Activity	Melanoma Breast

[a] Adapted from Refs. 8, 10 and 27. See these for further details.

of severe toxic side effects. The major dose-limiting toxicity is renal in nature with changes varying from abnormal urinalyses to fatal renal dysfunction [12]. Of 32 cancer chemotherapeutic agents reviewed for nephrotoxicity [13], cisplatin represented the highest risk. Interestingly enough, *trans*-$[PtCl_2(NH_3)_2]$ has been found not to be nephrotoxic [14]. The circumvention of this toxicity by hydration and simultaneous administration of a diuretic such as D-mannitol was first demonstrated in animals [15] and its extension to human trials [16] made the clinical use of cisplatin possible, and infusion in this manner allows tolerable doses of 50 to 100 mg/m^2. There is general agreement that this important breakthrough was fundamental for the continuing development of the complex as a clinically useful drug.

Other diuretics such as furosemide may be used [17] and administration in hypertonic saline was also shown to limit the nephrotoxicity [18, 19] and allow higher doses. The rationale in the latter case was that high chloride concentration would inhibit hydrolysis and the complex would pass through the kidneys in the unreactive chloride form. However, the mechanism by which hydration protects the kidneys is not fully understood and may be due to a lesser concentration of cisplatin and a shorter exposure in the renal tubules [20].

A number of other agents, all sulfur nucleophiles, have been shown to inhibit cisplatin nephrotoxicity and the structures of a number of these are shown in Figure 2.2. The fact that sulfur compounds, such as thiosulfate [21], sulfiram and its metabolite diethyldithiocarbamate (DDTC) [22], thiourea (TU) and methionine [23], penicillamine [24], reduced glutathione [25], and the radioprotectant WR-2721 (*S*-2,3-aminopropylaminoethylphosphorothioic acid) [26] affect the nephrotoxicity is of importance mechanistically. Three of these compounds (thiosulfate, diethyldithiocarbamate, and WR-2721) are in clinical trials. The

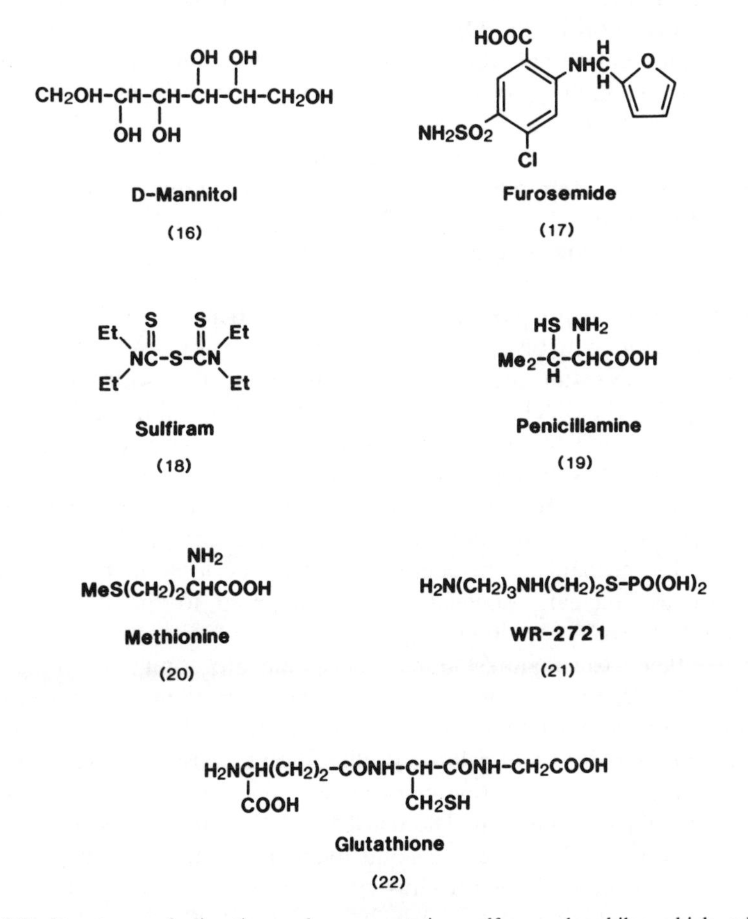

Fig. 2.2. Structures of diuretics and representative sulfur nucleophiles which relieve platinum-induced nephrotoxicity.

timing of administration of these sulfur nucleophiles is critical as they may also diminish the antitumour effect. Thiosulfate and WR-2721 must be administered prior to or concurrent with cisplatin to be effective in protection against nephrotoxicity while diethyldithiocarbamate may be administered later.

The diminution of antitumour activity by sulfur nucleophiles may be due to the sequestration of active metabolites or to interaction with Pt—DNA adducts. Some agents such as DDTC do not affect the antitumour activity of cisplatin [28], whereas other agents such as TU do so. Thiourea can remove Pt—DNA crosslinks and inhibits the cytotoxicity [29]. Indeed, thiourea has been of great utility in allowing examination of the mechanism of formation of Pt—DNA adducts (see Chapter 4.5). Depletion of intracellular glutathione (GSH) content, on the other hand, results in enhanced platinum cytotoxicity [30], and GSH can also quench Pt monoadducts [31]. Thus, all sulfur nucleophiles do not behave in the same manner with platinum complexes and the fundamental aspects of these interactions and their relationships to platinum complex metabolism are now being explored (Chapter 3.7). Some aspects of the biochemical mechanisms relating to these points are summarized in recent volumes [27, 91].

The mechanism of platinum nephrotoxicity may be similar to that of mercury and may involve depletion of SH groups of the renal tubules and in this case the sulfur agents would act in competition with the renal SH groups [32]. The ATPase enzyme is critical for kidney function (see Chapter 12.1) and this has also been proposed as the site of action [33, 34], although it has been pointed out that the inhibiting concentrations are high and unlikely to be achieved *in vivo* [35]. Other urinary enzymes have also been examined for inhibition by platinum complexes [28, 36]. In general, the mechanisms of action related to the toxic manifestations of cisplatin are unclear, especially when compared to the details of the proposed Pt—DNA reaction (Chapter 4).

The other toxic manifestations of cisplatin are common among antitumour agents. Since most of these drugs exert their major effect by inhibition of DNA synthesis at some time in the cell cycle, those normal tissues with a high rate of cellular proliferation are also affected adversely. These tissues are mainly bone marrow elements, gastrointestinal epithelial cells, hair follicles, and skin. The nausea and vomiting caused by cisplatin is very severe, and may cause some patients to desist from treatment. Standard antiemetic drugs are thus employed in treatment [37]. With the application of higher doses, which is a relatively recent feature of chemotherapeutic regimens, other toxic side effects are being reported. The hearing loss, or ototoxicity, has been reviewed; younger children appear to be more susceptible [38, 39]. Also, at high or prolonged doses neurotoxicity becomes a factor. Recent clinical toxicology findings have been summarized [40]. A summary of preventive clinical interventions to reduce nephrotoxicity has been published [41] and once the nephrotoxicity is overcome the major dose-limiting toxicities are considered to be anorexia and peripheral neuropathy.

Alteration of this activity–toxicity spectrum requires a daunting amount of both clinical and chemical research, much of it presently underway in various centers. Both disease-oriented and drug-oriented research strategies have been proposed and summarized [42]. In the former, combination with other drugs, surgery or radiation is necessary to expand the clinical role. Combination with other drugs may be cumulative or synergistic, and in many cases gives significantly better results than the individual drugs. The combination with radiation treatment is also of considerable current interest, with very promising results emerging from clinical trials [43, 44] (see Chapter 8).

In drug-oriented terms, an increase in allowable dosage by limitation of toxic side effects and changes in the route of administration (intraperitoneal, intraarterial or intravesical rather than infusion) are all desirable. These aspects are demonstrated by the observations that at high doses (in clinical terms, 120 mg/m^2) some activity against advanced cancer was observed, whereas more normal doses showed no effect [18]. However, as stated, the increased toxicity observed renders this dose inapplicable. Similarly, new methods of administration could produce higher regional concentrations of the drug, and produce beneficial effects in localized tumours.

Besides bettering the clinical spectrum of cisplatin itself, some of these desired improvements could be incorporated into analogues, and an increasingly important aspect of the drug-oriented research involves development and clinical testing of 'second-generation' platinum complexes. In accordance with the requirements outlined above, the selection of potentially useful drugs involves the search for complexes which satisfy one, or all, of three basic criteria:

1. Development of new selectivity, including a broader spectrum of activity than cisplatin, and especially activity in cisplatin-resistant tumour lines.
2. Modification of the therapeutic index, either through greater clinical effectiveness or reduced toxicity, in the latter case with activity at least equal to cisplatin.
3. Modification of pharmacological properties such as solubility which would allow for other routes of administration.

2.3. Development of Cisplatin Analogues

Concurrent with the clinical development of cisplatin, an intense synthetic program to develop structure-activity relationships was initiated. Apart from the importance of this work in the understanding of the mechanism of action of cisplatin, the systematic search for analogues derived from a clinically-used compound provides an important source of new compounds [45]. For instance, of 22 new drugs introduced into clinical trials by the National Cancer Institute in the four-year period from 1975, eight were analogues of previously tested anti-cancer agents.

In this respect, the structure-activity studies which will be discussed in detail in Chapter 3 have been particularly useful. These relationships are empirical and are not of the 'Hansch' type, where activity is related to some physicochemical parameter [46]. Essentially, modification of the cisplatin structure by variation of the neutral amine and ionic chloride groups was studied. Of approximately 1200 complexes investigated by the NCI, nearly 200 have met minimal standards of activity, a greater 'hit rate' than for purely organic substances, although it must be remembered that the structural type is more limited. Of these, 14 presented a *T/C* value of greater than 150%, regarded as the minimal standard of activity and thus were worthy of further evaluation [1, 47]. In a parallel survey a comparative study of 74 analogues in L1210 leukemia revealed that twenty complexes (27%) showed equivalent activity to cisplatin [48].

The studies on development and comparison of the toxicity/activity spectrum of cisplatin analogues have been extensively documented from the three principal groups at NCI, Surrey and Bristol-Myers [7, 48—51]. The candidates for clinical trials were evaluated on activity in the primary screen (L1210 or ADJ/PC6) and the secondary screen (B16 melanoma). Further considerations of preliminary toxicology studies and chemical considerations such as stability and water solubility narrowed the field further. The structures of the principal platinum analogues which are undergoing, or have undergone, clinical trials in various centers are shown in Figure 2.3. It is of interest to examine their properties with respect to cisplatin [52] and the goals for development outlined in Section 2.2, and to examine the process by which complexes suitable for clinical testing were selected. A recent article has summarized some of these results [53].

An examination of the structures shows that they include complexes with NH_3 and substituted chloride groups, i.e. cyclobutanedicarboxylato, and complexes substituted on the amine, i.e. 1,2-diaminocyclohexane, as well as a Pt(IV) complex. The systematic IUPAC names, along with abbreviations and trivial names, have been summarized [54]. This neces-

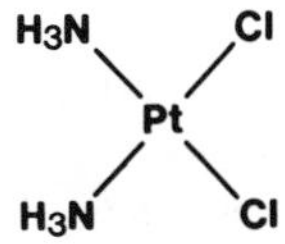

Fig. 2.3. Structures of principal second-generation platinum complexes: (23) CBDCA {diammine(1,1-cyclobutane-dicarboxylato)platinum(II)}; (24) CHIP {*cis*-dichloro-*trans*-dihydroxy-*cis*-bis(isopropylamine)platinum(IV); (25) 4-carboxyphthalato(1,2-diaminocyclohexane)platinum(II); (26) TNO-6 {(aquo)sulfato(1,1-bis-aminomethyl)cyclohexane)platinum(II); (27) MALEN malonato(ethylenediamine)platinum(II).

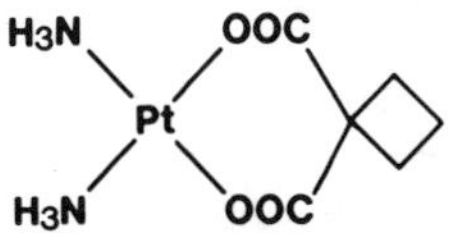

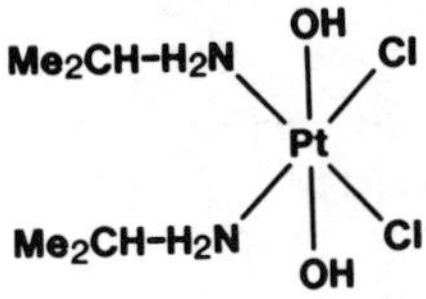

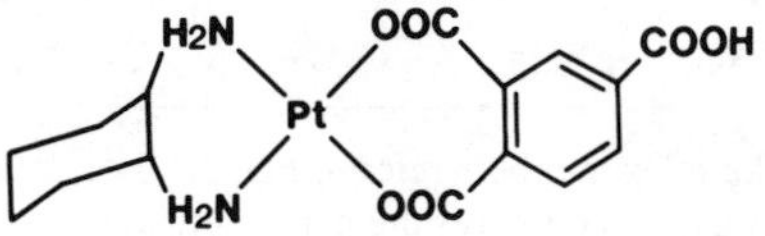

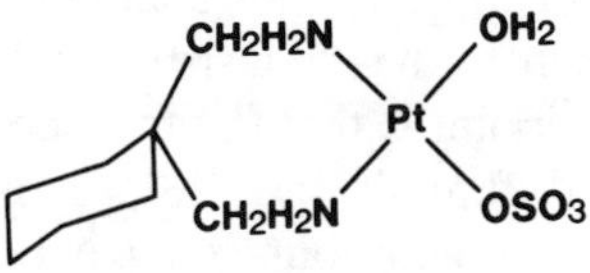

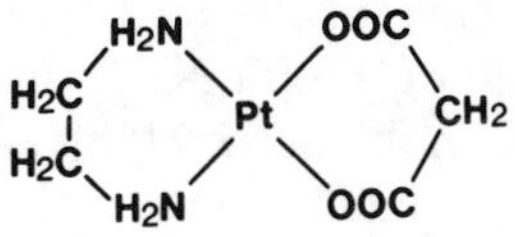

sarily brief summary cannot emphasize individual properties of the many other analogues studied but is intended to give a general picture. No major trends are discernible in their development. Indeed, if for instance only lack of nephrotoxic effects were considered, the range of potential candidates would also be unwieldy and complexes tend to be chosen for individual reasons.

The aqueous solubility and acute toxicity data are summarized in Table

TABLE 2.V
Aqueous solubility and acute toxicity for cisplatin and analogues.

Complex[a]	Solubility (mg/ml)	LD_{50} (mg/kg)	Comparative Ratio[b]		Cross-Resistant[c]
			L1210	B16	
(12)	1.0	14	1	1	—
(23)	17	130	0.76	0.95	Yes
(24)	10–20	52	0.95	0.83	Yes
(25)	25 ($NaHCO_3$)	77	1.21	0.80	No
(26)	10–20	11–20	1.17	0.83	No
(27)			0.78	0.86	Yes

[a] Complex numbers refer to Fig. 2.3. For details see Refs. 48 and 52.
[b] The ratio compares the maximum effects of each complex versus cisplatin.
[c] Refers to activity against lines made resistant to cisplatin.

2.V along with a comparison of their activity versus L1210, cisplatin-resistant L1210, and B16 melanoma. The data show that while some complexes may be better than cisplatin in the primary L1210 screen none are more active in the next screen, B16. Indeed, only one complex (the sodium salt of *cis*-[Pt(NH_3)$_2$(hydroxymalonato)]) has consistently been more active in this screen [48]. An interesting point is that the two screens rank the platinum complexes somewhat differently [7, 48], and no complexes are superior in both screens. Most compounds have better aqueous solubility than the original compound and have lower toxicity, as measured by acute toxicity values. Only one complex (Complex 26) has an LD_{50} equivalent to cisplatin and, in one survey, actually slightly less [55].

In the L1210 assay, three of the compounds, containing 1,2-diaminocyclohexane (dach) and 1,1-diaminomethylcyclohexane (damch), show greater activity than cisplatin and are also active against the cisplatin-resistant line. However, none of the compounds is better in the next screen, B16 melanoma. In general, with rare exceptions [51], complexes containing 1,2-diaminocyclohexane show lack of cross resistance with cisplatin (i.e. are active in cisplatin-resistant lines), a property shared by some other chelating diamines whose structures are shown in Figure 2.4. This, and their good activity in the primary screen, are reasons for their potential development. The question of resistance and lack of resistance is a complicated one, because it involves questions of selective uptake and elimination as well as 'target' interactions [27]. An interesting point here is that L1210 cell lines may be inherently more sensitive to 1,2-diamino-

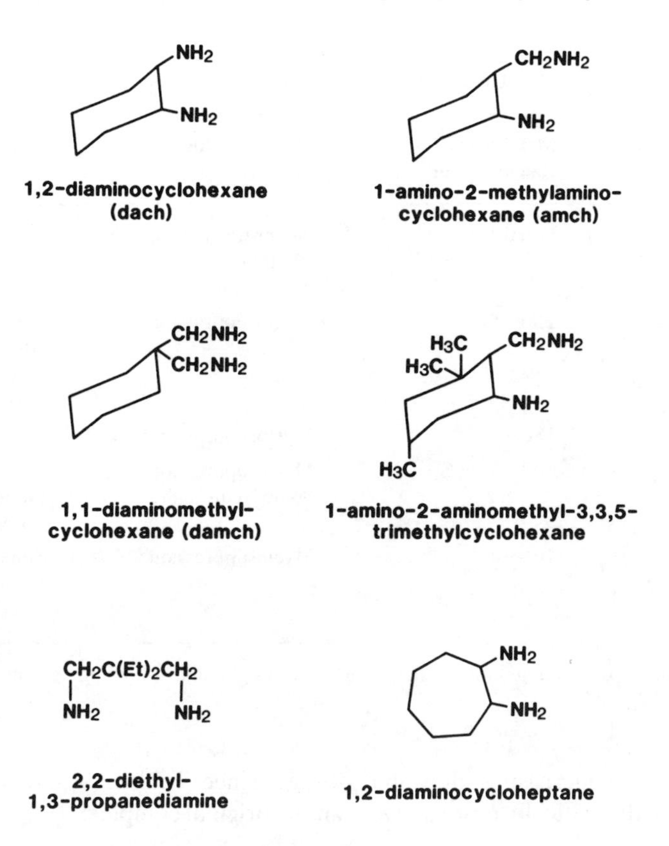

Fig. 2.4. Structures of amines whose platinum complexes have shown lack of cross-resistance with *cis*-$[PtCl_2(NH_3)_2]$.

cyclohexane complexes and, although clinical trials of complexes active in resistant lines are very desirable, these preliminary results may be misleading and the final clinical efficacy may not be as promising.

The early recognition that more inert platinum complexes may reduce nephrotoxicity, perhaps by being excreted in their unreactive forms, led to emphasis on dicarboxylate ligands derived from oxalate and malonate, and Complexes 23 and 27, although not as active as the parent complex, have been studied (see below) for their reduced toxic side effects. In clinical trials, Phase I studies define the toxicology of the drug in man and allow dose-limiting toxicities and optimal doses to be defined. The toxicity spectrum of these second-generation complexes compared to cisplatin is shown in Table 2.VI. The results show the emphasis on reducing the

TABLE 2.VI
Toxicity spectrum of cisplatin analogues.[a]

Complex	Max. Dose Range (mg/m^2)	Dose-Limiting Toxicity	Nephrotoxicity
(12)	Usual 120 (High > 120)	Nephrotoxicity (Peripheral Neuropathy)	Yes
(23)	250—520	Myelosuppression	No (But some at highest doses)
(24)	180—270	Myelosuppression	No
(25)	600—800	Myelosuppression Some Neuropathy	Some (at highest doses)
(26)	10—40	Myelosuppression	Some (at highest doses)

[a] All complexes exhibit emesis, that of complexes 23 and 24 being less than that of complex 12.

toxicity and increasing allowable dosage, since Table 2.V indicates that none are dramatically more active than the original complex.

2.3.1. CLINICAL TRIALS OF CISPLATIN ANALOGUES

The predictive power of animal screens is in the end only as good as the clinical results. The Phase II and Phase III studies currently in progress are already beginning to delineate the trends in the analogues. Naturally, the clinical trials take some time before trends emerge but it is clear at this stage that Complex 23 is most promising and may be considered the foremost 'second-generation' complex.

A timely review on Complex 23 has been published [57]. The complex entered clinical trials in 1981 and is showing a very similar activity profile to cisplatin, with good response in ovarian, small cell lung, head and neck, and testicular cancers. The advantage, therefore, over the parent complex is clearly in its reduced toxic side effects. Some indications of activity in cisplatin-resistant lines, perhaps due to use of higher doses, are also emerging. The lack of nephrotoxicity means that no hydration is required and in fact the drug may be administered in an outpatient setting. A detailed account of the development of Complex 23 is found in the first articles of Reference 57, and again it is of interest to note that in the initial screening on the primary L1210 screen the complex barely meets minimal standards of activity.

The 1,2-diaminocyclohexane complexes which have received clinical trials have been disappointing. Complex 25 has problems with stability and is difficult to formulate for pharmaceutical purposes [58]. The presence of three carboxylato groups can give isomers, which also complicates matters. Furthermore, the initial results do not show any significant improvements. Because of the lack of cross resistance the complexes with this amine ligand are of considerable interest, as stated, and many other dach complexes have received attention; the structures of some proposed candidates for further development include oxalato, malonato, isocitrato, and glutarato as leaving group. The use of carboxylate ligands in this system obviously stems from their use in the Pt—NH_3 system, but these complexes do not appear to be as stable as the ammine analogues. The chemical reasons for this are not clear. One possible way of avoiding these stability problems could be use of the Pt(IV) derivative, [$PtCl_4$(dach)], (tetraplatin) [59].

The use of Pt(IV) complexes is also exemplified by Complex 24, which at this stage does not appear to have a significantly different profile from Complex 23 [60]. The Pt(IV) derivative of cisplatin, *cis,trans,cis*-[$PtCl_2(OH)_2(NH_3)_2$] has been studied in the USSR [61].

The clinical findings, therefore, do uphold the preclinical data since the spectrum of activity is little different from that of the parent compound, the advantages being lack of toxicity. Further platinum complexes must be considerably more active or structurally significantly different to merit the investment in resources and time for clinical development. The broadening of the range of tumours susceptible to platinum treatment, therefore, is somewhat distant. In this respect, application of results from mechanistic studies to new classes of complexes is important.

2.4. Biochemical Mechanism of Action of Platinum Complexes

A full description of the mechanism of action of any drug requires the study of the rates of absorption, distribution, metabolism, and excretion, besides the study of the interactions with the proposed target molecule. The concentration of any drug at its active site, and thus its efficacy, is controlled by the above factors. A summary of the pharmacokinetic data for platinum complexes is beyond the scope of this work but much of this is to be found in other volumes and reviews [2, 52, 62].

Tissue binding involving sulfur sites has also been demonstrated in a number of cases [27, 30, 34, 63, 64]. Binding of platinum to plasma proteins *in vitro* is complete within 24 h and binding has been demonstrated to albumin and globulin [65]; and binding to low molecular weight compounds such as glutathione has also been demonstrated [66, 67]. The chemistry of platinum complexes related to their uptake and activity will be summarized in Section 3.7. Clearly, molecules can undergo various transformations *in vivo* and, indeed, there are many examples where metabolism is required for activation, and a description of drug—target effects must take this into consideration.

The complicated and involved nature of cellular processes warns against attempts to ascribe one unique mechanism to the biological action of any set of compounds, but at the same time a framework is required for the explanation of these effects and for further structured and rational development. The early biological observations which led to the development of platinum complexes are, as outlined earlier, indicative of an ability to affect replicative processes and, indeed, the ability to selectively kill rapidly dividing cells also argues for a mechanism involving inhibition of nucleic acid synthesis. Most antibiotics and drugs are believed to exert their action to a large extent by this mechanism and considerable evidence has been accumulated on both their biological and physical (i.e. binding) effects on nucleic acids [68].

In the case of platinum complexes many gross biological effects have been attributed to nucleic acid interaction and since the vast majority of mechanistic work predicates the role of DNA as a target it is useful to summarize these data here.

2.4.1. BIOLOGICAL EFFECTS IMPLYING DNA AS TARGET

The biological effects of cisplatin implicating DNA as the primary target *in vivo* have been summarized [69—73, 29]. Briefly, these include:

(1) *Induction of Filamentous Growth in Bacteria.* This effect is almost certainly due to the selective inhibition of DNA synthesis since cell growth (i.e. RNA and protein synthesis) occurs normally.
(2) *Induction of Lysogeny.* As outlined, the growth of phage from lysogenic bacteria is induced by agents which interact with DNA.
(3) *Mutagenesis.* The platinum complexes have been shown to be mutagenic in both bacterial (prokaryotic) and human (eukaryotic) cells. The *cis*-isomer is more mutagenic than the *trans*-isomer, implying differences in their DNA binding. Further, repair-deficient mutants are more sensitive than those proficient in repair.

(4) *Inactivation of Viruses.* The ability of platinum complexes to eliminate the infectious activity of extracellular papovavirus SV40 indicates inactivation of the viral DNA.

(5) *Inactivation* of transforming DNA and DNA-containing bacteriophages.

To these biological effects a number of biochemical effects also are shown to be DNA-related:

(6) *Inhibition of DNA Synthesis.* Compared to its effect on RNA and protein synthesis, cisplatin consistently and uniquely inhibits DNA synthesis both *in vitro* and *in vivo*.

(7) *Inactivation of Template for DNA Polymerase.* The decrease in DNA synthesis was shown to be due to reaction on the DNA template, rather than inhibition of DNA polymerase.

(8) *Selective Binding to DNA.* Assessment of platinum binding to macromolecules from cultured cells at known levels of cell killing and adjustment for molecular weight differences showed that more platinum was bound to DNA, and interestingly that at doses which gave equal toxicity more of the *trans*-isomer was bound than the *cis*-isomer.

(9) *Different Platinum Complexes Show Similar Levels of Binding.* The amounts of platinum bound to DNA from implanted tumour cells for equitoxic doses of Complexes 12, 23 and 24 (Figure 2.3) are very similar, indicating similar effects, despite different concentrations. The binding of Complex 23 to DNA has been shown recently to be identical to that of Complex 12 [74].

(10) *The Correlation Between Antitumour Activity and Growth Inhibition.* The ratio of the binding (expressed as r_b, or Pt/nucleotide) of the *trans*-isomer compared to the *cis*-isomer at equitoxic concentrations *in vitro* correlates well with that obtained for equivalent antitumour activity *in vivo* [75].

The overall biological effects indicative of interference with replication could, of course, have direct or indirect causes. The summarized biochemical effects argue for a direct effect. For more details see the references quoted above. A recent provocative study argues against quantitative correlations between antitumour activity and inhibition of DNA synthesis [73]. Certainly, other platinum complexes can inhibit DNA synthesis but do not produce antitumour effects (see Chapter 3). This difference, one of the great fascinations of the field, has focused attention on the nature of the binding (lesions) of the *cis*-isomer to DNA, in order to explain this unique specificity.

In this latter respect the overall role of repair processes is receiving increasing attention [29, 71]. The physical binding of any drug to DNA is eventually translated into cell death by an inability to replicate, which also results from an inability of the cell to repair that lesion. The nature of that specific lesion for cisplatin and the explanation of why tumour cells are less capable of its repair (since the complex must be somewhat selectively toxic to tumour cells) lie, then, at the heart of the understanding of the mechanism of action.

2.4.2. BIOLOGICAL EFFECTS NOT INVOLVING DNA

The emphasis on the primary role of DNA in cisplatin-induced cytotoxicity is reflected in the smaller number of studies on other aspects, although it should be stressed that any alternative explanation must account for the differing biological effects of the *cis*- and *trans*-isomers.

The induction of giant nuclei in *P. polycephalum* occurs at concentrations which do not inhibit synthesis [73], and the authors compare these results with previous suggestions that perturbation of the mitotic machinery, rather than gross inhibition of DNA synthesis, could cause cell death [76, 77]. The translation of these results to a mammalian situation is, however, problematic. The preferential inhibition of the activity of a stimulatory protein of eukaryotic transcription has been reported [78].

Consideration of membrane as a target for chemotherapeutic drugs has been reviewed and relevant studies with cisplatin summarized [79]. The amino acid uptake mechanism in L1210 cells is affected by cisplatin [80] and platinum complexes inhibit plasma membrane phosphatase activity in ascites cells [81]. Microtubule protein polymerization is also affected adversely [82]. Effects on mitochondrial functions and properties have been examined [83—86], along with studies on inhibition of sulfhydryl-containing enzymes [87—90].

2.5. Summary

The platinum complexes cisplatin, *cis*-$[PtCl_2(NH_3)_2]$, and its 1,1-cyclobutanedicarboxylato analogue, $[Pt(CBDCA)(NH_3)_2]$ have good clinical utility in the treatment of certain cancers. Initial studies on the antibacterial effects of platinum complexes led to the discovery of their antitumour potential. Toxic side effects may now in general be overcome by suitable clinical manipulation. The early dose-limiting nephrotoxicity is now circumvented routinely and sulfur nucleophiles are particularly effective in obviating this toxicity.

A number of further 'second-generation' analogues are undergoing evaluation for their clinical utility. Alteration of the amine may lead to complexes that are non-cross-resistant with cisplatin and the introduction into the clinic of such a complex is a high priority. A large quantity of biological and biochemical data indicates that the mechanism of action of cytotoxicity involves DNA binding with subsequent effects on the ability of the cell to replicate.

References

1. B. Rosenberg: *Interdisciplinary Science Reviews* **3**, 134 (1978) Reprinted in *Nucleic Acid-Metal Ion Interactions* (v. 1, Ed. T. G. Spiro), p. 1. Wiley, New York (1980).
2. B. Rosenberg: *Cisplatin, Current Status and New Developments* (Eds. A. W. Prestayko, S. T. Crooke, and S. K. Carter), p. 9. Academic Press, London (1980).
3. B. Rosenberg, L. Van Camp, and T. Krigas: *Nature (London)* **205**, 698 (1965).
4. B. Rosenberg, L. Van Camp, E. B. Grimley, and A. J. Thomson: *J. Biol. Chem.* **242**, 1347 (1967).
5. S. Reslova-Vasilukova: *Platinum Coordination Complexes in Cancer Chemotherapy* (Eds. T. A. Connors and J. J. Roberts), p. 98. Springer-Verlag (1974).
6. B. Rosenberg, L. Van Camp, J. E. Trosko, and H. V. Mansour: *Nature (London)* **222**, 385 (1969).
7. M. K. Wolpert-DeFilippes: *Cisplatin, Current Status and New Developments* (Eds. A. W. Prestayko, S. T. Crooke, and S. K. Carter), p. 183. Academic Press, London (1980).
8. J. R. Durant: *Cisplatin, Current Status and New Developments* (Eds. A. W. Prestayko, S. T. Crooke, and S. K. Carter), p. 317. Academic Press, London (1980).
9. A. W. Prestayko: *Cancer and Chemotherapy*, v. III, p. 133 (1981).
10. P. J. Loehrer and L. H. Einhorn: *Ann. Intern. Med.* **100**, 704 (1984).
11. R. E. Drasga, L. H. Einhorn, and S. D. Williams: *CA.* **32**, 66 (1982).
12. U. Schaeppi, I. A. Heyman, R. W. Fleischman, H. Rosenkrantz, V. Ilievski, R. Phelan, D. A. Cooney, and R. D. Davis: *Toxicol. App. Pharmacology* **25**, 230 (1973).
13. R. B. Weiss and D. S. Poster: *Cancer Treatment Revs.* **9**, 37 (1982).
14. J. B. Leonard, E. Ecclestone, D. Jones, P. Todd, and A. Walpole: *Nature (London)* **234**, 43 (1971).
15. E. Cvitkovic, J. Spaulding, V. Bethune, J. Martin, and W. F. Whitmore: *Cancer* **39**, 1357 (1977).
16. D. M. Hayes, E. Cvitkovic, R. B. Golbey, E. Scheiner, L. Helson, and I. Krakoff: *Cancer* **39**, 1372 (1977).
17. J. M. Ward, M. E. Grabin, E. Berlin, and D. M. Young: *Cancer Res.* **37**, 1238 (1977).
18. R. F. Ozols, B. J. Corden, J. Colins, and R. C. Young: *Platinum Coordination Complexes in Cancer Chemotherapy* (Eds. M. P. Hacker, E. B. Douple, and I. H. Krakoff), p. 321. Martinus Nijhoff, Boston (1984).
19. R. F. Ozols, B. J. Corden, J. Jacobs, M. N. Wesley, Y. Ostchega, and R. C. Young: *Ann. Intern. Med.* **100**, 19 (1984).
20. M. F. Pera, B. C. Zook, and H. C. Harder: *Cancer Res.* **39**, 1269 (1979).
21. S. B. Howell and R. Taetle: *Cancer Treat. Rep.* **64**, 611 (1980).
22. R. F. Borch, D. L. Bodenner, and J. C. Katz: *Platinum Coordination Complexes in*

Cancer Chemotherapy (Eds. M. P. Hacker, E. B. Douple, and I. H. Krakoff), p. 154. Martinus Nijhoff, Boston (1984).
23. J. H. Burchenal, K. Kalaher, K. Dew, L. Lokys, and G. Gale: *Biochimie* **60**, 961 (1978).
24. T. F. Slater, M. Ahmed, and S. A. Ibrahim: *J. Clin. Hematol. Oncol.* **7**, 534 (1977).
25. F. Zunino, O. Tofanetti, A. Besati, E. Cavalatti, and G. Savi: *Tumori* **69**, 105 (1983).
26. J. M. Yuhas and F. Culo: *Cancer Treat. Rep.* **64**, 57 (1980).
27. *Biochemical Mechanisms of Platinum Antitumour Drugs* (Eds. D. C. H. McBrien and T. F. Slater), IRL Press, Oxford (1986).
28. P. C. Dedon, R. Qazi, and R. F. Borch: In Ref. 27, pp. 199—225.
29. L. A. Zwelling: *Biological Consequences of Platinum-DNA Cross-links in Mammalian Cells* (Platinum, Gold, and Other Metal Chemotherapeutic Agents, Ed. S. J. Lippard), pp. 27—50. ACS Symposium Series 209, Washington (1983).
30. C. L. Litterst, F. Bertolero, and J. Uozumi: In Ref. 27, pp. 227—254.
31. K. Micetich, L. A. Zwelling, and K. W. Kohn: *Cancer Res.* **43**, 3609 (1983).
32. M. W. Weiner and C. Jacobs: *Fed. Proc. Fed. Am. Soc. Exp. Biol.* **42**, 2974 (1983).
33. A. M. Guarino, D. S. Miller, S. T. Arnold, J. B. Pritchard, R. D. Davis, M. A. Urbanek, T. J. Miller, and C. L. Litterst: *Cancer Treat. Rep.* **63**, 1475 (1979).
34. P. T. Daley-Yates and D. C. H. McBrien: *Cisplatin Nephrotoxicity* (Nephrotoxicity: Assessment and Pathogenesis, Eds. P. H. Bach, F. W. Bonner, J. W. Bridges, and E. A. Lock). Wiley, New York (1982).
35. C. L. Litterst: *Agents and Actions* **15**, 520 (1984).
36. C. Litterst, J. H. Smith, M. A. Smith, J. Uozumi, and M. Copley: *Uremia Investigation* **9**, 111 (1985—6).
37. L. J. Seigel and D. L. Longo: *Ann. Intern. Med.* **95**, 352 (1981).
38. M. J. Moroso and R. L. Blair: *J. Otolaryngol.* **12**, 365 (1983).
39. L. Koegel Jr.: *Am. J. Otol.* **6**, 190 (1985).
40. J. J. Roberts, W. J. F. van der Vijgh, J. B. Vermonken, and E. B. Douple: *Cancer Chemotherapy* (Eds. H. M. Pinedo and B. A. Chabner), p. 118, Elsevier, New York (1985).
41. R. S. Finley, C. L. Fortner, and W. R. Grove: *Drugs Clin. Pharm.* **19**, 362 (1985).
42. S. K. Carter: *Platinum Coordination Complexes in Cancer Chemotherapy* (Eds. M. P. Hacker, E. B. Douple and I. H. Krakoff), p. 359. Martinus Nijhoff, Boston (1984).
43. De Wit: *Int. J. Radiat. Oncol. Biol. Phys.* **13**, 403 (1978).
44. E. B. Douple: *Plat. Met. Rev.* **29**, 118 (1985).
45. S. K. Carter: *Cancer Chemother. Pharmacology* **1**, 69 (1978).
46. C. Hansch: *Acc. Chem. Res.* **2**, 232 (1969).
47. P. J. Sadler, M. Nasr, and V. L. Narayanan: *Platinum Coordination Complexes in Cancer Chemotherapy* (Eds. M. P. Hacker, E. B. Douple, and I. H. Krakoff), p. 290. Martinus Nijhoff, Boston (1984).
48. W. T. Bradner, W. C. Rose, and J. B. Huftalen: *Cisplatin, Current Status and New Developments* (Eds. A. W. Prestayko, S. T. Crooke, and S. K. Carter), p. 171. Academic Press, London (1980).
49. W. C. Rose and W. T. Bradner: *Platinum Coordination Complexes in Cancer Chemotherapy* (Eds. M. P. Hacker, E. B. Douple, and I. H. Krakoff), p. 228. Martinus Nijhoff, Boston (1984).
50. A. H. Calvert, S. J. Harland, K. R. Harrap, E. Wiltshaw, and I. E. Smith: *ibid.*, p. 240.
51. K. R. Harrap, M. Jones, C. R. Wilkinson, H. McD. Clink, S. Sparrow, B. C. V.

Mitchley, S. Clarke, and A. Veasey: *Cisplatin, Current Status and New Developments* (Eds. A. W. Prestayko, S. T. Crooke, and S. K. Carter), p. 193. Academic Press, London (1980).
52. F. H. Lee, R. Canetta, B. F. Isell, and L. Lenaz: *Cancer Treat. Rev.* **10**, 39 (1983).
53. C. F. J. Barnard, M. J. Cleare, and P. C. Hydes: *Chem. in Britain*, 1001 (1986).
54. J. Reedijk: in *Platinum Coordination Complexes in Cancer Chemotherapy* (Eds. M. P. Hacker, E. B. Douple, and I. H. Krakoff), p. 3. Martinus Nijhoff (1984).
55. J. B. Vermorken, W. W. ten Bokkel Huinink, J. G. McVie, W. J. F. van der Vijgh, and H. M. Pinedo: in *Platinum Coordination Complexes in Cancer Chemotherapy* (Eds. M. P. Hacker, E. B. Douple, and I. H. Krakoff), p. 330. Martinus Nijhoff, Boston (1984).
56. W. C. Rose, J. E. Schurig, J. B. Huftalen, and W. T. Bradner: *Cancer Treat. Rep.* **66**, 135 (1982).
57. Paraplatin (Carboplatin): Current Status and Future Prospects (Eds. S. K. Carter and K. Hellman): *Cancer Treatment Rev.* **12**, Supplement A (1985).
58. D. P. Kelsen, H. Scher, and J. Burchenal: *Platinum Coordination Complexes in Cancer Chemotherapy* (Eds. M. P. Hacker, E. B. Douple, and I. H. Krakoff), p. 310. Martinus Nijhoff, Boston (1984).
59. W. K. Anderson, D. A. Quagliato, R. D. Haugwitz, V. L. Narayanan, and M. K. Wolpert-DeFilippes: *Cancer Treat. Rep.* **70**, 997 (1986).
60. P. Lelieveld, W. J. F. van der Vijgh, R. W. Veldhuizen, D. van Velzen, L. M. van Putten, G. Atassi, and A. Danguy: *Eur. J. Cancer* **20**, 1087 (1984).
61. M. A. Presnov, A. L. Konovalova, A. M. Kozlov, V. K. Brovtsyn, and L. F. Romanova: *Neoplasma* **32**, 73 (1985).
62. See *Platinum Coordination Complexes in Cancer Chemotherapy* (Eds. M. P. Hacker, E. B. Douple, and I. H. Krakoff), Martinus Nijhoff, Boston (1984) and J. Clin. Hematol. Oncol., 7 (1977).
63. R. F. Borch and M. E. Pleasants: *Proc. Natl. Acad. Sci USA* **76**, 6611 (1979).
64. B. Leyland-Jones, C. Morrow, S. Tate, C. Urmacher, C. Gordon, and C. W. Young: *Cancer Res.* **43**, 6072 (1983).
65. R. C. Manaka and W. Wolf: *Chem.-Biol. Inter.* **22**, 353 (1978).
66. C. L. Litterst, S. Tong, Y. Hirokata, and Z. H. Siddik: *Cancer Chemother. Pharmacol.* **8**, 67 (1982).
67. P. T. Daley-Yates and D. C. H. McBrien: *Biochem. Pharmacol.* **32**, 181 (1983).
68. M. J. Waring: *Ann. Rev. Biochem.* **50**, 159 (1981).
69. L. A. Zwelling and K. W. Kohn: *Cancer Treatment Rep.* **63**, 1439 (1979).
70. J. J. Roberts and A. J. Thomson: *Progress in Nuc. Acid Res. and Mol. Biol.* **22**, 71 (1979).
71. J. J. Roberts: *Adv. Inorg. Biochem.* (v. 3, Eds. G. L. Eichhorn and L. G. Marzilli), p. 273. Elsevier, New York (1981).
72. J. J. Roberts and M. F. Pera Jr.: in *Platinum, Gold and Other Metal Chemotherapeutic Agents* (ACS Symposium Series, v. 209, Ed. S. J. Lippard), Washington, p. 1 (1983).
73. J. P. Macquet, J. L. Butour, N. P. Johnson, H. Razaka, B. Salles, C. Vieussens, and M. Wright: in *Platinum Coordination Complexes in Cancer Chemotherapy* (Eds. M. P. Hacker, E. B. Douple, and I. H. Krakoff), p. 27. Martinus Nijhoff (1984).
74. R. J. Knox, F. Friedlos, D. A. Lydall, and J. J. Roberts: *Cancer Res.* **46**, 1972 (1986).
75. B. Salles, J. L. Butour, C. Lesca, and J. P. Macquet: *Biochem. Biophys. Res. Comm.* **112**, 555 (1983).

76. H. C. Harder and B. Rosenberg: *Int. J. Cancer* **6**, 207 (1970).
77. J. A. Howle and G. R. Gale: *Biochem. Pharmacol.* **19**, 2757 (1970).
78. N. Matsumoto, K. Sekimizu, M. Horikoshi, M. Ohtsuki, Y. Kidani, and S. Natori: *Cancer Res.* **43**, 4338 (1983).
79. T. R. Tritton and J. A. Hackman: in *Experimental and Clinical Progress in Cancer Chemotherapy* (Ed. F. M. Muggia), p. 81. Martinus Nijhoff (1985).
80. K. J. Scanlon, R. L. Safirstein, H. Thies, R. B. Gross, S. Waxman, and J. B. Guttenplan: *Cancer Res.* **43**, 4211 (1983).
81. S. K. Aggarwal and I. Niroomand-Rad: *J. Histochem. Cytochem.* **31**, 307 (1983).
82. V. Peyrot, C. Briand, A. Crevat, D. Braguer, A. M. Chauvet-Monges, and J. C. Sari: *Cancer Treat. Rep.* **67**, 641 (1983).
83. A. Binet and P. Volfin: *Biochim. Biophys. Acta* **461**, 182 (1977).
84. A. Binet and P. Volfin: *Biochimie* **60**, 1052 (1978).
85. M. Beltrame, L. Sindellari, and P. Arslan: *Chem.-Biol. Inter.* **50**, 247 (1984).
86. C. L. Litterst, S. Tong, Y. Hirokata, and Z. H. Siddik: *Pharmacology* **27**, 46 (1983).
87. J. L. Aull, R. L. Allen, A. R. Bapat, H. H. Daron, M. E. Friedman, and J. F. Wilson: *Biochim. Biophys. Acta* **571**, 352 (1979).
88. J. L. Aull, A. C. Rice, and L. A. Tebbets: *Biochemistry* **16**, 672 (1977).
89. M. E. Friedman, H. B. Otwell, and J. E. Teggins: *Biochim. Biophys. Acta* **391**, 1 (1975).
90. M. E. Friedman and J. E. Teggins: *Biochim. Biophys. Acta* **350**, 263 (1974).
91. R. F. Borch: 'The Platinum Anti-Tumour Drugs' in *Metabolism and Action of Anti-Cancer Drugs* (Eds. G. Powis and R. A. Prough), p. 163. Taylor and Francis, London (1987).

CHAPTER 3

STRUCTURE–ACTIVITY RELATIONSHIPS OF PLATINUM–AMINE COMPLEXES

The intensive program of analogue synthesis initiated after the discovery of cisplatin has led to clinical trials for a number of second-generation complexes. Again it should be emphasized that analogue development in general has been responsible for many compounds of clinical usefulness and their study contributes to our better understanding of the parent complex.

The analogue syntheses were guided by a set of empirical structure—activity relationships that emerged early and that have been extensively tabulated [1—8]. More recent reviews summarize the range of complexes studied [9, 10]. These relationships reduced the number of possible complexes to a manageable number. It is worthwhile, in the light of continuing research, to examine the present status of these relationships, which may be stated as follows:

1. The complexes should be electrically neutral and exchange at least some ligands rapidly under biological conditions.
2. The complexes must be in the *cis* configuration, since the corresponding *trans*-isomers are inactive.
3. The platinum complexes may be in oxidation state (II) or (IV).
4. The leaving groups should be anionic, monodentate or bidentate, and moderately hard Lewis bases with a restricted range of lability.
5. The nonexchanging groups, the nature of which is critical to activity, should be relatively inert amines, preferentially primary.

The emphasis here will be on structural aspects; little attempt will be made to cover the vast amount of biological data. There is an increasing awareness that all complexes for biological testing must be rigorously characterized and purified — even apparently straightforward analogues do not give exactly the same chemistry as cisplatin, for instance, and one disturbing aspect of some early data is the lack of reproducibility and characterization of complexes.

While the neutral amine complexes clearly give the most active species the spectrum of complexes with demonstrated antitumour activity in animal models is in fact quite wide and allows scope for development as well as understanding of their mechanisms of action, especially where these may differ from cisplatin.

3.1. The Requirement for Neutrality

Complexes of type $[PtCl_2(NH_3)_2]$ may be considered as part of a series ranging from $[PtCl_4]^{2-}$ to $[Pt(am)_4]^{2+}$, (am = neutral amine), Figure 3.1. For NH_3, all these complexes have some form of biological activity, *in vitro* or *in vivo*. Table 3.I shows that, beside *cis*-$[PtCl_2(NH_3)_2]$, only one compound, derived from the $[PtCl_3(NH_3)]^-$ anion, has any antitumour activity [3]. This early study used $[Pt(NH_3)_4]^{2+}$ as counterion, and as this itself is inactive it may interfere with the result and allow for no firm conclusion. The activity of the amminetrichloroplatinate(II) anion as the water-soluble potassium salt, $K[PtCl_3(NH_3)]$, was, however, confirmed in a later study by Macquet and Butour using L1210 cells [11].

The anion is clearly not as active as the neutral compound but the result is noteworthy. In principle, the trend in activities in this series may be related to differing cellular absorption with subsequent different intracellular concentrations or to different reactivity with the proposed target molecule. Studies have shown that some of these complexes such as K_2PtCl_4 [12], *trans*-$[PtCl_2(NH_3)_2]$ [13] and $[Pt(dien)Cl]^+$ [14] do inhibit DNA synthesis, and indeed all bind to DNA (see Chapter 1); they are not unreactive but the inhibition does not lead to an antitumour effect. Biodistribution data *in vivo* have correlated the activity and toxicity of

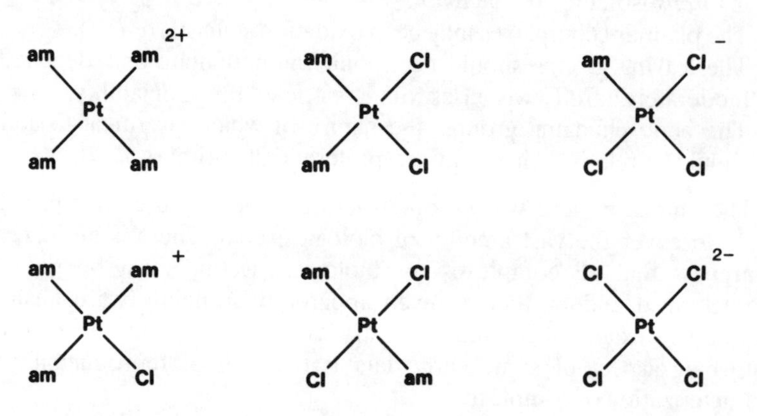

Fig. 3.1. Structures of platinum complexes $[Pt(am)_nCl_{4-n}]$, $n = 0$–4.

TABLE 3.I
Antitumour and toxicity data for the general series $[PtCl_n(am)_{4-n}]^{(2-n)+}$, (am = NH_3).[a]

Complex	ID_{50}(μM)	%*T/C*	Toxic Level (mg/kg)
$[PtCl_4]^{2-}$	48	110	40–50
$[Pt(am)Cl_3]^-$	7.3	146	~ 50
trans-$[PtCl_2(am)_2]$	67	116	40
cis-$[PtCl_2(am)_2]$	2.3	205	10
$[Pt(am)_3Cl]^+$	142	106	> 100
$[Pt(am)_4]^{2+}$	1796	108	> 200

[a] For fuller details see Refs. 3 and 11. The activity data are for L1210, both *in vitro* and *in vivo*. Toxic levels, rather than LD_{50} data, are reproduced from the same experimental data for consistency [3].

platinum complexes with physicochemical parameters such as stereochemistry, aquation constants, net charge, and partition coefficients [12].

The tissue distribution of *trans*-$[PtCl_2(NH_3)_2]$ roughly parallels that of the *cis*-isomer but more of the *trans*-isomer is absorbed and retained. Thus, membrane transport and tissue uptake are not limiting factors in the determination of antitumour activity. This is true, indeed, for the series $[PtCl_4]^{2-}$ to $[Pt(NH_3)_4]^{2+}$, as summarized in Table 3.II. Within this series the partition coefficients do parallel the order of retention. The antitumour activity is roughly paralleled by acute toxicity trends. These data clearly imply that other factors such as intrinsic reactivity or binding to intracellular target molecules are of fundamental importance in the antitumour action of the *cis*-isomer. Conversely, design of suitably active charged complexes or even *trans*-complexes with different ligands other than amine will not be limited by lack of uptake, and represent valid goals for drug development.

Neutral complexes of type $[(HL^+)PtCl_3]$, where L is a protonated diamine, have been evaluated for activity, the rationale being to increase the rate of chloride hydrolysis, using ligands of higher *trans* influence than chloride, whilst maintaining overall charge neutrality in complexes of structural type $[Pt(HL^+)Cl_3]$; a maximal *T/C* value of 148% was reported for 3-aminoquinuclidine [13]. Such complexes are, however, very insoluble in water, but do confirm that antitumour activity may be observed in complexes that are structurally different from cisplatin. Positively charged

TABLE 3.II
Comparison of biodistribution data for the general series $[PtCl_n(am)_{4-n}]^{(2-n)+}$, (am = NH_3).[a]

Complex	% Dose Retained[b] (Rel. Ret'n)	Relative Order of Partition Coefficient (K_d)[c]	$t_{1/2}$(B) days[d]
$[PtCl_4]^{2-}$	8.91 (2.60)	—	6.3
$[Pt(am)Cl_3]^-$	8.15 (2.37)	0.84	2.8
trans-$[PtCl_2(am)_2]$	9.54 (2.78)	6.2	> 7
cis-$[PtCl_2(am)_2]$	3.44 (1.0)	1.0	> 7
$[Pt(am)_3Cl]^+$	3.62 (1.05)	0.11	0.23
$[Pt(am)_4]^{2+}$	1.24 (0.36)	0.036	0.13

[a] Adapted from Ref. 12; see this reference for further details. Counter ions omitted for clarity.
[b] Equimolar doses, except for *trans*-$[PtCl_2(am)_2]$.
[c] The distribution coefficient is between *n*-octanol/saline at 37 °C.
[d] Biological half-time.

complexes derived from sulfoxides as leaving group (see Section 3.3) also show activity.

Recently, antitumour effects with $PtCl_4^{2-}$ salts of large, planar aromatic molecules such as ethidium [14] and rhodamine 123 [15] have been reported. These 2:1 salts show enhanced antitumour activity and reduced toxicity compared to the simple dyes that constitute the cations. The structure of the rhodamine species confirms the ionic nature [15]. The mechanism of action, and particularly the means whereby the acute toxicity of the parent dyes is reduced, is of some interest.

3.2. Complexes of Pt(IV)

The generally limited water solubility of $[PtCl_2(am)_2]$ complexes prompted the preparation of Pt(IV) derivatives, either as *cis*-$[PtCl_4(am)_2]$ or *cis,trans,cis*-$[PtCl_2(OH)_2(am)_2]$ in attempts to improve the aqueous solubility which might occur, for instance, due to the hydrophilic nature of the —OH group [5]. The Pt(IV) complexes were found to be active, confirming the original observations of Rosenberg [16]. No regular relationship exists between the solubility of the Pt(II) and Pt(IV) analogues, but the hydroxo complexes, prepared by oxidation with H_2O_2, are generally more soluble than the tetrachloro derivatives.

The greatly reduced toxicity for *cis,trans*-$[PtCl_2(OH)_2(NH_3)_2]$ results in a higher therapeutic index than that for the original cisplatin in one system, ADJ/PC6, but not in L1210 cells. Toxicity is not always reduced and the reverse is found for cyclohexylamine and cyclopentylamine derivatives, which was disappointing in view of the remarkably high indices of the parent Pt(II) complexes, although these were again only found in one system. No clear-cut correlations between toxicity, activity, and solubility were found in this study, but a suitable combination of these factors was found in the compound *cis*-dichloro, *trans*-dihydroxobis-(isopropylamine)platinum(IV), $[PtCl_2(OH)_2(i\text{-}PrNH_2)]$, (CHIP). The greatly increased water solubility counters the somewhat reduced efficacy but, as noted, the complex is undergoing clinical trials (Chapter 2.3.1). The tetrachloro derivative $[PtCl_4(dach)]$ also has greater solubility than its Pt(II) analogue and, in view of the high activity of complexes with this amine and its lack of cross resistance with cisplatin, the Pt(IV) derivative is also being considered for clinical trials (see also Chapter 2.3.1).

The activity of Pt(IV) species is considered to result from reduction to the Pt(II) complex. Biological reductants such as ascorbic acid and cysteine readily effect this reduction [9]; the reduction has been observed in tissue culture [17]. In accordance with these observations, *cis*-$[PtCl_2(i\text{-}PrNH_2)_2]$ has been identified as one of the metabolites of the corresponding Pt(IV) complex [18]. Similarly, there is a report surveying Pt(IV) complexes that concludes that only those with active reduction products had activity [19].

There is, however, increasing recognition that Pt(IV) complexes may interact with cellular targets, especially DNA, without the need for prior reduction. A report [20] that CHIP reacts with DNA by strand breakage, thus implying participation of free radicals and a different mode of action than that of the Pt(II) complexes, has not been substantiated and is now attributed to the presence of hydrogen peroxide as solvate in the preparation. The crystal structure of the perhydrate confirms this [21]. The preparation of the anhydrous form has been reported [21, 22] and this complex does not produce radical decomposition products [23]. Further studies on the interaction of Pt(IV) complexes with DNA showed no effect on the electrophoretic mobilities of PM2 DNA but, upon incubation with ascorbate or $Fe(II)(ClO_4)_2$, reaction was observed by the Pt(II) reduction products [21, 24]. On the other hand, $[PtCl_4(dach)]$ does interact with closed circular DNA under conditions which do not imply reduction, and where extraneous H_2O_2 is not a problem [25] and some Pt(IV) complexes are more efficient phage inducers than their Pt(II) counterparts [26].

Model studies with nucleobases such as 5′-GMP showed that Pt(II) products are formed [27] but results from oxidation of Pt(II)-cytosine complexes are indicating novel modes of reaction and binding for the oxidized species [28] (see also Chapter 4).

The caution which must be exercised in the purity of materials, especially where side products also have biological activity, is underlined by the observation of a surprisingly rapid isomerization between the isomers of $[PtCl_2(OH)_2(NH_3)_2]$ [29]. The structures of three isomers have been resolved, complex 30 being obtained from 28 in solution [29, 30]:

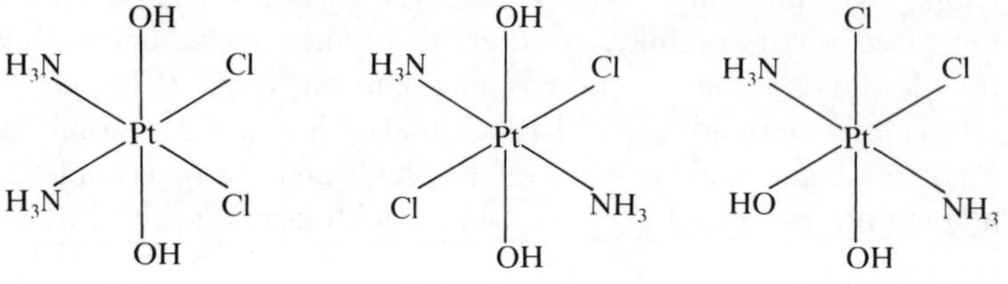

(28) (29) (30)

3.3. Nature of the Leaving Group

Examination of the second-generation complexes shows that leaving groups encompass a wide range of lability — from sulfato, which is displaced very readily in aqueous solution, ($t_{1/2}$ < 10 min), to inert dicarboxylates which undergo little aquation over a period of days. Monodentate carboxylates have been shown in some cases, e.g. chloroacetate, to have a rate of dissociation similar to that of chloride [9]. The significance of this wide range of reactivity in active species is unclear but it has been suggested that the initial period of uptake from blood may be a determining factor. Complexes with labile leaving groups may form chloro species with subsequent uptake problems. Certainly, no well defined window of lability is apparent.

The influence of leaving group lability with a range of amine complexes is notable for labile groups such as nitrate. Thus, while $[Pt(NO_3)_2(NH_3)_2]$ is very toxic, the nitrate derivatives for isopropylamine and 1,2-diaminocyclohexane maintain good activity without such toxic side effects, Table 3.III. Indeed for the former the nephrotoxic and emetic side effects are also greatly reduced and there does not seem to be any major difference between this complex and the Pt(IV) derivative, which, of course, is undergoing clinical trials. A further interesting point is that $[Pt(NO_3)_2(NH_3)_2]$ is only toxic when administered in water; in saline the complex behaves

TABLE 3.III
Comparative properties of $[PtX_2(am)_2]$, X = Cl, NO_3.[a]

Amine	X	Activity[b]	Toxic Level (mg/kg)	Ref.
NH_3	Cl	Active	9	3
NH_3	NO_3	Highly Toxic (Convulsions)	7	3
i-$PrNH_2$	Cl	Max %T/C = 171	Best Dose = 32 (LD_{50} = 33.5)	31
i-$PrNH_2$	NO_3	Max %T/C = 171	Best Dose = 64	31
(*R,R*)-dach	Cl	%T/C = 228 @ 3.12	Toxic Dose = 12.5	32
(*R,R*)-dach	NO_3	%T/C = 187 @ 6.25	Toxic Dose = 50	32

[a] Because these data are taken from various sources no overall comparison is intended or possible. For each amine, however, identical experimental conditions were employed for both chloride and nitrate. The tumours used were Sarcoma 180 (NH_3); L1210 (*i*-$PrNH_2$) and P388 (*R,R*-dach).

[b] Activity refers to maximal T/C value at stated dose, where applicable.

normally. Cisplatin itself is also more toxic when not administered in physiological saline [33, 34], which may reflect the toxicity of charged aqua species and hydroxo bridged oligomers (see Section 3.7).

The use of carboxylate and dicarboxylate leaving groups, independent of individual solubilities of the complexes, does seem to have a uniform effect in greatly reduced toxicity. Enzymatic activation has been proposed as a mechanism of action for these inert complexes, although the possibility of direct interaction with DNA may also be considered as found recently for CBDCA, (Complex 23, Figure 2.3) [35]. The discrepancy between the oxalate derivative of ethylenediamine, [Pt(en) (ox)], a strong neuromuscular toxin, and the good pharmacological properties of the malonate congener [Pt(en) (mal)], a second-generation candidate for clinical development is noteworthy. Considerable emphasis has been placed on carboxylates because the second generation CBDCA contains such a group; it is not clear whether factors such as stability and reactivity are constant for a range of amines. The malonato ligand in [Pt(en) (malonato)] has in fact been shown to be quite labile even in the presence of dilute chloride ion [36]. The boat conformation for the malonato ligand has been confirmed in at least three structural studies [36—38].

For the glutarate ligand, a crystal structure of $[Pt(NH_3)_2(glutarate)]$ shows a bridging structure rather than an eight-membered chelate ring

[39] whereas the analogous complex with dach is indicated by molecular weight measurements to be a monomeric chelate [40, 41]. The nature of the NH_3 complex in solution may not, of course, be the same as that of the most stable precipitated product, which may represent only a fraction of the species present, but these apparent discrepancies underline the need for more chemical studies on dicarboxylate complexes. It would be interesting to relate the conformational properties of proposed seven and eight membered rings to lability. Despite much effort, no one combination for the ligand of most interest, 1,2-diaminocyclohexane, has given sufficiently stable complexes; however, the chemical details are not as abundant as the biological data.

3.3.1. BISPLATINUM COMPLEXES CONTAINING BRIDGING DICARBOXYLATES

The use of aromatic dicarboxylates as leaving groups led to studies on tri- and tetra-carboxylates and some interesting bisplatinum complexes containing two $Pt(am)_2$ groups bridged by the polycarboxylate [42, 43]. Thus, for a 1,2,4,5-tetracarboxylate a general structure is:

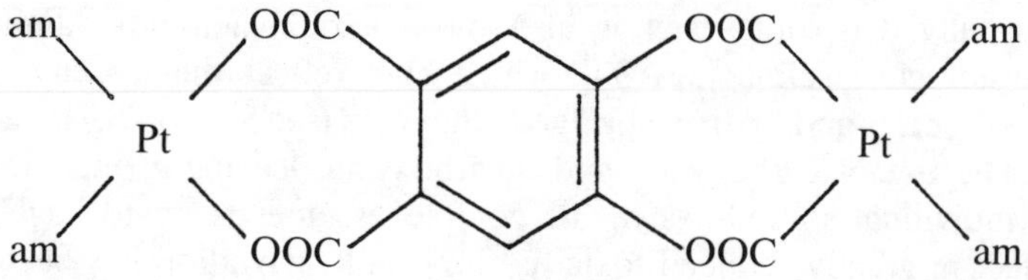

am OOC OOC am
Pt Pt
am OOC OOC am

These complexes represent an interesting attempt to obtain good localized concentrations upon hydrolysis of the $Pt(am)_2$ moiety. For some polycarboxylates, however, the possibility of isomers could complicate purification and structural assignment.

3.3.2. SULFOXIDES AS LEAVING GROUPS

An approach to expanding the range of leaving group has been made using dimethylsulfoxide (DMSO) and its congeners [44]. The lability of DMSO is similar to that of chloride [45] and complexes such as [Pt(1,2-dach) $(DMSO)_2]SO_4$ were prepared. These complexes were much more water soluble than the chloride and one sulfoxide ligand is readily lost because of the mutual labilization of the DMSO groups, which has been studied kinetically [44, 46]. These dipositive complexes had little activity, perhaps because of the charge, which could result in low intracellular concentrations, or the difficulty of displacing the second sulfoxide. Development of the use of sulfoxides led to the series $[PtCl(diam)(R_1R_2SO)]^+$ (diam = 1,2-dach or damch) where the geometry of the ligand could be altered to introduce more sterically hindered ligands and thus increase lability. These cationic complexes have good antitumour activity and are of particular interest because unsymmetrical sulfoxides are in fact chiral and a clear effect of chirality on biological activity has been noted where R_1R_2SO = methyl(*p*-tolyl)sulfoxide (Table 3.IV). This example represents the first demonstration of the effect of the chirality of the leaving group on biological activity, as well as the first use of a sulfur-containing leaving group [47]. Initial results indicate that, as would be expected, complexes differing only in handedness of sulfoxide react at different rates with DNA although the final products (i.e. cross links formed by displacement of sulfoxide) may be the same. This fact, coupled with the stereospecific effects of optically active amines (see Section 3.4.1) could be used to design very specific molecules.

A further very well-defined cationic series with antitumour activity is the triamine series *cis*-$[PtCl(NH_3)_2(py)]^+$, where py is a substituted pyridine ligand, purine or pyrimidine [120]. Unlike the sulfoxide complexes, mechanistic studies indicate that the pyridine ligand is not lost upon reaction with DNA and thus there is the possibility that these complexes

TABLE 3.IV
Biological activity of $[PtCl(diam)(R_1R_2SO)]NO_3$ in L1210 leukemia.*

diam	R_1R_2SO	Dose (mg/kg)	%*T/C***
damch	(−)MTSO	50 × 3	205 (1/6)
damch	(+)MTSO	50 × 3	229 (2/6)
R,R-dach	(−)MTSO	50 × 3	244 (2/6)
R,R-dach	(+)MTSO	50 × 3	129
S,S-dach	(−)MTSO	50 × 3	190 (1/6)
S,S-dach	(+)MTSO	50 × 3	139

MTSO = methyl(*p*-tolyl)sulfoxide. dach = 1,2-diaminocyclohexane. damch = 1,1-diaminomethylcyclohexane. The sign of the sulfoxide refers to the sign upon complexation.
* Injections on a 1,5,9 d schedule. All complexes dissolved in 0.9% saline. 60 d survivors in parentheses.
** *T/C* values calcd. at 60 d, exclusive of survivors.

produce a monodentate lesion but still exhibit good cytotoxicity, and thus react by a different overall mechanism.

3.4. Nature of the Nonleaving Group

The predominant feature of the structure–activity relationships must clearly be that the primary requisite for antitumour activity is the presence of an inert amine group, with the order of activity generally decreasing in the series $RNH_2 > R_2NH > R_3N$. Indeed, as has been pointed out, the basic skeleton describing active compounds is probably best summarized as follows:

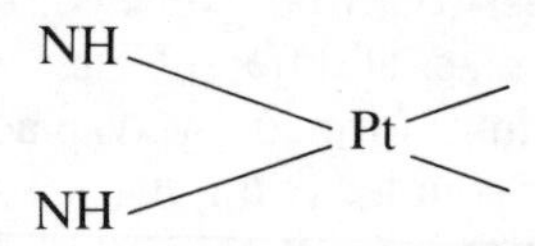

The range of amines includes normal and branched-chain alkylamines (e.g. isopropylamine), diaminoalkanes (ethylenediamine, propanediamine and their substituted derivatives), heterocyclic (ethyleneimine), alicyclic (cyclohexylamine), disubstituted cycloalkane diamines (1,2-diaminocyclohexane) and aromatic diamines (*o*-phenylenediamine), as well as 2-substituted (amino and aminomethyl) pyridines and analogues [9].

The amines dictate a considerable range in toxicity, perhaps because of solubility properties, although there is no correlation between partition coefficient and activity for a range of structurally similar amines [7]. There is some discrepancy in response to different tumour lines: the cyclohexylamine complex is very active against an ADJ/PC6 line but not against L1210 [5]. The high therapeutic indices for cyclohexyl- and cyclopentylamines in this study appear to be unique to this system and no conclusions for their general wide spectrum efficacy should be drawn. Some activity has been claimed for $[PtCl_2py_2]$ in Ehrlich Ascites but in no other line [48]. The aromatic ligands, however, may be positioned in a sterically unfavorable manner, and this may affect their binding.* Other reported complexes not containing NH are a pyridine derivative *cis*-$[PtCl_2(DMF)(2,4\text{-lutidine})]$ [127] and some bipyridine complexes based on $[PtCl_2(bipy)]$ [128, 129]. Note that these also contain nitrogen donors – no complexes with donors other than nitrogen have shown good activity.

The structure–activity relationships may only be valid for amine complexes. The limited mechanistic studies on other amine complexes besides NH_3 generally assume a mode of action similar to that of the parent complex. Evidence that an antibody raised to DNA damaged by cisplatin recognized DNA damaged by $[PtCl_2(dach)]$ supports the view that the types of damage are similar [49]. Comparative studies with en and dach complexes, however, indicated different extents of reaction with pBr322 DNA, which is surprising in view of the close similarity of these complexes [50].

More importantly the amines are also responsible for activity in cisplatin-resistant cells (Figure 2.4). This has important clinical implications and the demonstration of the lack of cross resistance in 1,2-diaminocyclohexane complexes [51] is responsible for much of the interest in developing these particular complexes. This ligand is also of interest from a mechanistic point of view because of the demonstration that different stereoisomers also have different biological properties.

3.4.1. STEREOSPECIFIC EFFECTS IN AMINE COMPLEXES

The fact that the stereoisomers of 1,2-diaminocyclohexane gave complexes with different activities was first demonstrated by Kidani and coworkers [32, 52]. The amine may be resolved into geometric (*cis, trans*) and optical isomers $\{$*trans-d*-(1S, 2S) and *trans-l*-(1R, 2R)$\}$:

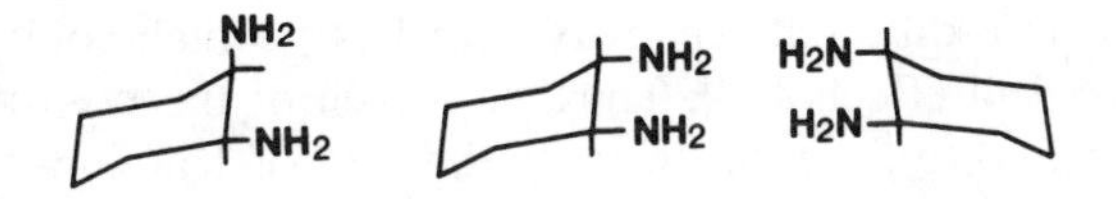

Other optically active amines studied are based on substituted ethylenediamine $H_2NCH(R)CH(R')NH_2$. The most clear-cut differences are found for dach complexes but the data are not consistent; thus while there is a distinct trend in chloride complexes with the *trans*-isomers being more active than the *cis*-isomer, these differences are not reflected in more water-soluble complexes with sulfate, nitrate or glucoronato. The activity of representative complexes is collected in Table 3.V. Note that, since the individual complexes vary in toxicity, overall therapeutic indices may not follow this trend. For 1,2-diaminocyclohexane the *trans-l* isomer gives consistently better results than the *trans-d* form, which itself is roughly equivalent to the *cis*-isomer. Some variation is found upon changing the leaving group.

The crystal structures of complexes of the stereoisomers $\{[PtCl_2($*cis*-dach)] [53], *trans*-$[PtCl_2($*trans-l*-dach$)_2]$ [Pt(*trans-l*-dach$)_2]Cl_4$ [54], [Pt(oxalato) (*trans-l*-dach)] and its malonate analogue [55], and $[Pt(acetato)_2($*trans*-dach)] [56]$\}$ confirm the differences in molecular structure. Notably, in the case of the *cis*-isomer [53] the plane of the ring (in the δ conformation) is

TABLE 3.V
Variation in activity for stereoisomers of chelating diamines in $[PtCl_2(diamine)]$.[a]

Diamine	Isomer	Dose (mg/kg)	%*T*/*C*
dach	*cis*	6.25	206
dach	*trans-d*	6.25	228
dach	*trans-l*	3.12	228
amch	*cis-d*	6.25	242
amch	*cis-l*	6.25	240
amch	*trans-d*	12.5	226
amch	*trans-l*	12.5	197

[a] Adapted from Ref. 32, which should be consulted for full details.

perpendicular to the coordination plane whereas in the *trans*-isomers [54, 55] this plane (λ-*gauche* for the 1*R*,2*R* form and δ-*gauche* for the 1*S*,2*S* form), is essentially coplanar with the coordination plane. These differences could indicate a different reactivity with DNA which could explain the different biological effects [32]. Thus, approach of the *cis*-form could be hindered compared to the *trans*-form. In the case of 1-amino-2-methylaminocyclohexane (amch) (see Figure 2.4), the *cis*-isomers of the dichloro complexes are in fact more active and this has been attributed to the greater flexibility of this ring compared to that of dach, thus diminishing the steric effect [32, 57]. Recent studies on the conformation of the adducts of the Pt—dach complexes with the dinucleoside monophosphate d(GpG) (Appendix 1) attempt to delineate these differences further (Figure 3.2) [58]. The difference in geometry of the dach ring cannot explain the differences in biological activity of the optical isomers but models shown in Figure 3.2 do show a different pattern of hydrogen bonding for these two isomers. These differences for the optical isomers are not as marked as for the geometric isomers and this has been confirmed over a range of amines [57, 59].

Studies on the interaction of $[PtCl_2(\text{chiral amine})_2]$ with model bases do not show large differences between the isomers studied, and NMR and CD studies suggest rigid structures for these adducts [60]. However, the studies with model bases have the bases in a head-to-tail arrangement rather than the head-to-head arrangement of the dinucleotide (see Chapter 4 for a fuller discussion of this and of conformational aspects). The overall

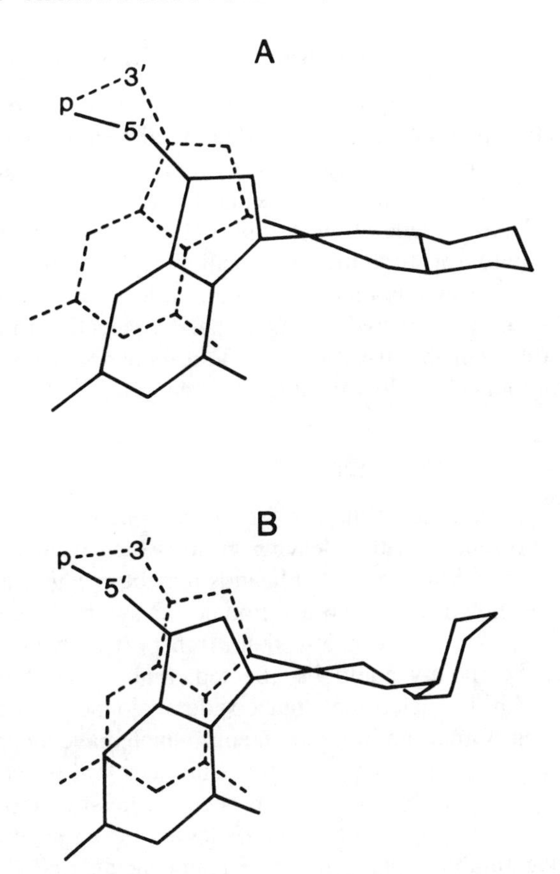

Fig. 3.2. Schematic representation of the interaction of Pt—dach isomers with a dinucleotide d(GpG). The interaction of the optically active *trans*-(*R*,*R*) form is shown in A and of the inactive *cis* form in B. From Reference 58.

shape of these molecules may also be important and it is difficult to extrapolate data from fused systems (cyclohexane rings) to 'open' systems such as 2,3-butanediamine [60]. The importance of this is underlined by studies on 2- and 3-aminomethylpiperidine complexes where differences in activity were also found [61].

It is also of interest to see whether optical isomers have equivalent uptake by cells. Early studies on the resolved, optically pure chelates $[M(phen)_3]^{2+}$ by Dwyer and co-workers did, in fact, show differences in both uptake and elimination [62]. This aspect in general needs to be explored more fully.

3.5. Novel and Targeted Approaches to $[PtX_2(am)_2]$ Complexes

The previous sections have outlined the basic approaches to synthesis of antitumour active platinum complexes. This section intends to give some more recent and novel examples of synthetic approaches and to discuss complexes synthesized within the standard structure—activity relationships. A possible 'third-generation' complex should also have the property of some selectivity for tumour, or specific tissue. If we consider the 'second-generation' stage as that of obtaining a less toxic derivative, then development of more selective agents, either by selective tumour uptake or radically different biodistribution, could enhance considerably the spectrum of tumours clinically treatable.

3.5.1. TERNARY COMPLEXES

The term 'ternary complexes' here refers to complexes which contain the *cis*-amine function along with a leaving group which itself has biological activity or relevance. Many potential ligands may be recognized by perusal of structures of known drugs and a rationale for synthesis would be that, since the ligand is known to be absorbed, transport or controlled release of the *cis*-$Pt(am)_2$ moiety could be affected. Figure 3.3 shows some of these ligands, such as nucleosides, nucleoside analogues and carboxylate-containing drugs. Other antitumour agents which have been platinated include doxorubicin [67], along with other quinones [68] (see also Chapter 7). In general, this approach has not been successful and no unique complex with distinctly superior characteristics has been recognized.

The selective tumour uptake of some amino acids, and the enhanced requirement for such nutrients, has prompted their utilization as carriers for a range of antitumour agents. In the case of platinum, some selective uptake was claimed for a tripeptide complex (containing ethylglycylglycinate) although the complex was not particularly active [69]. Complexes of the type *cis*-$[PtCl_2(t\text{-}BuNH_2)$ (amino acid)] have also been reported but are only moderately active and, in fact, none were more active than the monoanion-containing starting material $K[PtCl_3(t\text{-}BuNH_2)]$, [70]. The 'typical' amino acid binding mode of the N(amine)—O(acid) chelate does not, as expected, give active species.

This type of targetting approach would certainly be of use in making hitherto insensitive tumours more susceptible to platinum attack, but the finely-tuned and, perhaps, fortuitous combination of pharmacological properties of both independent species do not so far appear amenable to alteration in this manner.

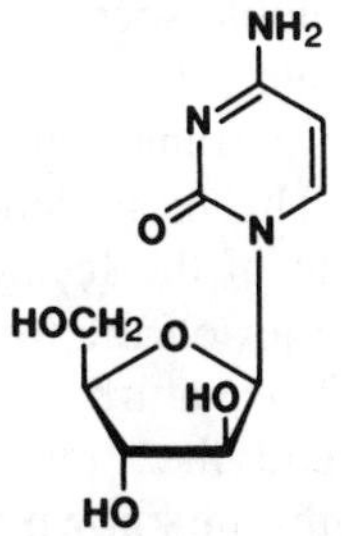

Ara-C
(31)

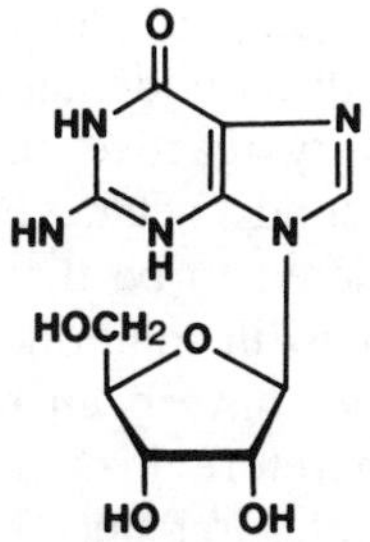

Guanosine
(32)

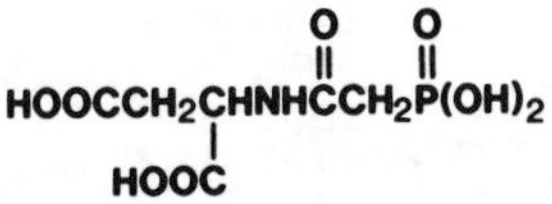

PALA
(33)

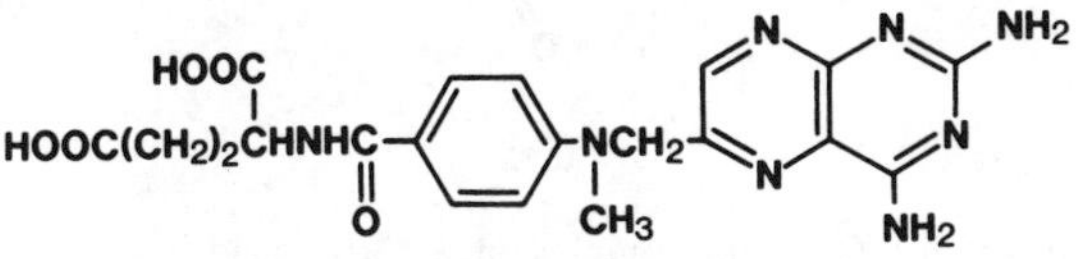

Methotrexate
(34)

Fig. 3.3. Compounds with antitumour and biological activity used as leaving groups for platinum complexes.

The use of ascorbate deserves special mention because of the unique structures that have emerged. Ascorbic acid is a weak, dibasic acid and in principle, therefore, can form uninegative and dinegative ligands [71]:

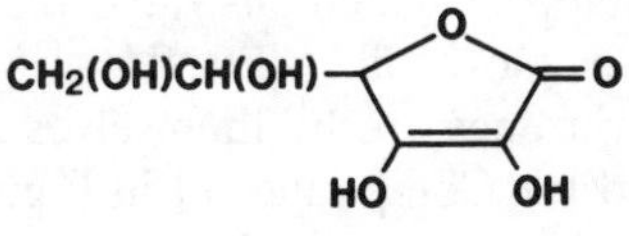

(35)

The ascorbate complexes give highly water soluble species and were first synthesized with monodentate anionic ascorbate. The coordinated ligand is, however, very reactive and the complex rearranges in solution to a unique O,C-bound chelate, Figure 3.4 [72]. The monodentate O-bound form is considered to be the most active form of the ligand but will not apparently persist in solution. Mechanistic studies on the formation of the C-bound species have been reported [121]. The nature of active species reported is therefore in some doubt [73], and the biological data are impossible to interpret since no correlation with the known chemistry was made. The observed chemistry does underline the need to examine carefully the stability of apparently active complexes and for full characterization of these complexes. All complexes should be characterized as fully as possible, which in many instances has not been the case.

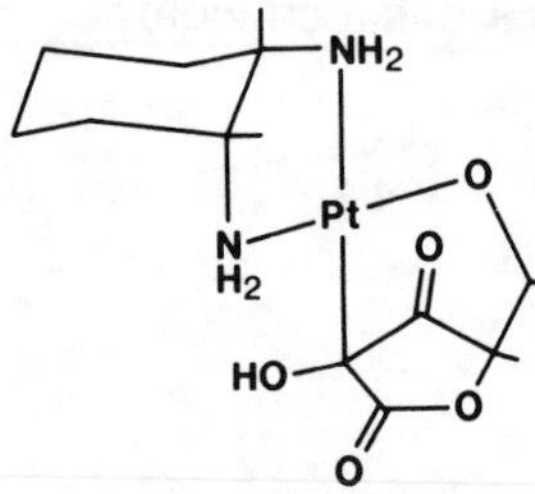

NH2
Pt
O
OH
N
H2
O
HO
O
O

(36)

Fig. 3.4. Structure of the C-bound complex of ascorbic acid bound to the (*cis*-dach)Pt moiety.

3.5.2. AMINES AS CARRIER LIGANDS

A corollary of the approach using biologically active leaving groups is to use ligands with biological activity or relevance as the neutral portion of the $[PtX_2(am)_2]$ molecule, where the am ligands present two appropriate amine functions. Figure 3.5 gives some examples.

The role of polyamines such as spermine in DNA-binding led to synthesis of the [Pt(spermine)] chelate but it demonstrated only marginal activity and, in fact, may behave more like $[Pt(NH_3)_4]^{2+}$ than *cis*-$[PtCl_2(NH_3)_2]$ [74]. The alkylating triazenes are by themselves antitumour active, but the structure as proposed for Compound 39 in Figure 3.5 [75] may be quite reactive as the tendency for neutral triazenes is to give bridging structures, in this case with elimination of HCl. The amino-

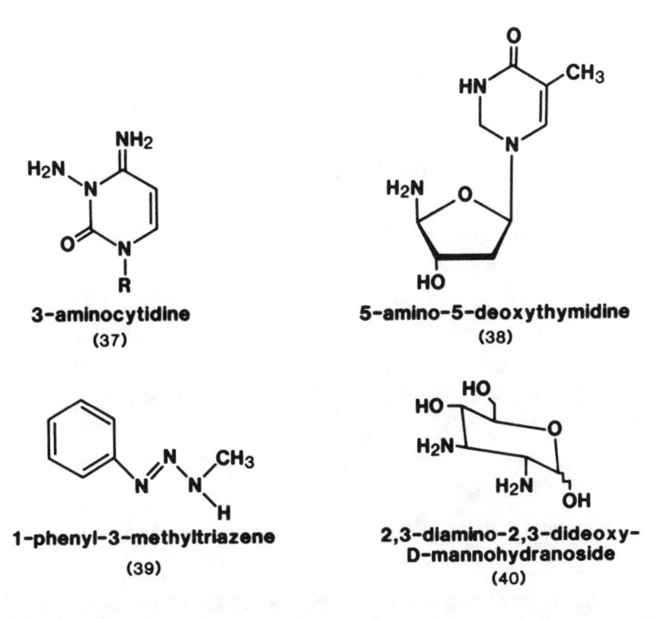

Fig. 3.5. Compounds used as neutral groups for antitumour platinum complexes.

nucleoside complexes are of interest [76, 77] in attempts to increase selectivity, as is the use of aminosugars as carriers [78]. Interestingly, in the latter case the sugar plane is not coplanar with the coordination plane and thus resembles the *cis*-isomer of dach.

The use of the 1,2-diarylethylenediamine estrogen receptors as the neutral ligand attempts to improve the susceptibility of mammary tumours [79]. In this respect, steroids have also been used to carry Pt for selectivity to hormone-dependent tumours [80]. The steroid is functionalized and linked via an *ortho*-catecholate group bound to platinum, in this system the neutral groups were in fact triphenylphosphine [80].

The interaction of intercalating agents with platinum complexes is of interest for a number of reasons. The platinum intercalation agents (see Chapter 1) have been studied for their possible antitumour effects but appear to be only marginally active [81] despite the fact that intercalators as a class give many active complexes. The activity of intercalator salts has been mentioned earlier [14] and complexes such as 8 and 9, Figure 1.8, should be of interest for their antitumour effects. The use of intercalators may also be summarized as attempting to increase the affinity of platinum for the target molecule DNA by use of DNA binding ligands, and represents an important mechanistic point worthy of further investigation.

Whether complexes can be designed which bind to DNA better than *cis*-[$PtCl_2(NH_3)_2$], and whether these would show increased antitumour activity, are clearly significant points. Further, given the detail with which Pt—DNA adducts are known, the design of suitable complexes with ligands capable of binding to DNA and producing specific (intrastrand) lesions should be possible. In this respect the interesting bisplatinum complexes containing two *cis*-[$Pt(NH_3)_2$] units linked together by a variable length diamine backbone in [{*cis*-[$PtCl_2(NH_3)$]}$_2H_2N(CH_2)_nNH_2$] have enhanced DNA-binding and cytotoxic properties [122]. These bisplatinum complexes are also non-cross-resistant with the parent cisplatin.

Finally, in this section, we should mention the complexing to immunoglobulins which enhances inhibition of DNA synthesis, not only of linked cisplatin but also for $PtCl_4^{2-}$, in comparison with the activity of the free compounds [82]. Again, the possibility exists of activating hitherto inactive or weakly active structures. The use of a metalloimmunoassay to target metal complexes is a novel approach which has also been described [83, 84].

3.6. Quantitative Structure—Activity Relationships

As stated, the development of complexes has been dictated by systematic changes based on empirical observations — the *cis*-$Pt(am)_2$ unit works. A more quantitative approach takes into account a mathematical equation with electronic and steric factors and such parameters as lipophilicity. The idea that biological response was a function of chemical composition was first advanced in 1869 and a review of these developments is available in Albert's book (Chapter 1, Ref. 1). An account of the quantitative approach which correlates concentration to obtain a given response with electronic factors (e.g. Hammett constants) and lipophilicity (partition coefficients) has been given by Hansch [85]. Few QSAR relationships have been applied to the platinum complexes.

There is no direct relationship between activity of amine complexes and lipophilicity [7]. An INDO—SCF calculation has been used to compare electrostatic potential functions with toxicity and activity. The electrostatic potential map, as expected, showed a relatively positive region around the amine groups and a negative region around the electronegative chlorine atoms. Various receptor sites, assuming DNA as the target, were then chosen on the molecule. The toxicity, LD_{50}, varied with the stability of the molecule, while the smaller the molecule the more likely it was to be active. The conclusions do not explain mechanistic details, but may be able to predict less toxic compounds [86]. Other MO calculations complement this study [87].

A further study attempting to relate the antitumour properties to reactivity involved the study of the kinetic lability of amines in the series *cis*-$[Pt(am)_2(DMSO)Cl]Cl$, where am is varied from cyclopropyl to cyclooctylamine. Apart from the high activity of this particular *cis*-$[PtCl_2(am)_2]$ series [2] (although, as noted, in only one tumour line) further reasons for the choice of this particular group of complexes were that it represents a homologous series where the basicity changes are a result of bond angle strain at the carbon adjacent to the nitrogen (without the necessity of introducing polar substituents with subsequent solvation changes), and steric effects are relatively constant where the pK_a changes occur [88]. The use of DMSO was prompted by the desire to have a strong *trans*-labilising group in order to study the leaving group effect of the amine. The basic reactions, then, are as presented below. The results have been summarized [5]:

$$cis\text{-}[PtCl_2(am)(DMSO)] + am \rightarrow cis\text{-}[PtCl(am)_2(DMSO)]^+ + Cl^-$$

and

$$[PtCl_3(DMSO)]^- + am \rightarrow trans\text{-}[PtCl_2(am)(DMSO)] + Cl^-$$

In the series *cis*-$[Pt(am)_2(DMSO)Cl]^+$, the rate of displacement of amine is dependent on the basicity of the amine, the least basic amine giving the most reactive substrate. This effect is made up of a leaving group effect and a *cis* effect, possibly because weaker amines would increase the electrophilicity of the platinum. The equilibrium constants for the reactions above are also sensitive to amine basicity. The study showed that the cationic species is less stable (with respect to loss of amine) than the neutral monoamine species. For the reactions of *trans*-$[PtCl_2(am)(DMSO)]$, the leaving group effect was also observed [89]. These observations are relevant because, since chloride replacement is the normal mode of reaction of *cis*-$[PtCl_2(am)_2]$, substitution by a group with a strong *trans* effect could labilize the amines and thus represent a pathway for complex inactivation. This aspect is of increasing importance as more work is being done on the metabolism of platinum complexes.

In the series $[PtCl_2(am)_2]$ the most active complexes are indeed the most stable, where displacement is unlikely and the antitumour activity is not related to amine loss.

3.7. Chemistry of Platinum–Amine Complexes

In a sense, the aqueous chemistry of the platinum amines with respect to biological relevance may be divided into activation, deactivation and toxicity; the interaction with DNA can be considered activation whereas reactions with other cellular components, metabolism and toxicity are related to deactivation. The relevant aqueous chemistry has been thoroughly reviewed [90, 91].

The chemistry will be initially concerned with dissociation of the chloride ligands. Many of the studies on platinum complexes relate to their interactions with DNA and its constituent bases, a topic which is covered separately (Chapter 4), but it is relevant here to summarize the aqueous reaction chemistry as it relates to antitumour activity. In a biological system, with many possible ligands, the reactions of Pt—aqua complexes with species other than DNA should also be considered. As noted in Chapter 2, there is much less known about the chemistry of platinum complexes in this 'intermediate' phase, i.e. after uptake but before activation by DNA binding (but see below). A general hydrolysis scheme for $[PtCl_2(am)_2]$ is presented in Figure 3.6. Use of the rates of hydrolysis and acidity constants, and knowledge that the chloride concentration of blood plasma is 103 mM compared to 4 mM in the cytoplasm, allow for the relative proportions of species present at the biological pH 7.4 to be calculated for am = NH_3 [90—92]:

Plasma:	24	1	1	0.0178	1.12	1.41
Cytoplasm:	0.915	1	1	0.46	2.90	3.63
	(i)	(ii)	(iii)	(iv)	(v)	(vi)

Detailed distribution curves have also been plotted [90]. The fact that the complex will be essentially undissociated in plasma is one reason for

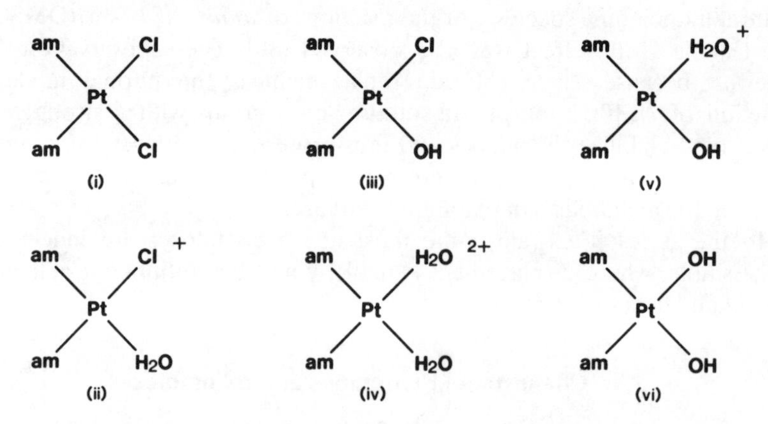

Fig. 3.6. Structures of hydrolysis products of platinum—amine complexes. Exact distribution of these species *in vivo* will depend on Cl^- concentration in the medium and pH.

supposing that the mechanism of cellular uptake is passive. The transport of a drug across a membrane may be passive or active. In the former, the drug molecules move from a region of higher concentration to a region of lower concentration and the rate of transfer is proportional to the concentration gradient, with equilibrium being achieved when the concentrations on both sides of the membrane are equal. A consequence of this is that where a species exists in both neutral and charged forms the neutral form is transferred across the membrane. Active transport, on the other hand, involves participation of some membrane carrier and the complex may be transported against a concentration gradient. The observation that the amino acid uptake process is inhibited by *cis*- and *trans*-$[PtCl_2(NH_3)_2]$ implies some contribution to the overall uptake by the active process [93], as also does the activity of intact charged species. The presence of an active system for Pt transport needs to be detailed further as this point could be important for design of newer drugs.

The reactive species in cytoplasm will clearly be a combination of species with aqua and hydroxo ligands. The hydroxo ligand is usually considered to be inert but may become involved via protonation to the labile aqua form [92]. The first pK_a for deprotonation of an aqua ligand is 5.6 and in concentrated solutions (> 0.1 mM) dimeric and trimeric species are formed through hydroxide bridging. These complexes were first structurally characterized for the NH_3 species [94–96], Figure 3.7. The trimeric dach and damch complexes have also been structurally characterized [97, 98]. The formation of these bridged species has been studied in solution by means of ^{195}Pt NMR [99–102], an extremely powerful tool in this respect [103, 117]. Their properties are of interest, but at biologically relevant concentrations they may not be of significance. The reactions of the diaqua cations with anions such as phosphate and

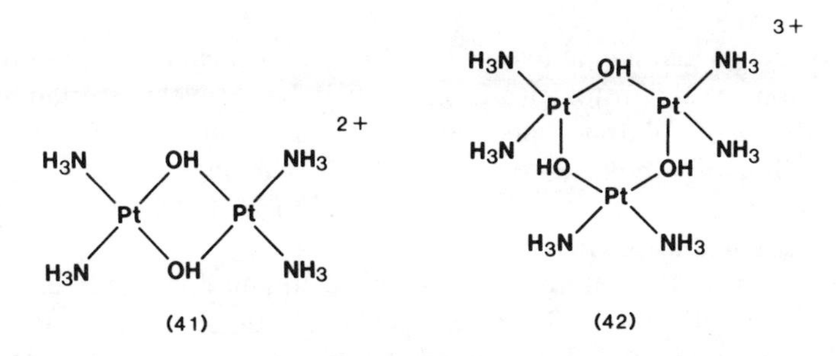

Fig. 3.7. Structures of the hydroxo-bridged dimers and trimers formed in the hydrolysis of *cis*-$[PtCl_2(NH_3)_2]$.

acetate have also been followed by multinuclear (^{195}Pt and ^{15}N) NMR, and the formation of bridged species and also oligomeric blue forms (see Chapter 5) have been shown to occur [104]. Recent chemical shift and coupling correlations have also been compiled for both amine and related systems [105, 106].

3.7.1. REACTIONS WITH BIOMOLECULES OTHER THAN DNA

Substitution of aqua ligands by nitrogen donors such as the nucleic acid bases leads to mono- and disubstituted complexes whose properties, of course, have been extensively studied (Chapter 4). Once inside the cell, however, only a small proportion of the platinum is complexed to the DNA. Metabolism of the platinum complexes therefore takes place with other biomolecules besides DNA. Amino acid reactions may take place which may sequester the complex. Further, in the case of substitution by nucleophiles with strong *trans* influence, especially those of sulfur, displacement of amine and deactivation may occur. Interestingly, amine labilization *trans* to a thioether in *cis*-$[Pt(NH_3)_2(Me_2S)_2]^{2+}$ has been observed [107]. *In vivo*, thioethers such as methionine could effect the same chemistry and the characterization of $[Pt(methionine)_2]^{2+}$, assigned the structure as shown, as a metabolite of cisplatin [108] confirms this view:

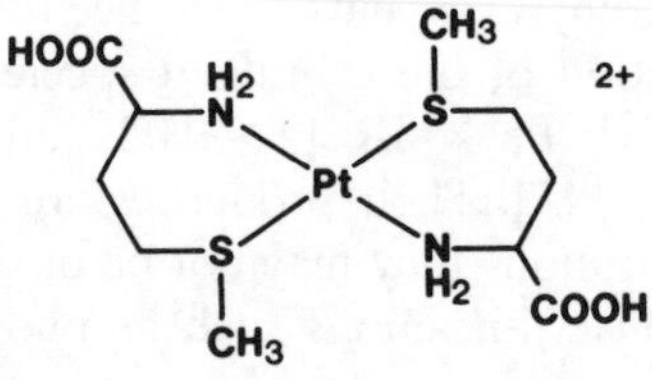

HOOC, H_2N, CH_3, S, Pt, S, CH_3, NH_2, COOH, 2+

(43)

This is evidence that amine loss can result in detoxification and the role of this reaction in nephrotoxicity is implicated but unclear. Certainly, the strong affinity of sulfur for platinum makes this reaction a feasible one. Indeed, platinum complexes such as K_2PtCl_4 and $[PtCl_2(am)_2]$ have been used as heavy metal markers in protein crystallography and some of these uses have been tabulated [91].

A general role for a thiol reaction with platinum complexes as deactivation cannot at present be stated, although tissue binding involving sulfhydryl groups is well documented and biochemical aspects have been discussed (Chapter 2.4). The reactions of Pt complexes with methionine have been studied [109], as well as reactions with glutathione [110]. Glutathione (Compound 22, Figure 2.2) is the predominant intracellular thiol [111] and in concentrations of 0.5—10 mM may react with many drugs. A recent brief review summarizes some of these reactions as they pertain to therapeutic efficacy of drugs but mention of Pt complexes was totally omitted [112]. The interaction of GSH with platinum—amines is of increasing interest, since there may also be correlations with platinum resistance in cells. It is known that GSH can quench Pt monoadducts on DNA [113] and that, in general, platination of isolated DNA is inhibited by GSH [114]. The role of GSH levels in resistant cells has been a topic of recent discussion [115, 116] and a Pt—purine—GSH adduct has been tentatively identified from a Pt—DNA incubation in the presence of GSH [114]. No definite conclusions may be drawn at this stage and the interrelationships between platinum toxicity, activity and thiol binding require more clarification. These aspects have been reviewed [123].

As stated, sulfur nucleophiles are expected to labilize amines in a *trans* position and the isolation of the platinum—methionine metabolite is further evidence that these reactions occur *in vivo*. Evidence for amine displacement, even of chelated ethylenediamine, has also been obtained from solution studies with methionine-containing peptides and RNase A [117]. In connection with these results the dissociation of doubly labelled $[PtCl_2(en)]$ (on the Pt and C) *in vivo* has been confirmed, negating earlier results [118].

Thus, although the chemistry of $[PtCl_2am_2]$ will be dictated by the formation of reactive species through chloride loss, many deactivating reactions are possible with intracellular nucleophiles, before reaction with cellular targets. The intracellular concentration of phosphates, for example, has recently been claimed to be sufficient to displace chloride also [119].

Considerable advances in the detection of intracellular metabolites have been made and a detailed description is available for complexes containing 1,2-diaminocyclohexane as carrier ligand [124, 125]. In these studies the widespread interaction with amino acids has been demonstrated and the hydrolysis of [Pt(malonato)(dach)] occurs without the requirement of enzymatic activation, as originally hypothesized. Biotransformation of platinum complexes has also been demonstrated by NMR studies [126]. These studies are an important complement to those of the DNA interactions discussed in Chapter 4, in terms of the overall objective of describing the intracellular chemistry of platinum complexes as fully as possible.

3.8. Summary

A wide range of structures are now known which violate the empirical structure—activity relationships for platinum complexes. The basic structure based on the *cis*-[Pt(amine)$_2$] unit does give the most active species and, indeed, few closely related second-generation analogues have demonstrated greater activity than *cis*-[$PtCl_2(NH_3)_2$] in a range of murine tumours. The identification of structurally novel complexes may lead to elucidation of new mechanistic pathways. Development of structurally novel complexes is important, not only from a mechanistic point of view, but also because new complexes may have a different spectrum of activity with activity also against cisplatin-resistant tumours.

Stereospecific effects on both nonleaving group (1,2-diaminocyclohexane) and leaving group (chiral sulfoxide) have been shown for both biological activity and DNA binding.

Upon entry into a cell platinum complexes undergo hydrolysis. The reactive aqua species may undergo oligomerization through hydroxo bridges, although this may not be relevant biologically. Activation of platinum complexes is through DNA-binding while metabolism of complexes occurs through binding to amino acids and tissue binding involving sulfhydryl groups is well documented.

Note: Recent results in the authors laboratory have given the unusual result that the *in vitro* activity of *trans*-[$PtCl_2(py)_2$] in L1210 leukemia is significantly greater than its *cis* isomer and of the order of cisplatin. The *trans*-pyridine derivative is not, however, active *in vivo* but the results do point out that the geometric differences noted so far may not dictate antitumour activity when the neutral ligands are other than primary or secondary amines.

References

1. M. J. Cleare: *Coord. Chem. Rev.* **12**, 479 (1974).
2. W. C. J. Ross, P. D. Braddock, A. R. Khokhar, and M. L. Tobe: *Chem.-Biol. Interact.* **5**, 415 (1972).
3. M. J. Cleare and J. D. Hoeschele: *Bioinorg. Chem.* **2**, 187 (1973).
4. P. D. Braddock, T. A. Connors, M. Jones, A. R. Khokhar, D. H. Melzack, and M. L. Tobe: *Chem.-Biol. Interact.* **11**, 145 (1975).
5. M. L. Tobe and A. R. Khokhar: *J. Clin. Hematol. Oncol.* **7**, 114 (1977).
6. M. J. Cleare: *J. Clin. Hematol. Oncol.* **7**, 1 (1977).
7. J. J. Roberts and A. J. Thomson: *Prog. Nuc. Acid Res. and Mol. Biol.* **22**, 71 (1979).
8. M. J. Cleare and P. C. Hydes: *Metal Ions in Biol. Syst.* **11**, 1 (1980).

9. M. J. Cleare, P. C. Hydes, D. R. Hepburn, and B. W. Malerbi: *Cisplatin, Current Status and New Developments* (Eds. A. W. Prestayko, S. T. Crooke, and S. K. Carter), p. 149. Academic Press, London (1980).
10. P. C. Hydes: *Platinum Coordination Complexes in Cancer Chemotherapy* (Eds. M. P. Hacker, E. B. Douple, and I. H. Krakoff), p. 216. Martinus Nijhoff, Boston (1984).
11. J. P. Macquet and J. L. Butour: *J. Natl. Cancer Inst.* **70**, 899 (1983).
12. J. D. Hoeschele, T. A. Butler, and J. A. Roberts: *Inorganic Chemistry in Biology and Medicine* (ACS Symposium Series v. 140, Ed. A. E. Martell), p. 181. ACS (1980).
13. D. B. Brown, A. R. Khokhar, M. P. Hacker, J. J. McCormack, and R. A. Newman: in *Platinum, Gold and Other Metal Chemotherapeutic Agents* (ACS Symposium Series 209, Ed. S. J. Lippard), p. 265. ACS (1983).
14. N. Farrell, J. Williamson, and D. J. McLaren: *Biochem. Pharmacol.* **33**, 961 (1984).
15. M. J. Abrams, D. H. Picker, P. H. Fackler, C. J. L. Lock, H. E. Howard-Lock, and R. Faggiani: *Inorg. Chem.* **25**, 3980 (1986).
16. B. Rosenberg, L. Van Camp, J. E. Trosko, and H. V. Mansour: *Nature (London)* **222**, 385 (1969).
17. M. Laverick, A. H. W. Nias, P. J. Sadler, and I. M. Ismail: *Br. J. Cancer* **43**, 732 (1981).
18. L. Pendyala, J. W. Cowens, S. Madajewicz, and P. J. Creaven: *Platinum Coordination Complexes in Cancer Chemotherapy* (Eds. M. P. Hacker, E. B. Douple, and I. H. Krakoff), p. 114. Martinus Nijhoff (1984).
19. E. Rotondo, V. Fimiani, and A. Cavallaro: *Tumori* **69**, 31 (1983).
20. S. Mong, D. C. Eubanks, A. W. Prestayko, and S. T. Crooke: *Biochemistry* **21**, 3174 (1982).
21. J. F. Vollano, E. E. Blatter, and J. C. Dabrowiak: *J. Am. Chem. Soc.* **106**, 2732 (1984).
22. P. C. Hydes and D. R. Hepburn: *U.K. Patent Appl. 2085440A* (1981); *C.A.* **97**, 61004 (1982).
23. R. J. Brandon and J. C. Dabrowiak: *J. Med. Chem.* **27**, 861 (1984).
24. E. E. Blatter, J. F. Vollano, B. S. Krishnan, and J. C. Dabrowiak: *Biochemistry* **23**, 4817 (1984).
25. N. Farrell and K. A. Skov: unpublished results.
26. R. K. Elespuru and S. K. Daley: *Platinum Coordination Complexes in Cancer Chemotherapy* (Eds. M. P. Hacker, E. B. Douple, and I. H. Krakoff), p. 58. Martinus-Nijhoff (1984).
27. J. L. van der Veer, A. R. Peters, and J. Reedijk: *J. Inorg. Biochem.* **26**, 137 (1986).
28. H. Schöllhorn, R. Beyerle-Pfnur, U. Thewalt, and B. Lippert: *J. Am. Chem. Soc.* **108**, 3680 (1986).
29. R. Kuroda, S. Neidle, I. M. Ismail, and P. J. Sadler: *Inorg. Chem.* **22**, 3620 (1983).
30. R. Faggiani, B. Lippert, C. J. L. Lock, and B. Rosenberg: *Can. J. Chem.* **60**, 529 (1982).
31. W. T. Bradner, W. C. Rose, and J. B. Huftalen: *Cisplatin, Current Status and New Developments* (Eds. A. W. Prestayko, S. T. Crooke, and S. K. Carter), p. 171. Academic Press (1980).
32. M. Noji, K. Okamoto, Y. Kidani, and T. Tashiro: *J. Med. Chem.* **24**, 508 (1981).
33. R. H. Earhart: *Cancer Treat. Rep.* **62**, 1105 (1978).
34. C. L. Litterst: *Toxicol. Appl. Pharm.* **61**, 99 (1981).
35. R. J. Knox, F. Friedlos, D. A. Lydall, and J. J. Roberts: *Cancer Res.* **46**, 1972 (1986).

36. S. D. Cutbush, R. Kuroda, S. Neidle, and A. B. Robins: *J. Inorg. Biochem.* **18**, 213 (1983).
37. M. A. Bruck, R. Bau, M. Noji, K. Inagaki, and Y. Kidani: *Inorg. Chim. Acta* **92**, 279 (1984).
38. C. G. van Kralingen, J. Reedijk, and A. L. Spek: *Inorg. Chem.* **19**, 1481 (1980).
39. M. V. Caparelli, D. M. L. Goodgame, R. T. Riley, and A. C. Skapski: *Inorg. Chim. Acta* **67**, L9 (1982).
40. J. D. Hoeschele, N. Farrell, W. R. Turner and C. D. Rithner: *Inorg. Chem.* **27**, 4106 (1988).
41. D. G. Cracienescu, R. Maral, G. Mathé: *Platinum Coordination Complexes in Cancer Chemotherapy* (Eds. M. P. Hacker, E. B. Douple, and I. H. Krakoff), p. 256. Martinus Nijhoff (1984).
42. F. D. Rochon and P. C. Kong: *Can. J. Chem.* **64**, 1894 (1986).
43. P. J. Andrulis, Jr. and P. Schwartz: U.S. Patent 4,565,884; C. A. 102 P: 1790652.
44. N. Farrell: In *Platinum, Gold and Other Metal Chemotherapeutic Agents* (ACS Symposium Series 209, Ed. S. J. Lippard), p. 279. ACS (1983).
45. R. Romeo and M. Cusumano: *Inorg. Chim. Acta* **49**, 167 (1981).
46. S. Lanza, D. Minniti, R. Romeo, and M. L. Tobe: *Inorg. Chem.* **22**, 2006 (1983).
47. N. Farrell, unpublished results.
48. J. A. Howle, G. R. Gale, and A. B. Smith: *Biochem. Pharmacol.* **21**, 1465 (1972).
49. S. J. Lippard, H. M. Ushay, C. M. Merkel, and M. Poirier: *Biochemistry* **22**, 5165 (1983).
50. I. Husain, S. K. Mauldin, A. Sancar, and S. G. Chaney: *Proc. Am. Assoc. Cancer Res.* **27**, 287 (1986).
51. J. Burchenal, K. Kalaher, K. Dew, L. Lokys, and G. Gale: *Biochimie* **60**, 961 (1978).
52. Y. Kidani, K. Inagaki, R. Saito, and S. Tsukagoshi: *J. Clin. Hematol. Oncol.* **7**(1), 197 (1977).
53. C. J. L. Lock and P. Pilon: *Acta Cryst.* **B37**, 45 (1981).
54. K. P. Larsen and F. Toftlund: *Acta Chemica Scand.* **A31**, 182 (1977).
55. M. A. Bruck, R. Bau, M. Noji, K. Inagaki, and Y. Kidani: *Inorg. Chim. Acta* **92**, 279 (1984).
56. F. D. Rochon, R. Melanson, J.-P. Macquet, F. Belanger-Gariepy, and A. L. Beauchamp: *Inorg. Chim. Acta* **108**, 17 (1985).
57. K. Okamoto, M. Noji, T. Tashiro, and Y. Kidani: *Chem. Pharm. Bull.* **29**, 929 (1981).
58. K. Inagaki and Y. Kidani: *Inorg. Chem.* **25**, 1 (1986).
59. M. Gullotti, A. Pasini, R. Ugo, S. Filippeschi, L. Marmonti, and F. Spreafico: *Inorg. Chim. Acta* **91**, 223 (1984).
60. M. Gulotti, G. Pacchioni, A. Pasini, and R. Ugo: *Inorg. Chem.* **21**, 2006 (1982).
61. K. Inagaki, K. Tajima, Y. Kidani, T. Tashiro, and S. Tsukagoshi: *Inorg. Chim. Acta* **37**, L547 (1979).
62. A. Shulman and F. P. Dwyer: *Metal Chelates in Biological Systems* (Chelating Agents and Metal Chelates, Eds. E. P. Dwyer and D. P. Mellor), pp. 383—431. Academic Press (1964).
63. A. B. Robins: *J. Clin. Hematol. Oncol.* **7**(1), 266 (1978).
64. R. E. Cramer, P. L. Dahlstrom, M. J. T. Seu, T. Norton, and M. Kashiwagi: *Inorg. Chem.* **19**, 148 (1980).
65. S. J. Meischen, G. R. Gale, and M. B. Naff: *J. Clin. Hematol. Oncol.* **12**, 67 (1982).
66. Pt compound NSC 356135, NCI antitumour screening data.
67. A. Pasini, G. Pratesi, G. Savi and F. Zunino: *Inorg. Chim. Acta* **137**, 123 (1987).

68. S. Yolles, R. M. Roat, M. F. Sartori, and C. L. Washburne: *ACS Symposium Series* **186**, 223 (1982).
69. W. Beck, B. Purucker, M. Girnth, H. Schoenenberger, H. Siedenberger, and G. Ruckdeshel: *Z. Naturforsch., Teil 3*: **31**, 832 (1976).
70. E. Bersanetti, A. Pasini, G. Pezzoni, G. Pratesi, G. Savi, R. Supino, and F. Zunino: *Inorg. Chim. Acta* **93**, 167 (1984).
71. A. E. Martell: *Chelates of Ascorbic Acid. Formation and Catalytic Properties* (Ascorbic Acid, Chemistry, Metabolism and Uses, Eds. P. A. Seib and B. M. Tolbert), pp. 153–178. ACS Symposium Series 200 (1982).
72. L. S. Hollis, A. Amundsen, and E. Stern: *J. Am. Chem. Soc.* **107**, 274 (1985).
73. M. P. Hacker, A. R. Khokhar, D. B. Brown, J. J. McCormack, and I. H. Krakoff: *Cancer Res.* **45**, 4748 (1985).
74. K. C. Tsou, K. F. Yip, K. W. Lo, and S. Ahmad: *J. Clin. Hematol. Oncol.* **7**(1), 322 (1978).
75. M. Julliard, G. Vernin, and J. Metzger: *Synthesis* **49** (1982).
76. M. Maeda, N. Abiko, H. Uchida, and T. Sasaki: *J. Med. Chem.* **27**, 444 (1984).
77. T.-s. Lin, R.-X. Zhou, K. J. Scanlon, W. F. Brubaker Jr., J. J. S. Lee, K. Woods, C. Humphreys, and W. H. Prusoff: *J. Med. Chem.* **29**, 681 (1986).
78. T. Tsubomura, S. Yano, K. Kobayashi, T. Sakurai, and S. Yoshikawa: *J. Chem. Soc. Chem. Comm.* 459 (1986).
79. H. Schonenberger, B. Wappes, M. Jennerwein, and M. Berger: *Cancer Treat. Reviews* **11** (Supp. A) 125 (1984).
80. O. Gandolfi, J. Blum, and F. Mandelbaum-Shavit: *Inorg. Chim. Acta* **91**, 257 (1984).
81. W. D. McFayden, L. P. G. Wakelin, I. A. G. Roos, and V. A. Leopold: *J. Med. Chem.* **28**, 1113 (1985).
82. E. Hurwitz, R. Kashi, and M. Wilchek: *J. Natl. Cancer Inst.* **69**, 47 (1982).
83. M. Cais: *Methods Enzymol.* **92**, 445 (1983).
84. M. Cais, S. Dani, Y. Eden, O. Gandolfi, M. Horn, E. Isaacs, Y. Josephi, Y. Saar, E. Slovin, and L. Snarsky: *Nature (London)* **270**, 534 (1977).
85. C. Hansch: *Acc. Chem. Res.* **2**, 232 (1969).
86. P. G. Abdul-Ahad and G. A. Webb: *Int. J. Quantum Chem.* **21**, 1105 (1982).
87. T. P. Carsey and E. A. Boudreaux: *Chem.-Biol. Interactions* **30**, 189 (1980).
88. R. Romeo and M. L. Tobe: *Inorg. Chem.* **13**, 1991 (1974).
89. P. D. Braddock, R. Romeo, and M. L. Tobe: *Inorg. Chem.* **13**, 1170 (1974).
90. R. B. Martin: *Hydrolytic Equilibria and N7 Versus N1 Binding in Purine Nucleosides of cis-Diamminedichloroplatinum(II): Palladium(II) as a Guide to Platinum(II) Reactions at Equilibrium* (Platinum, Gold, and Other Metal Chemotherapeutic Agents, ACS Symposium Series 209, Ed. S. J. Lippard), pp. 231–244 (1983).
91. M. E. Howe-Grant and S. J. Lippard: *Metal Ions in Biological Systems* **11**, 63 (1980).
92. M. C. Lim and R. B. Martin: *J. Inorg. Nucl. Chem.* **38**, 1915 (1976).
93. K. J. Scanlon, R. L. Safirstein, H. Thies, R. B. Gross, S. Waxman, and J. B. Guttenplan: *Cancer Res.* **43**, 4211 (1983).
94. R. Faggiani, B. Lippert, C. J. L. Lock, and B. Rosenberg: *J. Am. Chem. Soc.* **99**, 777 (1977).
95. J. A. Stanko, L. S. Hollis, J. A. Schriefels, and J. D. Hoeschele: *J. Clin. Hematol. Oncol.* **7**(1), 138 (1977).
96. R. Faggiani, B. Lippert, C. J. L. Lock, and B. Rosenberg: *Inorg. Chem.* **17**, 1941 (1978).
97. J.-P. Macquet, S. Cros, and A. L. Beauchamp: *J. Inorg. Biochem.* **25**, 197 (1985).

98. J. C. Dabrowiak, M. S. Balakrishnan, J. Clardy, G. D. van Duyne, and L. Silviera: *NMR and Crystallographic Studies on a Second Generation Cisplatin Analogue* (Platinum Coordination Complexes in Cancer Chemotherapy, Eds. M. P. Hacker, E. B. Douple, and I. H. Krakoff), p. 63. Martinus-Nijhoff (1984).
99. B. Rosenberg: *Biochimie* **60**, 859 (1978).
100. D. S. Gill and B. Rosenberg: *J. Am. Chem. Soc.* **104**, 4598 (1982).
101. C. J. Boreham, J. A. Broomhead, and D. P. Fairlie: *Aust. J. Chem.* **34**, 659 (1981).
102. M. Chikuma and R. J. Pollock: *J. Magn. Res.* **47**, 324 (1982).
103. P. S. Pregosin: *Coord. Chem. Rev.* **44**, 247 (1982).
104. T. G. Appleton, R. D. Berry, C. A. Davis, J. R. Hall, and H. A. Kimlin: *Inorg. Chem.* **23**, 3514 (1984).
105. T. G. Appleton, J. R. Hall, and S. F. Ralph: *Inorg. Chem.* **24**, 4685 (1985).
106. L. G. Marzilli, Y. Hayden, and M. D. Reily: *Inorg. Chem.* **25**, 974 (1986).
107. G. Annibale, L. Canovese, L. Cattalini, G. Marangoni, G. Michelon, and M. L. Tobe: *Inorg. Chem.* **23**, 2705 (1984).
108. C. M. Riley, L. A. Sternson, and A. J. Repta: *Anal. Biochem.* **130**, 203 (1983).
109. P. Melius and M. E. Friedman: *Inorg. Perspect. Biol. Med.* **1**, 1 (1978).
110. B. Odenheimer and W. Wolf: *Inorg. Chim. Acta* **66**, L41 (1982).
111. A. Meister and M. E. Anderson: *Ann. Rev. Biochem.* **52**, 711 (1983).
112. B. A. Arrick and C. F. Nathan: *Cancer Res.* **44**, 4224 (1984).
113. K. Micetich, L. A. Zwelling, and K. W. Kohn: *Cancer Res.* **43**, 3609 (1983).
114. A. Eastman and V. M. Richon: *Chem.-Biol. Inter.* **61**, 241 (1987).
115. P. A. Andrews, M. P. Murphy, and S. B. Howell: *Cancer Res.* **45**, 6250 (1985).
116. A. Eastman and E. Bresnick: *Biochem. Pharmacol.* **30**, 2721 (1981).
117. I. M. Ismail and P. J. Sadler: *^{195}Pt and ^{15}N-NMR Studies of Antitumour Complexes* (Platinum, Gold and Other Metal Chemotherapeutic Agents, Ed. S. J. Lippard), pp. 171—190. ACS Symposium Series 209, Washington (1983).
118. A. B. Robins and M. O. Leach: *Cancer Treat. Rep.* **67**, 245 (1983).
119. E. Segal and J.-B. lePecq: *Cancer Res.* **45**, 492 (1985).
120. L. S. Hollis, A. R. Amundsen, and E. W. Stern: *J. Med. Chem.* **32**, 128 (1989).
121. L. S. Hollis, E. W. Stern, A. R. Amundsen, A. V. Miller, and S. L. Doran: *J. Am. Chem. Soc.* **109**, 3596 (1987).
122. N. Farrell, S. G. de Almeida, and K. A. Skov: *J. Am. Chem. Soc.* **110**, 5018 (1988).
123. R. F. Borch: 'The Platinum Anti-Tumour Drugs' in *Metabolism and Action of Anti-Cancer Drugs* (Eds. G. Powis and R. A. Prough), p. 163. Taylor and Francis, London (1987).
124. S. K. Maudlin, I. Husain, A. Sancar, and S. G. Chaney: *Cancer Res.* **46**, 2876 (1986).
125. S. K. Maudlin, F. A. Richard, M. Plescia, S. D. Wyrick, A. Sancar, and S. G. Chaney: *Anal. Biochem.* **157**, 129 (1986).
126. J. D. Bell, R. E. Norman, and P. J. Sadler: *J. Inorg. Biochem.* **31**, 241 (1987).
127. F. D. Rochon and P. C. Kong: *J. Clin. Hematol. Oncol.* **12**, 39 (1987).
128. A. J. Canty and E. A. Stevens: *Inorg. Chim. Acta* **55**, 157 (1981).
129. L. Kumar, N. R. Kandasamy, T. S. Srivastava, A. J. Amonkar, M. K. Adwanka, and M. P. Chitnis: *J. Inorg. Biochem.* **23**, 1 (1985).

CHAPTER 4

INTERACTIONS OF PLATINUM COMPLEXES WITH DNA COMPONENTS

The identification of DNA as a primary target for metal-based drugs, especially cisplatin, has focused attention on the interactions of metal complexes with nucleic acid constituents, which include the simple purine and pyrimidine bases and their nucleoside and nucleotide derivatives. The structures, with abbreviations, are represented in Appendix 1. Simple complexes can represent models for cross links in DNA, which can be studied in more detail with small polynucleotides, from the simpler dinucleotides to oligonucleotides and this topic is covered in Section 4.4. There has been extensive use of substituted purines and pyrimidines as models for the DNA bases and in the examination of steric and electronic effects. The structures of many of these analogues are also collected in Appendix 1.

The principal aspects may be divided into two major areas — structural, concentrating on the sites of binding, both in solution and in the solid state, and reactivity, which means how metal binding affects the chemistry of the purine or pyrimidine moiety. These interests are of course generally relevant for all metal systems in terms of the same effects as are summarized in Table 1.I, and an extensive literature exists. The emphasis here will be on studies related to the metal complexes of chemotherapeutic interest. Therefore, other relevant studies and structures will be mentioned in this context although fuller details may be obtained in particular from the cited review references.

4.1. Structural Aspects Of Metal—Nucleobase Binding

The purines and pyrimidines present multiple binding sites and the particular choice for a metal will be affected by kinetic and thermodynamic considerations. A number of reviews have appeared since the early ones of Izatt *et al.* [1] and Eichhorn [2], and these have summarized the binding in solution [3—6] as well as purely solid state aspects [7—9].

4.1.1. BINDING IN SOLUTION

The factors which affect metal binding in solution may be summarized as:

4.1.1.1. *Basicity and pH*

The order of decreasing basicity, and therefore *possible* metal ion preference, of the endocyclic nitrogens of the common unsubstituted bases as judged by pK_a data are [3]:

$$\text{Thy } N_3\ (9.8) > \text{Ura } N_3\ (9.4) \sim \text{Gua } N_1\ (9.4) > \text{Cyt } N_3\ (4.6) > \text{Ade } N_1\ (4.1) > \text{Gua } N_7\ (1.9) > \text{Ade } N_7\ (\sim -1.6)$$

The values in parentheses are the pK_a values of the free bases; the nucleoside preference follows the same order.

The effect of pH of the solution is also apparent from these comparative values — at biological pH, the first three listed sites will be protonated and the metal ion is not considered to compete strongly for these centers. Thus, at neutral pH the order of preference may be predicted as: Cyt N_3 > Ade N_1 > Gua N_7 > Ade N_7. The order for CH_3Hg^+ binding for these four sites is, however, Cyt N_3 > Gua N_7 > Ade N_1 > Ade N_7 [10]. The species distribution has been discussed [3, 10] with respect to competitive binding to the N_1 and N_7 sites of guanine. The 'promotion' of Gua N_7 is especially relevant to studies on platinum. A similar promotion has been demonstrated for Cu(II) [11]. A study with *trans*-$[Ru(H_2O)(SO_3)(NH_3)_4]$ also established a strong preference for guanine binding ($K_{Gua}/K_{Ade} = 95$) [12].

The pH of the solution can be responsible for linkage isomerism (see below) and is also a critical feature for metal—aqua complexes, where deprotonation of the coordinated water group can lead to formation of hydroxo complexes whose potential oligomerization can lead to complications and certainly difficulties in interpretation.

4.1.1.2. *Hard-Soft Relationship of Donor Atom and Metal Ion*

The basicity data mentioned above refer only to the ring nitrogens. When exocyclic binding sites are considered there is an abundance of possibilities for cross linking. In this case the nature of the donor atom (oxygen or nitrogen) and the metal ion (hard or soft) is of importance. The hard alkali and alkaline earth metals would favor O-donors, as would also transition metals ions such as Cr^{3+}, Fe^{3+} and Mn^{2+}. The intermediate case can be found for Mn^{2+}, again, and Fe^{2+}, Zn^{2+}, and Cu^{2+} while those ions preferring N donors will be Ag^+, Hg^{2+}, Pd^{2+} and Pt^{2+}. As usual, there are exceptions to these trends. As has been pointed out [3], these definitions of metal—ligand combinations do not follow a rigid hard/soft delineation, since none of the potential donor ligand atoms can really be considered as soft.

4.1.1.3. *Steric Effects including Hydrogen Bonding*

In a nucleic acid, as distinct from the individual bases, the presence of hydrogen bonding will render some possible binding sites less favorable. Further, a combination of interligand hydrogen bonding and nonbonded repulsions can afford selectivity in particular bases. This has been exploited for cobalt complexes containing hydrogen bond donating and accepting ligands [13]. A further relevant example is that of rhodium acetate (see Chapter 6.1.2), which does not bind to doubly stranded DNA [14] but reacts in a highly specific manner with adenine based nucleosides and nucleotides [15]. While it should be pointed out that this interaction may not be important for the antitumour effect of this particular complex, the remarkable specificity has been attributed to favorable hydrogen bonding interactions between the oxygen atoms of the acetate ligands and the exocyclic amino group of the adenine ring [16].

4.1.1.4. *Base Stacking and Chelation*

At higher concentrations (>0.02M) the purine bases stack, and so do the pyrimidine bases, although to a lesser extent. This stacking, which may be metal ion promoted, can offer conformations suitable for chelation. A further mode of chelation on the individual bases is with suitably placed groups such as the N_7 (or N_1)—C_6 carbonyl pairs of guanine and the N_7 (or N_1)—C_6 amino group pairs of adenine. The former has been a popular suggestion for platinum, as it would explain the difference between the reactivity of *cis*- and *trans*-$[PtCl_2(NH_3)_2]$, but no direct evidence for chelation of this type has been observed (see 4.2.1.1). Indirect chelates via a water molecule or weak axial binding are well documented in structural examples [17, 18] and are shown schematically in Figure 4.1. Such bonding has been observed in structures with tRNA. Chelation in nucleotide complexes involving the phosphate group and a base donor is known [4, 8].

4.1.2. BINDING OF PLATINUM COMPLEXES IN SOLUTION

Very useful reviews of relevant aqueous platinum chemistry are available

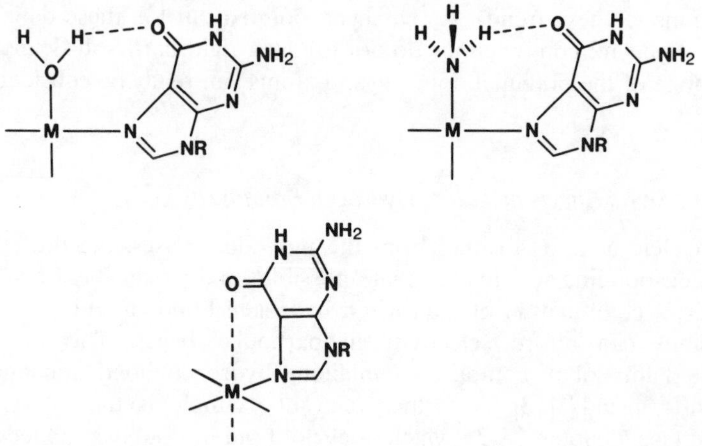

Fig. 4.1. Scheme for possible hydrogen bonding interactions and weak interactions between exocyclic groups and a metal center and aqua and ammine groups on the metal.

[19, 20]. The "promotion" of Gua N_7 as a preferred binding site occurs for Pt(II), in addition to the examples cited above. The exact trend for platinum has, however, been difficult to determine, in part due to kinetic factors. Since Pt(II) complexes reach equilibrium only slowly, there is always the question of whether the reaction analyzed and the complexes formed are under thermodynamic or kinetic control [3]. A general summary of experimental methods is available [6].

One approach has been to use the kinetically more labile palladium complexes as analogues of the platinum systems, although it has been emphasized that even if the reaction is 10^5 times faster than platinum, the palladium complexes still react slowly [21]. Using ^{1}H NMR and potentiometric studies on the tridentate diethylenetriamine chelate, $[Pd(dien)(H_2O)]^{2+}$, which allows substitution at only one site, an order of thermodynamic binding at pH 7 was obtained for nucleosides [22]:

$$\text{Guo } N_7 > \text{Guo } N_1 > \text{Urd } N_3 > \text{Thyd } N_3 > \text{Cytd } N_3 > \text{Ado } N_1 > \text{Ado } N_7$$

The relative orders for the two guanine possibilities, N_1 and N_7, confirm that in solution there is a competition between these two sites. Since metallation at N_7 in general renders the hydrogen at the N_1 site more acidic (see 4.2.1), the potential for binding at the second site increases upon initial N_7 binding. In the case of adenosine both sites are deprotonated at normal pH, and N_1 is favored. The tendency for binding at N_1 or N_7 has been considered in detail in the $[Pd(dien)(H_2O)]^{2+}$ system, and the intrinsic binding ratios calculated and compared to those for H^+ and CH_3Hg^+.

An interesting feature of the adenine system is that the presence of a 5′-phosphate group (i.e. nucleotide compared to nucleoside) favors metallation of N_7 by approximately 0.7 log units, and thus the relative order in AMP is $N_7 > N_1$. The extent of N_7 binding for both guanosine and 5′-GMP is much greater than could be expected from purely basicity considerations. In competition studies with mixtures of nucleotides, binding of 5′-GMP N_7 is favored over the pH range 5—9, but for other nucleotides is pH-dependent [21].

The results quoted above are claimed to be extendable to platinum [21] but may also be considered to represent the simplest and most favorable case for study: attainment of thermodynamic equilibrium at one substitution site. In the case of the $[Pd(en)_2]^{2+}$ cation, both N_7 and N_1 binding are observed for IMP, but the system is quite complicated [23].

A detailed examination of the reactions of both *cis*- and *trans*-$[PtCl_2(NH_3)_2]$ with 5′-GMP was carried out by Tobias and co-workers, and the species distribution studied as a function of concentration and pH [24]. The major species at neutral pH for both isomers was $[Pt(NH_3)_2(5'\text{-GMP})_2]^{2-}$, when the ratio of Pt complex to nucleotide (r) was <0.5. At higher ratios other reactions occurred. The binding is clearly N_7. In the presence of excess chloride, more than one species is present, and detailed NMR studies [25, 26] have confirmed their assignments as species containing only one 5′-GMP, as well as an interesting complex considered to contain a Pt bound to N_7 with another metal atom binding to N_1:

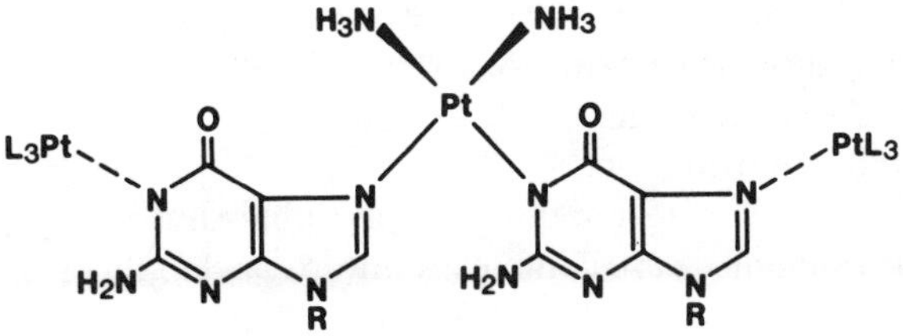

(45)

This binding situation is a consequence of the above mentioned deprotonation of N_1 caused by the first metal binding at N_7 and was first observed with studies on inosine [27]. Later results [25, 26] show convincingly that previous conclusions on this system are in error [28].

In a study of competitive reactions using a mixture of the common nucleotides, selective binding to 5′-GMP followed by 5′-AMP was found at low r values, $0.05 < r < 0.2$ [29]. Recent kinetic measurements gave second order rate constant values of 144 $M^{-1} s^{-1}$ for 5′-GMP as against 3.23 $M^{-1} s^{-1}$ for 5′-AMP [30]. A second step, with a rate constant of 23.8 $M^{-1} s^{-1}$, was noted for 5′-GMP.

A study of binding to ribonucleosides reported formation constants of $\log K = 3.7$, 3.6 and 3.5 for Guo, Ado and Cytd, respectively, implying very little thermodynamic selectivity [31], although these values have been questioned [3]. In these experiments the binding sites are difficult to assign, a general fault of UV-based studies. For adenine, an NMR study unequivocally demonstrated the N_1—N_7 binding mode at low Pt/base ratios [32]. The general feature of binding to bases appears to be that of kinetic control. Another feature of binding to nucleotides is that the presence of the phosphate group increases the rate of reaction with the metal complex. The strong discrimination for guanine binding, of course, is also reflected in the preferential binding on DNA where the ready accessibility in the major groove over the inward-directed adenine N_1 or cytosine N_3 contributes to the selectivity.

4.2. Reactivity of Platinum Metal—Base Complexes

The preceding discussion has emphasized studies of binding sites and thermodynamic/kinetic considerations. An area of interest, which is now receiving as much attention as the structural aspects, is the reactivity of the bases upon metallation. The principal points of nucleobase chemistry modified by metallation which have been studied are:

(1) Effect on acidity of nonbound nitrogens.
(2) Effect on C—H exchange.
(3) Susceptibility to oxidation.
(4) Reactivity of base—sugar and sugar—phosphate linkages.
(5) Reactivity of metal—base complexes toward nucleophiles.

These points all refer to alteration of the inherent chemistry of the bases, which, although not emphasized in relation to platinum metal complexes and their antitumour activity, is a recognized lesion which may lead to cell kill through inhibition of DNA synthesis. Further features of metal—nucleobase chemistry are the dynamic features of

(6) Linkage isomerism and
(7) Rotation around metal—base bonds.

4.2.1. EFFECT OF METALLATION ON ACIDITY OF NONCOORDINATED ATOMS

A major effect is that for guanine, for example, metallation of N_7 causes an increase in acidity at N_1. Table 4.I summarizes the principal published values [33]. Since the chemical shifts of these protons are also pH-dependent, metallation will also affect chemical shifts [36, 37] and we note here that this correlation of shift vs. pH is extensively used for assignment of binding sites, especially for oligonucleotides.

In view of the fact that the N_1 proton is involved in hydrogen bonding in the G—C base pair the metal-induced deprotonation is relevant, and could lead to mispairing. A unique case of unusual base pairing has been structurally demonstrated, Figure 4.2 [34].

The binding of the exocyclic 4-NH_2 group of cytosine has also been observed for both Pt and Ru. For platinum, deprotonation of this exocyclic group occurs even at pH 7, in contrast to the pK_a (12.8) of this group in the free ligand. The change is suggested to result from platination of N_3, the primary Pt binding site [38]. This deprotonation was observed in the reaction of the hydroxo-bridged dimer $[Pt(NH_3)_2(OH)_2(NH_3)_2Pt]^{2+}$ with cytosine, and results in the formation of a dimeric complex, with bridging via N_3 and 4-NH, similar to the type discussed for thymine and uracil (Chapter 5) [38]. The binding of Ru(III) to the exocyclic amines of adenosine and cytidine gives large increases in acidity for both the neighboring N_1 and N_3 atoms [39]. Thus, a pK_a of 9 is found for N_1 in the Ru(III) complex compared to a value of 1—3 for the Ru(II) analogue [39, 40].

TABLE 4.I
Influence of metallation on ring nitrogen basicity in guanine and its nucleoside and nucleotides.

Complex	pK_a (found)	ΔpK_a (shift)	Ref.
cis-$[Pt(NH_3)_2$(9-EtG)(1-MeC)]		1.6	34
$[Ru(NH_3)_5(Guo)]^{3+}$	7.4	2.1	35
$[Ru(NH_3)_5(Guo)]^{2+}$	8.7	0.8	35
$[Pt(dien)(Guo)]^{2+}$	8.3	1.2	33
cis-$[Pt(NH_3)_2(5'\text{-GMP})_2]^{2-}$	8.2	0.3	24
trans-$[Pt(NH_3)_2(5'\text{-GMP})_2]^{2-}$		0.6	24

4.2.1.1. *The Platinum-N_7, O_6 Chelate*

A possible binding mode which has received considerable attention is that of chelation to guanine N_7 and O_6. A major attraction of this postulate is that it would explain the difference between the *cis*- and *trans*-isomers, since the *trans*-isomer would not be capable of binding in such a manner. An early analysis of this binding mode [24] has been supplemented by more recent ones [25, 26, 41] and no clear evidence has been found to support such a hypothesis. All the evidence is in agreement that, after initial N_7 binding, the tendency is to bind to deprotonated N_1 [140].

4.2.2. EFFECT OF METALLATION ON C—H EXCHANGE

The C_8 hydrogens of purines and the C_5 hydrogens of pyrimidines are base-labile and exchange with, for instance, deuterium in D_2O. This process is catalyzed upon binding of a transition metal to the endocyclic nitrogens and, as a result, some unique structures have been observed. A kinetic analysis indicated that the exchange was still 10^5 times slower for platinated guanosine and inosine than for the protonated analogues but this is almost a 10^2 times greater difference than could be expected from considerations of pure basicities of the adjacent (N_7) nitrogens [41]. These results contradict previous estimates which implied a much more (by 10^4) rapid rate for this exchange [42]. The results with both $[Pt(en)(H_2O)_2]^{2+}$ and $[Pt(dien)(H_2O)]^{2+}$ [41] confirm earlier observations that the exchange is more rapid for a mono complex than a bis complex with two bound purines [43].

The exchange on uracil and cytosine complexes has also been noted [44], and the formation of substituted uracil from the Cl_2 oxidation of Pt(II)—uracil complexes to Pt(IV) also reflects the increased reactivity

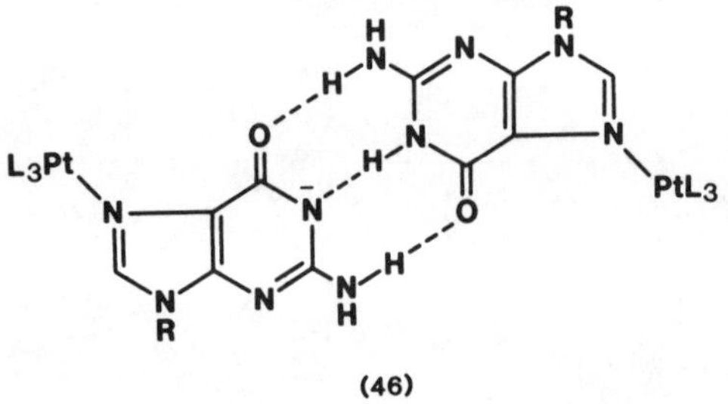

(46)

Fig. 4.2. The unusual hydrogen bonding scheme observed in the $[Pt(NH_3)_2(1\text{-MeCyt})(9\text{-EtGua})]^{2+}[Pt(NH_3)_2(1\text{-MeCyt})(9\text{-EtGua})H]^+$ complex. The two guanine rings form a platinated G—G pair in the solid state. From Reference 34.

[45]. In this latter case, reaction of *cis*-$[Pt(NH_3)_2(1\text{-}MeU)Cl]$ gave not only the expected product, $[Pt(NH_3)_2(1\text{-}MeU)Cl_3]$, but also similar products with the C_5-H displaced by Cl and HOCl, respectively [45].

The activation of the C_8-H bond upon metallation is really a mimic of the Mo-catalyzed enzyme xanthine oxidase which catalyzes the oxidation of xanthine to uric acid [46]:

$$\textbf{(47)} \xrightarrow[-(2e^-,\ 2H^+)]{+H_2O} \textbf{(48)}$$

Studies on the reactivity of the $[Ru(NH_3)_5(Guo)]^{2+}$ cation showed that one of the decomposition products, upon metal ion induced hydrolysis of the sugar—purine bond, was 8-hydroxyguanine, indicating oxidation of the bound ligand [47]. The propensity for Ru(III) to bind to this carbon (C_8) site, when no nitrogen ligands are available, has been demonstrated also in the caffeine and 4,5-dimethylimidazole complexes of $[Ru(NH_3)_5]^{2+}$ where the presence of a Ru—C bond has been structurally confirmed [48, 49].

The C_8-hydrogen exchange was also observed with the CH_3Hg cation [50], when C_8—Hg bond formation also occurred [51]. The fixation of Hg on C_5 of pyrimidines has been used for heavy metal labelling of nucleic acids and, in this respect, the formal epoxidation of the thymine double bond by OsO_4 for labelling purposes also falls within this topic, as an activation of the C—H bond. Finally, it is worthwhile mentioning that organopalladium intermediates have been used as a general synthetic route to C_5-substituted pyrimidines and pyrimidine nucleosides [52—54]. Indeed, the use of metal complexes for synthesis of methylated purines and pyrimidines could be of considerable use in producing these biologically relevant molecules in simpler syntheses than those presently employed. The alkylation of xanthine by N_7 methylation in a cobalt complex has been reported [55]. The use of the tricarbonylcyclohexadienyliron cation for alkylation of both purine and pyrimidine derivatives has also been studied [56].

4.2.3. SUSCEPTIBILITY TO OXIDATION OF METAL—BASE COMPLEXES

The xanthine oxidase type reactivity observed and discussed above is strictly an oxidation and more general oxidations may also be metal catalyzed. An interesting recent example is that of the susceptibility of Pt—cytosine complexes to H_2O_2 oxidation and their catalytic activity for decomposition of H_2O_2 [57]. In this work, platinum antitumour agents were reacted with DNA, RNA or their constituents, and their ability to decompose H_2O_2 measured. The activity was most marked with cytosine, with adenine showing minimal activity and the other common bases intermediate behavior. The *trans*-$[PtCl_2(NH_3)_2]$ is inactive, interestingly enough, as was the Pt(IV) analogue of cisplatin (Compound 14, Figure 2.1). There has been a suggestion that this activity may be relevant to the mechanism of antitumour action of platinum complexes, through effects on DNA leading to singly stranded, rather than doubly stranded helices. The complexes are, however, only slowly active at room temperature, especially with DNA and thus, although interesting, the theory does not seem likely at this stage to furnish alternative mechanisms for platinum activity. However, formally, the susceptibility could be relevant in radiosensitization (Chapter 8), where hydrogen peroxide is produced as a radiolysis product.

4.2.4. REACTIVITY OF NUCLEOSIDE AND NUCLEOTIDE LINKAGES

The cleavage of the base—sugar bond of nucleosides is catalyzed by alkylation of the purine ring at N_7 [58], and it might be thought that a similar process could prevail upon metallation. For the $[Pt(dien)(Guo)]^{2+}$ cation, results indicate that the linkage is in fact stabilized against acid decomposition [33].

A detailed examination of this aspect using the pentammine—ruthenium(III) system showed that the proton-catalyzed rate of cleavage was decreased by a factor of 30 over that of the free ligands, for both guanosine and deoxyguanosine [35]. However, the effects are not as straightforward as might appear and, if allowance is made for the reduced basicity of the purine ring on metallation, the protonated metal complex is actually more effective than protonated ligand alone [35]. The half-life for the decomposition of $[Ru(NH_3)_5(dGuo)]^{3+}$ is of the order of 4.5 days, with both guanine (base—sugar cleavage) and deoxyguanosine detected [59]. This is clearly too slow to result in biological effects leading to cell kill, and the same argument applies to Pt(II) systems [60].

4.2.5. REACTIVITY OF METAL—BASE COMPLEXES TOWARD NUCLEOPHILES

The observations that cyanide and sulfur nucleophiles such as thiourea can reverse Pt—DNA crosslinks have their parallels in model studies. In a survey of a series of metal—purine and pyrimidine complexes Lippert and Raudaschl-Sieber showed that the complexes varied in their reactivity to CN^- [135]. Since, at high binding ratios, not all the platinum is removed from DNA by cyanide (see Chapter 1.3.3.3) these lesions could be similar in structure to the model compounds unreactive toward the nucleophile [135]. Of interest is the fact that the Pt—pyrimidine complexes are most resistant, the bis(guanine) complexes reacting readily. Do these results imply that a Pt—pyrimidine linkage would be difficult to repair in a biological situation?

DDTC can inhibit platinum nephrotoxicity without affecting activity (Chapter 2.2). In this case, studies with model compounds showed that the bis(guanosine) complexes were not affected by DDTC (10 mM, 37 °C) whereas a mono(guanosine) complex and a bis(adenosine) complex reacted readily with the DDTC ligand [137]. These results confirm observations from studies on DNA, where very little platinum is removed by DDTC at low binding ratios, and allow the prediction that DDTC should not reverse the toxic lesion on DNA, as observed [137]. It would be interesting to correlate the various effects of the nephroprotective sulfur nucleophiles (Chapter 2.2) and their effects on antitumour activity of cisplatin with the behavior toward model compounds (see also Chapter 3.7).

4.2.6. LINKAGE ISOMERIZATION

Unsubstituted uracil and thymine, as their monoanions, exist in tautomeric forms with the single proton residing on either N_1 or N_3. Tautomeric complexes of uracil with the triammine—platinum cation, $[Pt(NH_3)_3Cl]^+$, were recognized by Inagaki and Kidani [61]. The protonation of 1-methylthymine platinated at N_3 in *cis*-$[Pt(1\text{-}MeT)_2(NH_3)_2]$ affords the rare iminol tautomer in an interesting reaction [62]. In cases where the platinum moiety is the *cis*-$[Pt(NH_3)_2]$ unit the formation of tautomer and linkage isomer complexes is complicated and factors affecting the formation of the individual species have been summarized [44, 63]. For example, in DMF the monouracil complex *cis*-$[Pt(NH_3)_2(HU)Cl]$ is N_1-bound but in mildly acidic aqueous medium the N_3-bound species predominates [63]. The formation of linkage isomers in solution in the $[Pt(NH_3)_3(HU)]^+$ system has also been observed [44].

Clear-cut examples of linkage isomerization have also been observed in the ruthenium(II,III) systems and these have been summarized [39, 60]. The synthesis of all three possible isomers (N_3, N_7 and N_9 bound) has been achieved [64]. The 7-methylhypoxanthine complexes undergo a facile, acid-catalyzed isomerization from N_3 to N_9.

The study of linkage isomerization, especially of the slow transformation of initially formed species, is relevant to the biological action of metal complexes because it is feasible that these transformations could occur slowly on DNA.

4.2.7. ROTATION AROUND METAL—PURINE BONDS

Another question related to the dynamics of platinum—nucleobase chemistry is the rotational freedom or otherwise of the metal—purine bonds. This point is relevant to the formation of, for instance, an intrastrand link upon binding of the platinum to the first base. Studies on the *cis*-$[Pt(NH_3)_2(Guo)_2]^{2+}$ cation indicated little restriction around the Pt—purine bond [65, 66] but with bulkier ligands (*N,N*-Me_2-en, *N,N,N′,N′*-Me_4-pn) restricted rotation is observed [65, 67].

The suggestion has been made that the guanine base is less sterically hindered than the other bases and forms the intrastrand link more easily. Very pertinent to this suggestion is the observation that *cis*-$[Pt(NH_3)_2(5'\text{-AMP})_2]$, where the 5′-AMP is bound by N_7 with the bases in a head-to-tail (htt) orientation, does exhibit restricted rotation, as evidenced by NMR [67]. The different exocyclic substituents in the 6-position (NH_2 for Ade, O for Gua) may be responsible for this difference between 5′-AMP and 5′-GMP [67].

The difference in guanine and adenine binding is further relevant because, in platinated adducts of DNA (Section 4.5), the predominant adduct formed is the d(GpG) complex while, in the second most important mixed guanine—adenine adduct, the d(ApG) sequence is favored over the alternative d(GpA) arrangement (see Section 4.4). The aspects of formation of the bidentate linkage from a monodentate (guanine) complex is being increasingly studied in detail to explain the sequence preferences on DNA.

4.3. Solid State Structural Studies

The determination of crystal structures for many platinum metal—nucleobase complexes has elucidated many of the binding features, although the foregoing discussion is evidence of the complexity of these mixtures in solution. The bases studied range from unsubstituted purines and pyrimidines to the nucleosides and nucleotides. One feature of this area is the general difficulty of obtaining suitable crystalline samples of the sugar (nucleoside) derivatives, and this has been circumvented in some way by use of 9-alkylated purines and 1-alkylated pyrimidines, thus blocking these potential nitrogen donors at the site of the sugar attachment and giving complexes which are more amenable to recrystallization.

The chemical reactivity summarized above is unlikely to be the principal feature of inhibition of DNA synthesis, despite the physical binding, and thus conformational changes which are repaired only with difficulty are a very likely source of the molecular mechanism of inhibition of cellular replication. The principal models for the proposed intrastrand link in DNA will be based on a bis(nucleobase) complex of the platinum—diamine unit. For reference, the principal structures for *cis*-Pt models are summarized in Table 4.II for alkylated bases, nucleosides and nucleotides, along with binding sites and comments. The emphasis on guanine binding is reflected in the number of structures for this base. A full summary of complexes of Pt(II) structurally characterized up to 1981 is to be found in [68], complementing the earlier reviews of metal complexes in general [7—9]. Selected examples have also been discussed [69].

TABLE 4.II

Selected structural parameters for complexes of general type *cis*-$[Pt(am)_2(nucleobase)_2]$.[a]

Complex	Dihedral Angles[b] *B/B'*	*B/M* *B'/M*	Ref.
$[Pt(NH_3)_2(9\text{-}EtG)_2]^{2+}$	68	75.4 49.2	73
$[Pt(NH_3)_2(Guo)_2]^{2+}$	74	74 70	74
$[Pt(en)(Guo)_2]^{2+}$	71	—	75
$[Pt(NH_3)_2(5'\text{-}IMP)_2]^{2-}$	40.7	61.8	76
$[Pt(NH_3)_2(3\text{-}MeAde)_2]^{2+}$	90.6	98.9 111.8	77
$[Pt(NH_3)_2(1\text{-}MeCyt)_2]^{2+}$	102.3	101.2 102.4	72, 78
$[Pt(NH_3)_2(1\text{-}MeU)_2]$	109	112.5 119.3	79

[a] Counteranions omitted for clarity. Abbreviations as in appendix.
[b] Dihedral angles as defined in text.

How do studies on model compounds help predict the distortion in DNA conformation upon metal binding? The binding sites for guanine clearly reflect the preference for N_7. Model studies also give conformational parameters for cross links of biological importance. The number of cross links statistically possible for cisplatin binding is high and, although most attention has been paid to the GG link, a full description of other possibilities, such as the GA, bis(pyrimidine) and mixed purine—pyrimidine complexes is necessary for comparison. The study of many of these has been undertaken, especially by Lippert, Lock and co-workers [70] and Marzilli and co-workers, and their relevance to DNA-binding modes examined [71, 72]. Besides the binding site, further aspects of solid state studies are the conformational parameters which describe the relative orientation of the bases and the description of their geometry with respect to the coordination plane; these factors describe the conformational alterations dictated for polymer binding.

The conventions for this analysis have been laid down in some detail by Marzilli and co-workers [71, 72]. The major conformational parameters for *cis*(bisnucleobase) complexes, where in fact the bases may be the same or two different bases, are defined as:

(a) The interbase dihedral angle (B/B') and
(b) The interbase/coordination plane angles (B/M and B'/M).

The first angle (B/B') describes the relative orientation of the two planar rings with respect to each other. The second angle (B/M or B'/M where M represents the ligating atoms PtN_4) measures the relation between the planar rings and the plane of the coordination sphere. A system for visualizing these angles has been explained [71, 72]. The major angles are summarized for selected complexes in Table 4.II. Some structures are shown in Figure 4.3. The variation in parameters and their implications have been thoroughly discussed [69—71].

In structural work, the exact conformation may be affected by a combination of intra- and inter-complex interactions such as counterion, medium of recrystallization, hydrogen bonding and steric effects of the particular substituted nucleobase studied [4, 5]. These effects have to be somewhat separated from 'inherent' effects such as the dihedral angles discussed above, although the interplay is complicated. The various parameters have been summarized for bis(purine), bis(pyrimidine) and mixed purine and pyrimidine complexes and for hydrogen bonding patterns. The reader is referred to the references [69—71] for detailed analysis of these points.

Firstly, it should be noted that with the exception of the first complex in

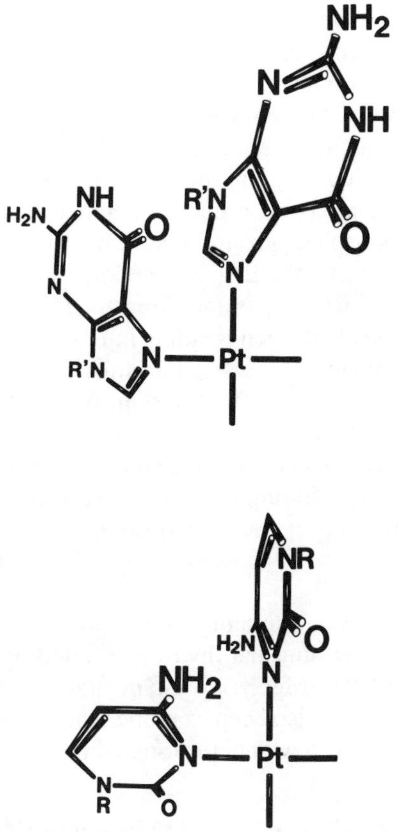

Fig. 4.3. Representation of orientation of two nucleobases in the complexes of *cis*-[Pt$(NH_3)_2Cl_2$] with (A) 9-EtGua and (B) 1-MeCyt. The guanine ligands are in the head-to-head configuration while the cytosine groups are head-to-tail. From References 72, 73, and 78.

Table 4.II, all structures have the head-to-tail orientation, rather than the head-to-head expected in DNA. The recent isolation of head-to-head isomers is of significance for comparative purposes, and as a 'real model' of the polynucleotide interaction [73]. The head-to-head arrangement has been extensively studied with various counteranions. In the case of the head-to-head isomers of the *cis*-[Pt$(NH_3)_2$(9-EtG)$_2$]$^{2+}$ cation, the dihedral angle between the bases is 72° and results in a short N_7—N_7 distance of 2.82 Å compared to 3.9 Å in DNA. However, the dihedral angles of the bases with the Pt coordination planes vary widely, depending on the counteranion. The steric effects are balanced in the O_6—O_6 separation which remains almost constant (varying from 3.37 to 4.10 Å) in comparison to the distance in DNA of 3.5—6 Å. This implies that there would not be a severe distortion of base stacking in the platinated DNA.

There is an interesting difference in going from the simple base to the nucleoside and nucleotide. Within similar complexes, the presence of a phosphate group in the molecule reduces the dihedral angles dramatically. The presence of the phosphate group in mononucleotide complexes allows for the study of hydrogen bonding interactions as, for example, in the complexes with IMP which are, in fact, nonstoichiometric [76]. The presence of an intracomplex hydrogen bonding interaction from O(phosphate) . . . H—O—H . . . O_6 (guanine) dictates this smaller *B*/*B*′ angle. This trend is continued in the dinucleotide complexes studied both in solution and in the solid state (see Section 4.4).

The cytosine structures also reflect thermodynamic tendencies, although, as stated, exocyclic amine binding has been observed. For bis(pyrimidine) complexes, the trends are in fact opposite to those observed for the bis(guanine) systems and result in severe crowding for the $[Pt(NH_3)_2(1\text{-}MeCyt)_2]$ cation.

Mixed ligand complexes containing one purine and one pyrimidine are few in number. The unusual hydrogen bonding of a [Pt(9-EtG)(1-MeCyt)]—[Pt(9-EtGH)(1-MeCyt)] system has been noted and the individual components have also been examined [80]. A mixed pyrimidine—Pt complex is that of 1-MeCyt with deprotonated thymine [81].

4.4. Structural Studies on Di- and Oligonucleotide Complexes

Over the last few years, significant advances in the depth of understanding of the conformational features of the polynucleotide—platinum complex interaction have come from both solution (CD, NMR) and solid-state (X-Ray) studies on oligonucleotide fragments, particularly dinucleotide complexes. Excellent summaries of the status of model studies and their relationship to the molecular mechanism of cisplatin can be found in references [82 and 141]. These complement the earlier reviews [83, 84]. The features outlined do in fact confirm some of those expected from model studies, although there has not been an extensive comparison with the *trans*-isomer.

4.4.1. STUDIES ON DINUCLEOTIDE COMPLEXES

The dinucleotide of guanine, d(GpG) and d(pGpG), is by far the best understood, and its platinated adduct has been the subject of a full conformational analysis [85] and X-ray crystal structure determination [86].

A full conformational analysis was carried out on the complex $[Pt(NH_3)_2\{d(GpG)\}]^+$ and, apart from the expected binding at N_7, with the guanines in a head-to-head arrangement, a major feature is that the sugar ring of the 5′-terminal adopts fully the N (C3′-*endo*) conformation, while the other sugar retains the normal S (C2′-*endo*) conformation [85]. This second sugar appears to exist in an N—S equilibrium. A model for the most abundant conformer in solution also indicates a dihedral angle of 53° between the bases and the platinum lies out of the guanine planes by 0.65 and 0.56 Å. For the remaining torsion angles, no large differences are found in comparison to B-DNA. There is considerable conformational freedom and thus, upon formation of one Pt—guanine link, the only major alteration necessary for chelation is the S to N conformational switch on the sugar. The results confirm earlier suggestions for both d(GpG) and d(pGpG) systems [36, 37].

The solid-state structure is shown in Figure 4.4. The compound contains four independent molecules with two distinct crystal types with resolution of 0.94 (initial R factor = 8.4) and 1.37 Å, respectively. The major features of the structure are base/base dihedral angles of from 76 to

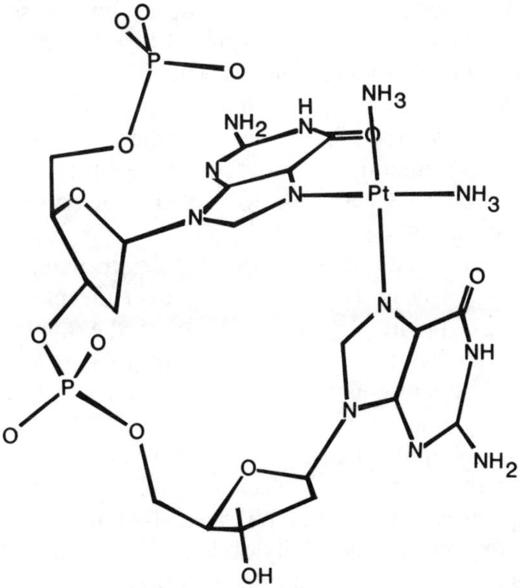

Fig. 4.4. Schematic representation of the structure of the adduct of *cis*-$[Pt(NH_3)_2Cl_2]$ with the dinucleotide d(pGpG). From Reference 86.

87°. The dihedral angles for the coordination plane are 111, 111, 80, and 77° for one set, and 86, 95, 58, and 60° for the other set. This is stated to preclude hydrogen bonding from NH(ammine) to O_6(guanine), although other contacts are present, especially a phosphate—ammine hydrogen bond. The 5′-sugars all adopt the N (C3′-*endo*) conformation and these sugars also contain the phosphate involved with the hydrogen bonding to the ammine. The other sugars appear to be more flexible, in a predominantly normal conformation. The conformational results are in good agreement with solution data, although the dihedral angles do differ. The base/base dihedral angles are somewhat larger than predicted from previous solution studies. The net result is that this conformational change could result in a kink in DNA, although the overall change in conformation was not as drastic as expected and confirms the idea of a localized denaturation on the polymer. CD and NMR studies have also been performed on Pt—dach complexes (see Section 3.4).

Studies have also been undertaken on other dinucleotides, containing other bases. For d(CpG) and d(pCpG), mixtures of N_3-*anti*—N_7-*anti* and N_3-*syn*—N_7-*anti* isomers were apparent [87]. These equilibrate by rotation of the cytosine around its glycosidic and Pt—N_3 bonds. For the d(GpC), sequence two sets of isomers are seen: G-*syn*—C-*anti*; G-*syn*—C-*syn* and G-*anti*—C-*anti*; G-*anti*—C-*syn*. The 5′ sugars in all cases adopt the N(C3′-*endo*) conformation and, as for the d(GpG) case above, the 3′ sugar is in equilibrium between the N and S forms. The two-step process starts with platination of the guanine N_7. The fact that the sugar conformations all change at the 5′ end may indicate the ability to induce similar conformational changes in DNA, as for the GG sequence. For the dinucleotide (GpA), mixtures of products were observed, one being the GN_7—AN_7 chelate [88]. Cross links were also detected for ApA and CpA by CD spectroscopy [89].

If the initial step in any dinucleotide complex formation is first binding to N_7 of guanine, why is d(GpG) favored over d(ApG) and why is the d(ApG) sequence favored dramatically over the alternative d(GpA)? Excellent NMR studies have shown that (ApG) reacts faster than (GpA) by a factor of 2, but subsequent chelation is in fact slower [142]. However the N_7 of the 5′-A in an (ApG) sequence is closer than that of the 3′-A in (GpA) (approx. 3 Å and 5 Å, respectively) [108] and this may be a major dictating factor in specificity. Molecular mechanics calculations on trinucleotide sequences have also emphasized the different hydrogen bonding patterns to explain the specificity of d(ApG) over d(GpA) [143]. These studies are contributing to the very detailed descriptions of the binding to DNA fragments.

4.4.2. STUDIES ON TRINUCLEOTIDES

The NMR approach has been extensively used for oligonucleotide complexes and allows the study of the effect of different sequences on the overall reaction. Thus, with the trinucleotide d(CpGpG), the GpG sequence adopts the configuration expected of the dinucleotide but the cytosine ring still stacks on the neighboring guanine, despite the fact that the sugar of the base has the N-conformation [90].

An interesting series of trinucleotides is formed by those in which two guanine residues are separated by another base of general sequence d(GpNpG). The interactions of these sequences with platinum complexes are relevant because platinated adducts with this sequence are obtained upon DNA digestion and also the fact that these chelates are implicated in the induction of mutation in *E. coli* [91]. For d(GpCpG), the chelation of both guanines is confirmed by NMR [92, 93]. The consequences of this binding is that the cytosine cannot stack and is 'pushed' or 'bulged' out, but the chelation does not appear to induce the conformational switch of the 5′-sugar, as in the dinucleotides, and thus the conformational changes are not the same. This unique binding could occur naturally in DNA. An examination of the interaction of *trans*-$[PtCl_2(NH_3)_2]$ with this trinucleotide also showed the formation, albeit much more slowly, of the GG chelate, a model for a 1,3-intrastrand link [144]. Similarly, in d(GpTpG) no distinct differences between *cis*- and *trans*-$[PtCl_2(NH_3)_2]$ binding have been observed [94]. Since any explanation of the antitumour activity must account for the differences in behavior of these isomers, the relevance of this sequence is in doubt, although the conformational distortion of this binding mode in an undecamer sequence is large (see below).

However, since a molecular explanation of the inherent cytotoxicity of *trans*-complexes cannot include an intrastrand link between adjacent bases, the 1,3-link as demonstrated by these studies presents a viable structural alternative. Is the difference in *cis* and *trans*, then, a reflection of a kinetic difference? The *trans*-isomer can form, in principle, an intrastrand link, but more slowly than the *cis*-isomer. The original monoadducts may be excised more efficiently or, indeed, the *trans*-complex itself may be sequestered more readily.

A structural determination of the cisplatin adduct of d(CpGpG) confirms the features of the dinucleotide structure [138]. Chelation to N_7 occurs, with the B/B' dihedral angle ranging from 80° to 84° and the sugar conformational switch observed. The intramolecular hydrogen bonding from the NH_3 ligands to guanine(O_6) and phosphate oxygen are also observed. The cytosine bases do not, however, stack on top of the neighboring guanine bases and, as a result of intermolecular hydrogen bonding interactions of the cytosine base, the conformation in the solid state does not match that in solution [138].

4.4.3. STUDIES ON OLIGONUCLEOTIDES

The study of higher order oligonucleotides has usually been undertaken to study the effect of sequence on the GG link. As pointed out from the restriction endonuclease studies, the sequence on either side of the GG link can be a determining factor in reactivity, since the sequence will dictate local conformation, etc. These studies are of particular interest now that the detailed geometry of the GG link is known and the results can be discussed in this context to give an understanding of the overall nature of the localized distortion on the DNA.

A number of tetranucleotides have been examined. In the case of d(CpCpGpG), for instance, the sugar conformational change is observed and no evidence for duplex stacking is noted [93]. Within the d(ApTpGpG) sequence, the internal guanine sugar again adopts the C3′-*endo* conformation [95]. An unusual temperature dependence for some of the resonances is noted, and this is not present in the dinucleotide, nor indeed in the complementary d(TpGpGpCpCpA) hexanucleotide [96], and presumably indicates conformational switching. Studies on d(TpGpGpT) with $[PtCl_2(en)]$ also expands the number of sequences with GG binding. Base stacking is observed on one side of the chelate, but not the other with the ^{31}P resonance shift corresponding to that in platinated DNA itself [97].

The study of higher oligonucleotides is also important because many of the smaller fragments are single stranded and the interaction with double stranded DNA may be different. Two studies have utilized complementary (i.e. double stranded) hexanucleotides. The specific chelation of the GG adduct was found in both cases, d(TpGpGpCpCpA) [96] and d(ApGpGpCpCpT) [98] and in one case the sugar conformational switch was observed [96]. Under conditions of duplex formation, intermediate complexes are noticed before conversion to the GG chelate. In both cases, a single stranded form for the final product (separated and purified by HPLC) is likely, providing further evidence for a local denaturing of the DNA upon platination.

Platinated oligonucleotides can, however, form duplexes with complementary strands. The octanucleotide d(GpApTpCpCpGpGpC) was platinated and shown to form a duplex with its complementary decanucleotide [99]. These conclusions are in agreement with studies on the decanucleotide d(TpCpTpCpGpGpTpCpTpC) which showed that all base pairs remain intact on platination, although the melting temperature is lowered [100, 101]. In contrast to this, in an undecamer sequence where the platination occurs in the d(GpTpG) sequence the melting temperature is also lowered, indicating a greater distortion than in a GG platination [102]. In this case, hydrogen bonding of both an AT (from the central T flanked by two Pt—G bonds) and GC pair is interrupted. The conformational distortion was considered to be greater for this sequence than the GG sequence [102].

Molecular mechanics calculations for the platinated double stranded oligonucleotides d(GpGpCpCpGpGpCpC)—d(CpCpGpGpCpCpGpG) and the decamer d(TpCpTpCpGpGpTpCpTpC)—d(GpApGpApCpCp-GpApGpA) had accurately predicted the features observed in the dinucleotide structure [103]. The model includes the presence of the phosphate—ammine hydrogen bond and also predicts the loss of one hydrogen bond involved in GC base pairing. The interplanar angle between G_5—G_6 of the decamer was predicted to be 50°. An interesting result, using the A-DNA conformation for the oligonucleotides, is that the bases on either side of the platinated bases also move in such a way as to 'accommodate' the kink, thus distributing the distortion among several base pairs. On consideration of the accepted flexibility of DNA, the implication from calculations and the solution studies may be that the final distortion again may be dependent on base sequence, somewhat similar to the results seen from molecular biology experiments. Further refinement of this 'kinked model' using high- and low-salt forms has been discussed and the model is shown to be energetically feasible [104] although the theoretical nature of these models has been noted.

4.4.3.1. *Structural Studies on Oligonucleotides*

A number of studies have reported binding to oligonucleotide fragments and tRNA. The incubation of cisplatin with the helical dodecanucleotide depicted below showed binding to N_7 positions at the inner guanine residues [105]:

Pt(30%) (at G_4); Pt(22%) (at G_{10})

$$5'\text{—}C_1\text{—}G_2\text{—}C_3\text{—}G_4\text{—}A_5\text{—}A_6\text{—}T_7\text{—}T_8\text{—}C_9\text{—}G_{10}\text{—}C_{11}\text{—}G_{12}$$

$$\text{G—C—G—C—T—T—A—A—G—C—G—C-5}'$$

Pt(61%) (at the G opposite C_9)

Not all guanine sites are occupied and those bound are indicated with relative occupancies. The binding is monodentate in all cases but there is an aqua bridge between the Pt atom and the guanine O_6. The immediate geometry around the platinum confirms expectations from the model studies, but there is evidence that the guanine ring moves out of the base pair stack into the major groove. This shift, of almost 0.5 Å in this case, could also destabilize the helix. The different reactivities of the guanines were rationalized in terms of decreased reactivity at the end of the helix and also the bending of the helix, where the sharpest bending occurs on the top half of the helix, thus restricting the major groove to Pt attack [105]. The destruction of the crystal pattern at higher substitution indicates that more severe distortions occur.

It is of interest here to mention the structural work done on tRNA because there is a good consistency in general results as the polymer is built up. The metal binding of tRNA has been summarized [106, 107]. A number of studies have reported on Pt—tRNA binding and although various differences are observed, depending on the incubation time and medium used, the relevant data are instructive and are presented in Table 4.III, along with the studies on a related Ru complex [108—110]. While

TABLE 4.III
Summary of binding sites and interactions for platinum and ruthenium complexes on tRNA.

Complex	Binding Sites (Residue)	H-Bonding Interactions	Ref.
cis-$[PtCl_2(NH_3)_2]$	G3—G4 C25—Gm26 G42—G43—A44—G45 A64—G65		108
cis-$[PtCl_2(NH_3)_2]$	N_7(G15)	NH_3—P14 NH_3—P15 H_2O—O_6(G15)	109
	N_7(G18)	H_2O—O_6(G18) H_2O—N_6(Am58)	
trans-$[PtCl_2(NH_3)_2]$	N_1(A73)	NH_3—O_6(G1) NH_3—N_3, O2(C74)	108, 110
	N_7(Gm34)	NH_3—O_6(Gm34) NH_3—P34	
$[Ru(NH_3)_5(H_2O)]^{3+}$	N_7(G15) N_7(G18)	NH_3—O(6)	109

caution must be exercised because not all binding regions are double-helical, the consistency of G and A binding is again confirmed. Again, binding of only one base—metal bond is observed. The hydrogen bonding interactions are of particular relevance, confirming the generality of this interaction, and both amine—phosphate and amine—exocyclic group hydrogen bonds are observed. Interestingly, a noncovalent mode of binding is observed for the Ru complex using a shorter incubation time. The experimental details usually require incubation for a considerable time with various counteranions and it is reasonable to suggest that this situation exists *in vivo*. In an intracellular situation, where the DNA is surrounded by positively charged histones and counterions, electrostatic interactions may be of critical importance in the affinity of metal complexes for DNA. In view of the previously demonstrated stabilization of DNA by free amines, this interaction may be important for DNA binding. One question that is of particular interest is: are hydrogen bonding interactions different for the two isomers? In at least one case, that of hydrogen bonding from NH_3 to adjacent P_{14} and P_{15} oxygens, this can only occur for the *cis*-isomer [109].

4.5. Platinum Adducts from DNA Degradation

The likely binding sites of platinum bound to DNA have also been confirmed from studies on degradation products from the Pt—DNA interaction (Table 4.IV). Acid hydrolysis cleaves the sugar bonds and so no sequence information is obtained, as for enzymatic degradation. The products have been summarized [82, 145].

The two major products are the GG and AG adducts. In the case of the latter the d(pApG) sequence is formed rather than d(pGpA) [111], a result also indicated in one of the crystallographic studies mentioned above [108]. Recently, evidence for a small amount of d(pGpA) adduct has been reported in a study of the action of T4 DNA polymerase on Pt—DNA restriction fragments [146].

An interesting aspect of adduct chemistry is the isolation of monoadducts containing only one Pt—purine bond and their subsequent conversion to bispurine complexes. The monoadducts may be trapped by a variety of reagents including thiourea, NH_4CO_3, and glutathione. These monoadducts disappear with time and are estimated to represent 10—20% of adducts in approximately the first two hours upon incubation of the platinum complex with DNA. In the case of *trans*-$[PtCl_2(NH_3)_2]$ the monoadducts represent up to 80% of all adducts. The purine base in all monoadducts formed is uniformly guanine and their transformation into bispurine adducts is interesting. This change occurs in the absence of free

TABLE 4.IV
Platinum complexes formed from DNA degradation.

DNA	Conditions[a]	Adducts	Method	Ref.
cis-$[PtCl_2(NH_3)_2]$:				
Salmon sperm	5 h, 50° $r_b < 0.1$	$(dpG)_{1,2}$ d(pGpG) d(pApG) d(pGpNpG)	Enzymatic degradation	111
Salmon sperm	36 h, RT $r_b < 0.4$	(Gua) $(Gua)_2$ (Gua—Ade)	Mild acid hydrolysis	112
Calf thymus	48 h, 37° $r_b < 0.1$	$(Gua)_2$ (Gua—Ade)	Mild acid hydrolysis	113
$[PtCl_2(en)]$:				
Calf thymus	16 h, 37° $r_b < 0.1$	$d(Guo)_2$ d(pGpG) d(pApG) d(pApNpG)	Enzymatic digestion	114
trans-$[PtCl_2(NH_3)_2]$:				
Salmon sperm	5 h, 50° $r_b < 0.1$	$d(pG)_{1,2}$	Enzymatic digestion	115
[Pt(dien)Cl]Cl:				
Salmon sperm	$r_b < 0.1$	(Gua)		116

[a] r_b is usually defined as the molar ratio of Pt bound per nucleotide.

complex — the monoadducts therefore being seen to undergo chemical reactions on the DNA [147]. Kinetic evidence for the transient monoadduct intermediate and its subsequent transformation has been reported [148] and the chemical reactivity of the monoadducts have been examined by reaction with exogenous guanosine to form the bis(guanosine) adducts [147].

4.6. On the Molecular Mechanism of Antitumour Action of Cisplatin

The current acceptance of DNA as the cellular target for platinum complexes is a major reason for the studies undertaken and outlined in this chapter. These results, taken together with biological results, can lead to elucidation of the molecular mechanism of action, and the assignment of the lesion(s) responsible for cytotoxicity. Particularly, two major questions have been addressed — what is the lesion responsible for cisplatin toxicity (if indeed there is only one unique lesion)? and what is the explanation for the differential cytotoxicity of the *cis*- and *trans*-isomers? There is still considerable debate on these questions and much remains to be resolved [82, 141]. In cellular studies, the correlation of DNA binding and cytotoxicity must also take into account the rates at which these phenomena occur.

The major adducts observed from studies *in vitro* are likely to be the same, and in the same proportion, as those that are formed in the DNA of whole cells [123]. Studies on nucleosomes and chromatin also confirm differential binding to DNA, with differences in the *cis*- and *trans*-isomers, and with some physical characteristics similar to those of model complexes [119—122]. DNA interstrand links account for less than 1% of the *cis* adducts, although they can increase to around 3% with time [123]. DNA—protein cross links have also been observed from cultured cells with an approximate three-fold frequency over interstrand links [124—126]. There are relatively more DNA—protein links for the *trans*-isomer, and also a certain proportion of these cross links increases with time [127]. Intrastrand links are extremely difficult to quantitate and immunochemical methods have been developed for this purpose [149, 150]. The contribution of these various lesions to cytotoxicity must account for their relative frequency and ease of chemical and biochemical repair.

The GG intrastrand link has received the most attention as the major adduct of cisplatin binding. While GG sequences may be the most prevalent sites for platination, kinetic studies show that the rate of reaction is dependent only on G content, and not on the statistical amount of GG [117, 118]. Since chelation occurs in a two-step manner the preponderance of GG links must be accounted for by a favored second step [142].

The differential cytotoxicity of the *cis*-isomer is attributed to its formation of the unique intrastrand link. This could produce a kink or, to account for its relatively minor perturbation of the local area as perceived by structural studies, could be more difficult to repair [132]. The theory that the *cis*-isomer differs from the *trans*-complex in the repair of their lesions on DNA [132] has been disputed recently, where results indicated that the different cytotoxic effects could result from the intrinsic effects of their DNA-bound adducts [139].

There is not, therefore, full agreement on the fundamental importance of the GG intrastrand link as the unique locus of DNA damage. Other possibilities to be considered include interstrand links and the slower forming intrastrand lesions. The demonstrated correlation of interstrand links with cytotoxicity in some systems [124, 125, 128] is not, however, general and inactivation of phage DNA by *cis*-$[PtCl_2(NH_3)_2]$ does not correspond with the degree of interstrand links [129–131]. For the *trans*-isomer, however, inactivation of phage DNA is accounted for on the basis of interstrand links [131] and, indeed, a combination of DNA–DNA interstrand and DNA–protein cross links must account for the cytotoxicity of the *trans*-isomer, since at least the GG intrastrand link is sterically forbidden. Again, it is worth remembering that the *trans*-complex is cytotoxic, despite the absence of the GG link.

Interstrand links for the *cis*-isomer are maximal after 12 h, while the maximum effect for the *trans*-isomer is achieved within 6 h [124, 127, 133]. This may mean faster repair of *trans* binding or slower forming *cis* linkages, perhaps by a slow two-step reaction. The principle adduct on the DNA is the GG intrastrand link but other linkages, such as AG may form with time [111]. The firmly bound Pt, which is not removable by CN^- (see Section 4.2.6), is not recognized by antibodies specific for the Pt–GG link [134] and this indicates that slow-forming adducts may occur. What is the importance of these adducts in cytotoxicity? Little monoadduct formation has been observed under the assay conditions employed [136], and the correlation of these adducts with cytotoxicity is at present unclear.

The structural studies outlined in this chapter contribute to the discussion on effectiveness of different binding modes as they are charaterized. Another interesting feature is the nature of the affinity of Pt–amine complexes toward DNA. The presence of amine–phosphate hydrogen bonds could lead to interpretation of the amines as a DNA-recognition portion of the molecule, especially *in vivo*. The assignment of a dual-function nature, whereby the amines recognize and might even orient the complex to DNA, with subsequent inactivating Pt–base bonding is tempting. This 'dual functionality' would, in fact, be analogous to that attributed to bleomycin (Chapter 7) where DNA binding and radical production are two distinct steps. A further point to emerge from analysis of the dinucleotide structure is that there may be considerable steric restraint (independent of hydrogen bonding) for four aromatic rings (2 pyridine, 2 purine) around the platinum, and may explain the lack of activity of these complexes compared to simple primary and secondary amines. It may, however, be possible to design compounds capable of an intrastrand link but with DNA-recognition components on the periphery of the molecule, rather than directly bound to the Pt. More relevant – if the end result of importance is the bending of DNA, rather than an intrastrand link *per se* – how can this be achieved by other structures? It is also of considerable interest to design more active *trans* complexes – what are the structural and chemical (kinetic) features required? The analysis of systems containing these compounds can benefit from the enormous amount of information gathered so far, as well as confirm some of the present theories about the mechanism of action of the clinically used complexes.

4.7. Summary

The interactions of platinum complexes with DNA, oligonucleotide sequences and constituent bases give a detailed description of the binding and contribute to the understanding of the molecular mechanism of action of cisplatin.

There is a well demonstrated kinetic effect for guanine N_7 and adducts isolated from DNA reflect this. Platination affects chemical properties of the bound bases as well as affecting conformations of nucleoside and nucleotide units. The adducts formed on DNA may be isolated by either chemical or enzymatic degradation. The predominant adduct is a chelate formed by binding to N_7 of two adjacent guanines (GG), the next most prominent being (AG) by binding to N_7 of a neighboring adenine/guanine sequence. The nature of the adducts, and their relative ratios, vary with time and this may be important with respect to cytotoxicity.

Physical studies, especially NMR and X-ray crystallography, have elucidated the structures of many oligonucleotide adducts and from this knowledge we can propose how Pt binding affects the conformation of DNA, and how this may be further responsible for inhibition of DNA synthesis. A major feature is a conformational switch of the sugar residues of the bound purine which leads to a kink or bend in the DNA. Studies on small portions of DNA may help elucidate the different binding of *cis*- and *trans*-isomers of $[PtCl_2(NH_3)_2]$ which may account for their different cytotoxicity.

References

1. R. M. Izatt, J. J. Christenson, and J. H. Rytting: *Chem. Rev.* **71**, 439 (1971).
2. G. L. Eichhorn: *Inorganic Biochemistry* (v. 2, Ed. G. L. Eichhorn), pp. 1191 and 1210. Elsevier, Amsterdam (1973).
3. R. B. Martin and Y. H. Mariam: *Metal Ions in Biol. Syst.* **8**, 57 (1979); R. B. Martin: *Acc. Chem. Res.* **18**, 32 (1985).
4. L. G. Marzilli, T. J. Kistenmacher, and G. L. Eichhorn: *Structural Principles of Metal Ion-Nucleotide and Metal Ion–Nucleic Acid Interactions* (Nucleic Acid–Metal Ion Interactions v. 1, Ed. T. G. Spiro), p. 179. Wiley, New York (1980).

5. L. G. Marzilli: *Adv. Inorg. Biochem.* **3**, 47 (1981).
6. L. G. Marzilli: *Prog. Inorg. Chem.* **23**, 255 (1977).
7. D. J. Hodgson: *Prog. Inorg. Chem.* **23**, 211 (1977).
8. R. W. Gellert and R. Bau: *Metal Ions in Biol. Syst.* **8**, 1 (1979).
9. V. Swaminathan and M. Sundaralingam: *CRC Crit. Rev. Biochem.* **245** (1979).
10. R. B. Simpson: *J. Am. Chem. Soc.* **86**, 2059 (1964).
11. A. M. Fiskin and M. Beer: *Biochemistry* **4**, 1289 (1965).
12. G. M. Brown, J. E. Sutton, and H. Taube: *J. Am. Chem. Soc.* **100**, 2767 (1978).
13. L. G. Marzilli and T. J. Kistenmacher: *Acc. Chem. Res.* **10**, 146 (1977).
14. A. Erck, L. Rainen, J. Whileyman, I. Chang, R. Howard, G. Serio, and A. P. Kimball: *Proc. Soc. Exp. Biol. Med.* **145**, 1278 (1975).
15. K. Das and J. L. Bear: *Inorg. Chem.* **15**, 2093 (1976).
16. N. Farrell: *J. Inorg. Biochem.* **14**, 261 (1981).
17. D. J. Szalda, T. J. Kistenmacher, and L. G. Marzilli: *J. Am. Chem. Soc.* **98**, 8371 (1976).
18. D. J. Szalda, T. J. Kistenmacher, and L. G. Marzilli: *Inorg. Chem.* **15**, 2783 (1976).
19. M. E. Howe-Grant and S. J. Lippard: *Metal Ions in Biol. Syst.* **11**, 63 (1980).
20. R. B. Martin: *Hydrolytic Equilibria and N_7 Versus N_1 Binding in Purine Nucleosides of cis-Diamminedichloroplatinum(II): Palladium(II) as a Guide to Platinum(II) Reactions at Equilibrium* (Platinum, Gold and Other Metal Chemotherapeutic Agents, ACS Symposium Series v. 209, Ed. S. J. Lippard), p. 231. ACS, Washington (1983).
21. P. I. Vestues and R. B. Martin: *J. Am. Chem. Soc.* **103**, 806 (1981).
22. K. H. Scheller, V. Scheller-Krattiger, and R. B. Martin: *J. Am. Chem. Soc.* **103**, 6833 (1981).
23. I. Sovago and R. B. Martin: *Inorg. Chem.* **19**, 2868 (1980).
24. G. Y. H. Chu, S. Mansy, R. F. Duncan, and R. S. Tobias: *J. Am. Chem. Soc.* **100**, 593 (1978).
25. F. J. Dijt, G. W. Canters, J. H. J. den Hartog, A. T. M. Marcelis, and J. Reedijk: *J. Am. Chem. Soc.* **106**, 3644 (1984).
26. S. K. Miller and L. G. Marzilli: *Inorg. Chem.* **24**, 2421 (1985).
27. R. S. Tobias, G. Y. H. Chu, and H. J. Peresie: *J. Am. Chem. Soc.* **97**, 5305 (1975).
28. G. M. Clore and A. M. Gronenborn: *J. Am. Chem. Soc.* **104**, 1369 (1982).
29. S. Mansy, G. Y. H. Chu, R. E. Duncan, and R. S. Tobias: *J. Am. Chem. Soc.* **100**, 607 (1978).
30. S. Eapen, M. Green, and I. M. Ismail: *J. Inorg. Biochem.* **24**, 233 (1985).
31. W. M. Scovell and T. O'Connor: *J. Am. Chem. Soc.* **99**, 120 (1977).
32. P.-C. Kong and T. Theophanides: *Inorg. Chem.* **13**, 1981 (1974).
33. J. P. Macquet, J. L. Butour, and N. P. Johnson: *Physicochemical and Structural Studies of the In Vitro Interactions between Platinum(II) Compounds and DNA* (Platinum, Gold, and Other Metal Chemotherapeutic Agents (ACS Symposium Series v. 209, Ed. S. J. Lippard), p. 75. ACS, Washington (1983).
34. R. Faggiani, C. J. L. Lock, and B. Lippert: *J. Am. Chem. Soc.* **102**, 5419 (1981).
35. M. J. Clarke and H. Taube: *J. Am. Chem. Soc.* **96**, 5413 (1974).
36. J. C. Chottard, J. P. Girault, J. Y. Lallemand, and D. Mansuy: *J. Am. Chem. Soc.* **102**, 5565 (1982).
37. J. P. Girault, G. Chottard, J. Y. Lallemand, and J. C. Chottard: *Biochemistry* **21**, 1352 (1982).
38. R. Faggiani, B. Lippert, C. J. L. Lock, and R. A. Speranzini: *J. Am. Chem. Soc.* **103**, 1111 (1981).
39. M. J. Clarke: *Metal Ions in Biol. Syst.* **11**, 231 (1980).

40. M. J. Clarke: *J. Am. Chem. Soc.* **100**, 5068 (1978).
41. B. Noszal, V. Scheller-Krattiger, and R. B. Martin: *J. Am. Chem. Soc.* **104**, 1078 (1982).
42. J. R. Jones and S. E. Taylor: *J. Chem. Soc. Perkin Trans.* **2**, 1773 (1979).
43. G. Y. Chu and R. S. Tobias: *J. Am. Chem. Soc.* **98**, 2641 (1976).
44. B. Lippert: *Inorg. Chem.* **20**, 4326 (1981).
45. G. Muller, J. Riede, R. Beyerle-Pfnur, and B. Lippert: *J. Am. Chem. Soc.* **106**, 7999 (1984).
46. S. J. N. Burgmayer and E. I. Steifel: *J. Chem. Ed.* **943** (1985).
47. M. J. Clarke, K. F. Coffey, H. J. Perpall: *Anal. Biochem.* **122**, 404.
48. R. J. Sundberg, R. F. Bryan, I. F. Taylor, and H. Taube: *J. Am. Chem. Soc.* **96**, 381 (1974).
49. M. J. Clarke and H. Taube: *J. Am. Chem. Soc.* **97**, 1397 (1975).
50. S. Mansy and R. E. Tobias: *Biochemistry* **14**, 7701 (1975).
51. E. Buncel, A. R. Norris, W. J. Racz, and S. E. Taylor: *Inorg. Chem.* **20**, 98 (1981).
52. D. E. Bergstrom and J. L. Ruth: *J. Am. Chem. Soc.* **98**, 1587 (1976).
53. J. L. Ruth and D. E. Bergstrom: *J. Org. Chem.* **43**, 2870 (1978).
54. C. F. Bigge, P. Kalaritis, and M. P. Mertes: *Tetrahedron Lett.* **19**, 1653 (1979).
55. L. G. Marzilli, L. A. Epps, T. Sorrell, and T. J. Kistenmacher: *J. Am. Chem. Soc.* **97**, 3351 (1975).
56. F. Franke and I. D. Jenkins: *Aust. J. Chem.* **31**, 595 (1978).
57. J. Turkevich and R. S. Miner Jr.: *J. Clin. Hematol. Oncol.* **13**, 65 (1983).
58. P. D. Lawley and P. Brookes: *Biochem. J.* **89**, 127 (1963).
59. M. J. Clarke and P. E. Morrissey: *Inorg. Chim. Acta* **80**, L69 (1982).
60. M. J. Clarke: *Ruthenium Anticancer Agents and Relevant Reactions of Ruthenium-Purine Complexes* (Platinum, Gold and Other Metal Chemotherapeutic Agents v. 209, Ed. S. J. Lippard), p. 335. ACS, Washington (1983).
61. K. Inagaki and Y. Kidani: *Bioinorg. Chem.* **9**, 157 (1978).
62. B. Lippert: *Inorg. Chim. Acta* **55**, 5 (1981).
63. B. Lippert: *Platinum(II) Complex Formation with Uracil and Thymine* (Platinum, Gold and Other Metal Chemotherapeutic Agents, ACS Symposium Series v. 209, Ed. S. J. Lippard), p. 147. ACS, Washington (1983).
64. M. J. Clarke: *Inorg. Chem.* **16**, 738 (1977).
65. A. T. M. Marcelis, J. L. van der Veer, J. C. M. Zwetsloot, and J. Reedijk: *Inorg. Chim. Acta* **78**, 195 (1983).
66. R. E. Cramer and P. L. Dahlstrom: *J. Am. Chem. Soc.* **101**, 3679 (1979).
67. M. D. Reily and L. G. Marzilli: *J. Am. Chem. Soc.* **108**, 6785 (1986).
68. B. de Castro, T. J. Kistenmacher, and L. G. Marzilli: *Agents Actions Suppl. 8 (AAS8)* **435** (1981).
69. B. L. Heyl, K. Shinozuka, S. K. Miller, D. G. van der Veer, and L. G. Marzilli: *Inorg. Chem.* **24**, 661 (1985).
70. A. P. Hitchcock, C. J. L. Lock, W. M. C. Pratt, and B. Lippert: *Platinum Complexes with DNA Bases, Nucleotides, and DNA* (Platinum, Gold, and Other Metal Chemotherapeutic Agents, ACS Symposium Series v. 209, Ed. S. J. Lippard), p. 209. ACS, Washington (1983).
71. T. J. Kistenmacher, J. D. Orbell, and L. G. Marzilli: *Conformational Properties of Purine and Pyrimidine Complexes of cis-Platinum: Implications for Platinum(II)–DNA Crosslinking Modes* (Platinum, Gold and Other Metal Chemotherapeutic Agents, ACS Symposium Series v. 209, Ed. S. J. Lippard), p. 191. ACS, Washington (1983).

72. J. D. Orbell, L. G. Marzilli, and T. J. Kistenmacher: *J. Am. Chem. Soc.* **103**, 5126 (1981).
73. H. Schöllhorn, G. Raudaschl-Sieber, G. Muller, U. Thewalt, and B. Lippert: *J. Am. Chem. Soc.* **107**, 5932 (1985).
74. R. E. Cramer, P. L. Dahlstrom, M. J. T. Seu, T. Norton, and M. Kashiwagi: *Inorg. Chem.* **19**, 148 (1980).
75. R. W. Gellert and R. Bau: *J. Am. Chem. Soc.* **97**, 7379 (1975).
76. T. J. Kistenmacher, C. C. Chiang, P. Chalilpoyil, and L. G. Marzilli: *J. Am. Chem. Soc.* **101**, 1143 (1979).
77. J. A. Orbell, C. Solorzano, L. G. Marzilli, and T. J. Kistenmacher: *Inorg. Chem.* **21**, 2630 (1982).
78. R. Faggiani, B. Lippert, and C. J. L. Lock: *Inorg. Chem.* **21**, 3210 (1982).
79. D. Neugebauer and B. Lippert: *J. Am. Chem. Soc.* **104**, 6596 (1982).
80. R. Faggiani, B. Lippert, C. J. L. Lock, and R. A. Speranzini: *Inorg. Chem.* **21**, 3216 (1982).
81. B. Lippert, R. Pfab, and D. Neugebaur: *Inorg. Chim. Acta* **37**, 1495 (1979).
82. J. Reedijk, A. M. J. Fichtinger-Schepman, A. T. van Oosterom, and P. van de Putte: *Structure and Bonding* **67**, 54 (1987).
83. A. Pinto and S. J. Lippard: *Biochim. Biophys. Acta* **780**, 167 (1985).
84. A. T. M. Marcelis and J. Reedijk: *Recl. Trav. Chim. Pays Bas* **102**, 121 (1983).
85. J. H. J. den Hartog, C. Altona, J. C. Chottard, J. P. Girault, J. Y. Lallemand, F. A. A. M. de Leeuw, A. T. M. Marcelis, and J. Reedijk: *Nuc. Acids Res.* **10**, 4715 (1982).
86. S. E. Sherman, D. Gibson, A. H. J. Wang, and S. J. Lippard: *Science* **230**, 412 (1985).
87. J. P. Girault, G. Chottard, J. Y. Lallemand, F. Huguenin, and J. C. Chottard: *J. Am. Chem. Soc.* **106**, 7227 (1984).
88. K. Inagaki and Y. Kidani: *Inorg. Chim. Acta* **80**, 171 (1983).
89. I. A. G. Roos, A. J. Thomson, and S. Mansy: *J. Am. Chem. Soc.* **96**, 6484 (1974).
90. J. H. J. den Hartog, C. Altona, G. A. van der Marel, and J. Reedijk: *Eur. J. Biochem.* **147**, 371 (1985).
91. J. Brouwer, P. van de Putte, A. M. J. Fichtinger-Schepman, and J. Reedijk: *Proc. Natl. Acad. Sci. USA* **78**, 7010 (1981).
92. A. T. M. Marcelis, J. H. J. den Hartog, and J. Reedijk: *J. Am. Chem. Soc.* **104**, 2664 (1982).
93. A. T. M. Marcelis, J. H. J. den Hartog, G. A. van der Marel, G. Wille, and J. Reedijk: *Eur. J. Biochem.* **135**, 343 (1983).
94. J. L. van der Veer, G. J. Ligtvoet, H. van den Elst, and J. Reedijk: *J. Am. Chem. Soc.* **108**, 3860 (1986).
95. J. M. Neumann, S. Tran-Dinh, J. P. Girault, J. C. Chottard, T. Huynh-Dinh, and J. Igolen: *Eur. J. Biochem.* **141**, 465 (1984).
96. J. P. Girault, J. C. Chottard, E. R. Guittet, J. Y. Lallemand, T. Huynh-Dinh, and J. Igolen: *Biochem. Biophys. Res. Comm.* **109**, 1157 (1982).
97. R. A. Byrd, M. F. Summers, G. Zon, C. Spellmeyer Fouts, and L. G. Marzilli: *J. Am. Chem. Soc.* **108**, 504 (1986).
98. J. P. Caradonna, S. J. Lippard, M. J. Gait, and M. Singh: *J. Am. Chem. Soc.* **104**, 5793 (1982).
99. B. van Hemelryck, E. Guittet, G. Chottard, J. P. Girault, T. Huynh-Dinh, J. Y. Lallemand, J. Igolen, and J. C. Chottard: *J. Am. Chem. Soc.* **106**, 3037 (1984).

100. J. H. J. den Hartog, C. Altona, J. H. van Boom, G. A. van der Marel, C. A. G. Haasnoot, and J. Reedijk: *J. Am. Chem. Soc.* **106**, 1528 (1984).
101. J. H. J. den Hartog, C. Altona, J. H. van Boom, and J. Reedijk: *FEBS Letters* **176**, 393 (1984).
102. J. H. J. den Hartog, C. Altona, H. van den Elst, G. A. van der Marel, and J. Reedijk: *Inorg. Chem.* **24**, 986 (1985).
103. J. Kozelka, G. A. Petsko, S. J. Lippard, and G. J. Quigley: *J. Am. Chem. Soc.* **107**, 4079 (1985).
104. J. Kozelka, G. A. Petsko, G. J. Quigley, and S. J. Lippard: *Inorg. Chem.* **25**, 1076 (1986).
105. R. M. Wing, P. Pjura, H. R. Drew, and R. E. Dickerson: *EMBO J.* **3**, 1201 (1984).
106. G. A. Petsko: *Methods Enzymol.* **114**, 147 (1985).
107. M. M. Teeter, G. J. Quigley, and A. Rich: *Metal Ions and Transfer RNA* (Nucleic Acid–Metal Ion Interactions Ed. T. J. Spiro), p. 145. Wiley, New York (1980).
108. J. C. Dewan: *J. Am. Chem. Soc.* **106**, 7239 (1984).
109. J. R. Rubin, M. Sabat, and M. Sundaralingam: *Nuc. Acids Res.* **11**, 6571 (1983).
110. A. Jack, J. E. Ladner, D. Rhodes, R. S. Brown, and A. Klug: *J. Mol. Biol.* **111**, 315 (1977).
111. A. M. J. Fichtinger-Schepman, J. L. van der Veer, J. H. den Hartog, P. H. M. Lohman, and J. Reedijk: *Biochemistry* **24**, 707 (1985).
112. N. P. Johnson, A. M. Mazard, J. Escalier, and J. P. Macquet: *J. Am. Chem. Soc.* **107**, 6376 (1985).
113. R. O. Rahn: *J. Inorg. Biochem.* **21**, 311 (1984).
114. A. Eastman: *Biochemistry* **22**, 3927 (1983).
115. A. M. J. Fichtinger-Schepman, P. H. M. Lohman, and J. Reedijk: *Nuc. Acids Res.* **10**, 5345 (1982).
116. N. P. Johnson, J. P. Macquet, J. L. Wiebers, and B. Monsarrat: *Nuc. Acids Res.* **10**, 5255 (1982).
117. N. P. Johnson and J.-L. Butour: *J. Am. Chem. Soc.* **103**, 7351 (1981).
118. N. P. Johnson, J. D. Hoeschele, and R. O. Rahn: *Chem-Biol. Inter.* **30**, 151 (1980).
119. S. J. Lippard and J. D. Hoeschele: *Proc. Natl. Acad. Sci. USA* **76**, 6091 (1979).
120. Z. M. Banjar, L. S. Hnilica, R. C. Biggs, J. Stein, and G. Stein: *Biochemistry* **23**, 1921 (1984).
121. C. Houssier, M. C. Depauw-Gillet, R. Hacha, and E. Fredericq: *Biochim. Biophys. Acta* **739**, 317 (1983).
122. L. G. Marzilli, M. D. Reily, B. L. Heyl, C. T. McMurray, and W. D. Wilson: *FEBS Letts.* **176**, 389 (1984).
123. J. J. Roberts and F. Friedlos: *Chem.-Biol. Inter.* **39**, 181 (1982).
124. L. A. Zwelling, T. Anderson, and K. W. Kohn: *Cancer Res.* **39**, 365 (1979).
125. G. Laurent, L. C. Erickson, N. A. Sharkey, and K. W. Kohn: *Cancer Res.* **41**, 3347 (1981).
126. L. A. Zwelling, S. Michaels, H. Schwartz, P. P. Dobson, and K. W. Kohn: *Cancer Res.* **41**, 640 (1981).
127. M. C. Strandberg, E. Bresnick, and A. Eastman: *Chem.-Biol. Inter.* **39**, 169 (1982).
128. L. C. Erickson, L. A. Zwelling, J. M. Ducore, N. A. Sharkey, and K. W. Kohn: *Cancer Res.* **41**, 2791 (1981).
129. K. Micetich, L. A. Zwelling, and K. W. Kohn: *Cancer Res.* **43**, 3609 (1983).
130. J. Filipski, K. W. Kohn, and W. M. Bonner: *Chem.-Biol. Inter.* **32**, 321 (1980).

131. K. V. Shooter, R. Howse, R. K. Merrifield, and A. B. Dobbins: *Chem.-Biol. Inter.* **5**, 289 (1972).
132. R. B. Cicarelli, M. J. Solomon, A. Varshavsky, and S. J. Lippard: *Biochemistry* **24**, 7533 (1985).
133. A. C. M. Plooy, M. van Dijk, and P. H. M. Lohman: *Cancer Res.* **44**, 2043 (1984).
134. S. J. Lippard, H. M. Ushay, C. M. Merkel, and M. Poirier: *Biochemistry* **22**, 5165 (1983).
135. G. Raudaschl-Sieber and B. Lippert: *Inorg. Chem.* **24**, 2426 (1985).
136. A. M. J. Fichtinger-Schepman, J. L. van der Veer, P. H. M. Lohman, and J. Reedijk: *Biochemistry* **24**, 707 (1985).
137. D. L. Bodenner, P. C. Dedon, P. C. Keng, and R. F. Borch: *Cancer Res.* **46**, 2745 (1986).
138. G. Admiraal, J. L. van der Veer, R. A. G. de Graaf, J. H. J. den Hartog, and J. Reedijk: *J. Am. Chem. Soc.* **109**, 592 (1987).
139. J. J. Roberts and F. Friedlos: *Cancer Res.* **47**, 31 (1987).
140. G. Raudaschl-Sieber, L. G. Marzilli, B. Lippert, and K. Shinozuka: *Inorg. Chem.* **24**, 989 (1985).
141. S. E. Sherman and S. J. Lippard: *Chem. Rev.* **87**, 1153 (1987).
142. T. Van Hemelryck, J.-P. Girault, G. Chottard, P. Valadon, M. Laoui, and J.-C. Chottard: *Inorg. Chem.* **26**, 787 (1987).
143. T. W. Hambley: *J. Chem. Soc. Chem. Comm.* 221 (1988).
144. D. Gibson and S. J. Lippard: *Inorg. Chem.* **26**, 2275 (1987).
145. A. Eastman: *Pharmacol. Ther.* **34**, 155 (1987).
146. A. Rahmouni, A. Schwartz, and M. Leng: *Platinum and Other Metal Coordination Compounds in Cancer Chemotherapy* (Ed. M. Nicolini) 127. Martinus-Nijhoff, Dordrecht (1988).
147. J.-L. Butour and N. P. Johnson: *Biochemistry* **25**, 4534 (1986).
148. W. Schaller, H. Reisner, and E. Holler: *Biochemistry* **26**, 943 (1987).
149. M. C. Poirier, S. J. Lippard, L. A. Zwelling, H. M. Ushay, D. Kerrigan, and C. C. Thill: *Proc. Natl. Acad. Sci. (USA)* **79**, 6443 (1982).
150. A. M. J. Fichtinger-Schepman, R. A. Baan, A. Luiten-Schuite, M. Van Dijk, and P. H. M. Lohman: *Chem.-Biol. Inter.* **55**, 275 (1985).

CHAPTER 5

THE PLATINUM–PYRIMIDINE BLUES

The platinum—pyrimidine blues, derived from cisplatin, are an interesting class of complexes with antitumour activity. They were first discovered as a result of studies on the interaction of *cis*-$[PtCl_2(NH_3)_2]$ and its aquated products with pyrimidines. The identification of DNA as the target for platinum attack naturally led to a systematic study of the reactions of the complex with nucleic acid constituents. The observation that with polyuridine no immediate reaction took place but that a blue color slowly developed in solution, from which a blue solid could be isolated, led to the preparation of a series of complexes containing substituted uracils and thymines.

The results of antitumour screening on these blue complexes showed them to have promising activity, characterized by high aqueous solubility and low toxicity [1]. The binding of a uracil blue to closed circular DNA has been reported [2], as has the radiosensitizing properties of a thymine blue [3]. Further development was, however, hampered by their lack of homogeneity and variability in synthesis, despite a consistent stoichiometry of one $Pt(NH_3)_2$ unit per pyrimidine. Extensive studies have elucidated the principal structural and chemical features of these species, a good example where interest in biological activity has led to development of new platinum coordination chemistry.

5.1. Properties of the Platinum–Pyrimidine Blues

Reaction of a solution of the diaqua cation, *cis*-$[Pt(NH_3)_2(H_2O)_2]^{2+}$, formed *in situ* by the reaction of *cis*-$[PtCl_2(NH_3)_2]$ with silver nitrate, with a 2,4-dihydroxopyrimidine, or its sugar derivative, over a period of days at approximately neutral pH allows isolation of the dark blue products as salts of stoichiometry $[Pt(NH_3)_2(\text{pyrimidine})](NO_3)_{1-2}$ [4]. Few blue complexes are obtained with other amines although a purple complex from cyclopropylamine with apparent stoichiometry of Pt/amine/pyrimidine = 3/2/3 (independent of any charge neutralisation) has been reported [5]. A blue species from $[PtCl_2(en)]$ has also been reported [6], and its structure reported [25] but the only extensively studied compounds contain NH_3 as the amine.

Mass spectral studies of platinum—thymine blue showed a distribution of cluster ions with weights ranging up to 3000, indicating the presence of a number of oligomeric species [7]. The presence of white components in this mixture was later confirmed by HPLC studies [8]. Earlier studies had shown that the visible spectra did not obey Beer's law, and also were time and pH dependent [4]. The paramagnetic nature of the blues was indicated by EPR studies, the signal again being time dependent [4]. The presence of discrete Pt—Pt distances was confirmed by EXAFS studies on the blue substance derived from uridine, a value of 2.93 Å being obtained [9].

The results cited above indicated the platinum—pyrimidine blues to be oligomeric, mixed-valent cations with metal—metal bonding, in line with opinions on the original 'platinblau', first described in 1908 from the reaction of platinum salts in the presence of amides [10]. The bridging, and thus chain extension, occurs through the pyrimidines and the remarkable propensity of platinum—ammine complexes to stack in species such as $[Pt(NH_3)_2Br_2Pt(NH_3)_2Br_4]$ [11] undoubtedly contributes to the chain extension and indeed, the platinum—pyrimidine blues are another example of this tendency.

5.2. Platinum Blues Not Derived from Pyrimidines

The general interest aroused by the antitumour activity of cisplatin and its interaction with biological substrates such as purines and pyrimidines are incentives for the understanding of the chemical nature of the platinum—pyrimidine blues and in this respect work on related and model species have contributed to this end. A number of other ligands have also given mixed-valent and blue species with the $[Pt(NH_3)_2(H_2O)_2]^{2+}$ cation. Ligands include phthalimide [12], phosphate [13], cytosine and derivatives [14, 15], tryptophan [16], succinamic acid [17], glutamine [18], asparagine [19], 1-Me-nicotinamide [20], guanosine [20], oxamic acid [21, 22] and cytidine [23]. Sometimes the simple salt K_2PtCl_4 can be used rather than the amine complex [17—23]. Few biological data have been presented for these species. With the exception of the unusual phosphate complexes all ligands contain an anionic amidate linkage.

5.3. X-Ray Structures of Platinum—Pyrimidine and α-Pyridone Blues

The understanding of the chemical nature of the platinum blues received a considerable impulse with the resolution of the first crystal structure of a platinum—α-pyridone blue [24]. Recently the structures of its ethylenediamine analogue [25] and that of the 1-methyluracil complex [26] have also been solved. The similarity in the pyrimidine and pyridone ligands produces very similar chemistry but the presence of an extra amidate linkage in the pyrimidines does obviously introduce complications such as further binding from the extra exocyclic oxo groups (see below) and which are relevant to a full understanding of the nature of the pyrimidine blues:

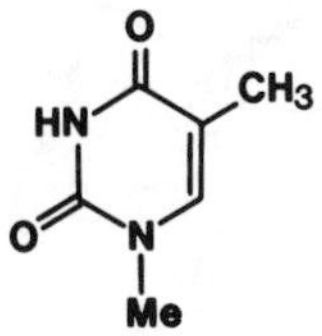

1-Me-Thymine

(52)

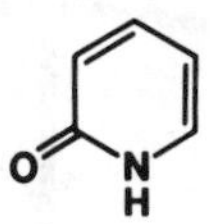

α-pyridone

(53)

Extensive chemical and structural studies have now elucidated the essential features of the platinum blues and these studies have utilized almost exclusively precursor complexes based on the ligands noted above. Before the details of their formation are examined a brief description of the blue structure is given.

Reaction of α-pyridone with *cis*-$[Pt(NH_3)_2(H_2O)_2]^{2+}$ at pH 7 and subsequent adjustment to low pH after a number of days give deep blue crystals with stoichiometry $[Pt_4(NH_3)_8(\alpha\text{-pyr})_4](NO_3)_5$, along with other products [27, 28]. The X-ray crystal structure (Figure 5.1) confirms the tetrameric nature of the product. For charge neutralization the compound is formally composed of three Pt(II) and one Pt(III) centers, thus giving an average oxidation state of 2.25 per platinum atom. The structure may be described as representing a dimer of dimers where each dimer contains two *cis*-$Pt(NH_3)_2$ units bridged by two α-pyridonate ligands in a head-to-head, rather than a head-to-tail, manner (see Section 5.4.2). Two major structural features are that within each dimer the two coordination planes are tilted at an angle of 27.4° relative to each other (because the bite of the amidate moiety, 2.3 Å, is shorter than the Pt—Pt distance of 2.78 Å), and also are twisted at an angle of 22° around the platinum—platinum bond axis. The dimers are connected by a slightly longer (2.885 Å) Pt—Pt bond with a Pt(1)—Pt(2)—Pt(2′) angle of 164.5° giving a slight zigzag effect to the chain. A number of hydrogen bonding interactions contribute to this linkage. Two weakly bonded nitrate ions cap the chain. These features have been confirmed in the more recent structures; the data are compared in Table 5.I.

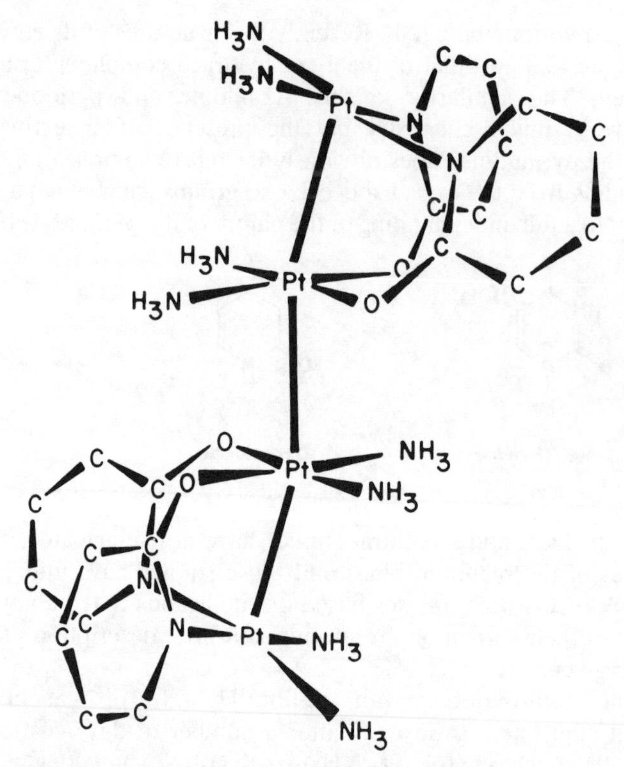

Fig. 5.1. Schematic representation of the structure of platinum—α-pyridone blue, $[Pt_4(NH_3)_8(\alpha\text{-pyr})_4]^{5+}$. From Reference 28.

TABLE 5.I
Structural and physical parameters for crystalline platinum blues

Complex	(Pt—Pt)[a]		∠Pt—Pt—Pt (deg.)	Dihedral angles[b]		μ[b]	λ_{max} (nm)
	outer	inner		τ	ω		
Pt—NH_3—α-pyridone	2.774	2.877	164.6	27.4	22.8	1.81	480,680
Pt—en—α-pyridone	2.830	2.906	164.3	32.1	24.3	1.934	532,745
Pt—NH_3—1-Me-Uracil	2.810	2.865 2.793	165.0 164.8	26.7	22.3	1.89	480,520 740

[a] Inner refers to the Pt—Pt bond forming the dimer of dimers.
[b] τ is the angle between adjacent platinum coordination planes and ω is the twist angle about the outer Pt—Pt vectors for the first two complexes and the inner vector in the third complex. See Ref. 26.

5.4. Formation of Platinum-Pyrimidine and α-Pyridone Blues

The formation of the blues may now be traced from monomeric to dimeric and on to tetrameric structures. This progression has not necessarily been chronological, being usually dependent on X-ray structural characterization, but it is perhaps useful to the interested chemist to describe the progression in this order. The solution chemistry is complicated and while selective and reproducible recrystallization has been achieved in most cases this does not mean that the species isolated are the only ones obtained from any particular reaction. A recent review covers many of the structures mentioned in the following discussion [29].

5.4.1. MONOMERIC COMPLEXES

Uracil or thymine may bind to a metal ion as the neutral ligand, the monoanion or the dianion (see Appendix 1). The multiplicity of binding sites of uracil has been discussed [30, 31]. The complexity of this binding has resulted in use of 1-methyl-substituted derivatives which, apart from resembling the nucleosides, facilitate studies by blocking one nitrogen and thus preventing formation of a ligand dianion. In the case of α-pyridone the neutral and monoanionic ligands need to be considered.

The reaction of *cis*-$[Pt(NH_3)_2(H_2O)_2]^{2+}$ with 1-MeU or 1-MeT anions in basic solution gives the bis(pyrimidine) complexes [32, 33] which with HCl yield the mono complexes [34] e.g.:

$$[Pt(NH_3)_2(H_2O)_2]^{2+} + 2\ \text{1-MeU} \rightarrow [Pt(NH_3)_2(\text{1-MeU})_2]$$
$$[Pt(NH_3)_2(\text{1-MeU})_2] + HCl \rightarrow [Pt(NH_3)_2(\text{1-MeU})Cl]$$

The analogous mono complexes with ethylenediamine have also been structurally characterized [35]. The corresponding bis-α-pyridone complex has been obtained from reaction with *cis*-$[PtCl_2(NH_3)_2]$ [36]: in fact this is a neutral 2-hydroxypyridine (α-pyrH) species with chloride counteranion, the minor tautomer being stabilized in the complex. The iminol form of 1-MeThymine has also been recognized upon protonation of *cis*-$[Pt(NH_3)_2(\text{1-MeT})_2]$ [33]. The stabilization of tautomeric forms of the unsubstituted pyrimidines has also been noted [31].

Concurrent ^{195}Pt NMR studies on the α-pyridone reaction also identified the aqua species $[Pt(NH_3)_2(\alpha\text{-pyr})H_2O]^{2+}$ as present in solution [37]. The mono chloro complexes all give aqua species on reaction with silver salts:

$$[Pt(NH_3)_2(L)Cl]^+ + AgNO_3 \rightarrow [Pt(NH_3)_2(L)(H_2O)]^{2+}$$

In some systems (1-MeU) the cationic aqua species may be isolated [38] and, in fact, there is crystallographic evidence for an aqua ligand in the related $[Pt(NH_3)_2(1\text{-Me-Cyt})H_2O]$ cation [39].

Binding of the metal to the ring nitrogen (N_3 in the case of the pyrimidines) increases the electron density on the exocyclic oxygens, and this property is responsible for much of the chemistry of these complexes, for they have a remarkable ability to self-associate or form dimeric complexes. In the case of uracil and thymine the stereochemical arrangement of the exocyclic oxygens is favorable for mixed-metal complex formation. This tendency has also been observed in other metal systems besides platinum [40, 41]. From the monomeric complexes, heteronuclear species of stoichiometry Pt,M, Pt,M_2 and Pt_2,M have been characterized [42, 43]. Further aggregation may occur and noteworthy in this respect is the crystallographic characterization of tridentate [44] and tetradentate [45] 1-MeUracil. A generalized scheme for complex formation has been written [42]. Dimerization may also occur via hydroxo bridge formation:

$$2[Pt(NH_3)_2(1\text{-MeU})H_2O]^+ \xrightarrow{-H^+,\,-H_2O} [\{Pt(NH_3)_2(1\text{-MeU})\}_2OH]$$

The extreme arrangements of this complex are shown in Figure 5.2 and represent possibilities of formation of intra- and interstrand links in a helix from a monobase, in this case a pyrimidine, aqua complex [31, 38]. The point is valid for other bases and certainly two metals on adjacent bases would seem to be a more difficult lesion for DNA to repair.

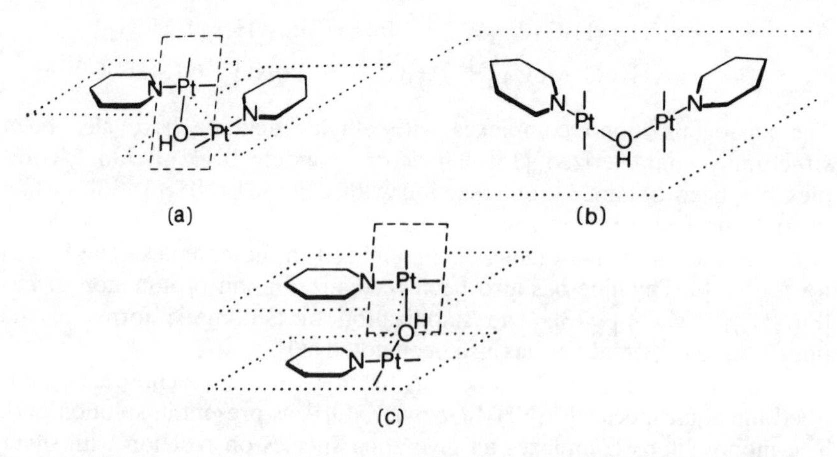

Fig. 5.2. Extreme arrangements for the interaction of a dinuclear platinum complex with DNA showing possible inter- (a and b) and intrastrand (c) links. From Reference 31.

5.4.2. DIMERIC COMPLEXES

Of more direct relevance to the formation of the platinum—pyrimidine blues is the tendency of the monomeric systems to give dimeric structures containing bridging pyrimidine or α-pyridone. This bridging is carried out by the exocyclic oxygen atoms and may occur in two ways. Where both bridging atoms on the same platinum are identical, e.g. both nitrogen or both oxygen, the head-to-head (hth) arrangement is obtained; the converse is the head-to-tail (htt) form. The structures of two head-to-tail complexes are represented in Figure 5.3. The synthetic routes used are outlined in Scheme 5.1 and in some cases can be selective (Consult References in Table 5.II for details).

Comparisons have been made between the complexes in this series. A number of features are of interest. Firstly, the hth dimer of α-pyridone further associates to give a tetramer and is therefore the Pt(II) analogue of the platinum—pyridone blue from which it thus differs in formal oxidation state by only one electron [37]. Indeed, addition of acid (pH 4.2—1.0) oxidizes this hth dimer to the blue, the htt isomer remaining in the Pt(II) state. As expected the Pt—Pt distances are longer in the Pt(II) complex than in the blue [2.877 (outer) and 3.129 Å (inner) versus 2.775 and 2.877 Å respectively] and the formation of the tetramer unit is facilitated by a number of stacking and hydrogen bonding interactions between the ammine hydrogen and the amidate oxygen atoms. The structure is essentially similar to that shown in Figure 5.1. Why the htt isomer is not

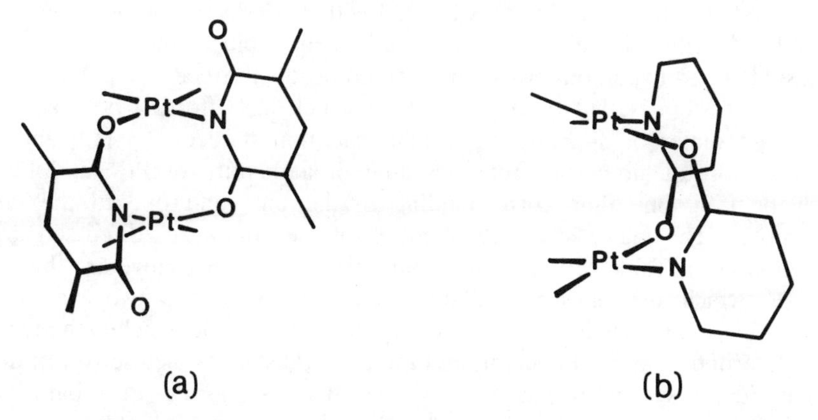

Fig. 5.3. Representation and comparison of the structures of the dimeric complexes formed from the reactions of *cis*-$[PtCl_2(NH_3)_2]$ with 1-MeT and α-pyridone. The head-to-tail configuration is shown in both cases. From References 33 and 36.

$M = Pt(NH_3)_2$

N—O = 1-MeT or 1-MeU *N—O = α-pyridone*

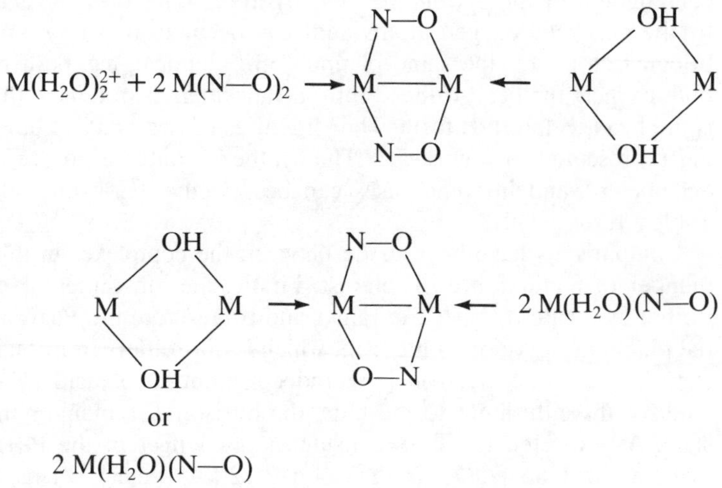

$$M(H_2O)_2^{2+} + 2\,M(N{-}O)_2 \rightarrow \text{M}\underset{\text{N—O}}{\overset{\text{N—O}}{\text{——}}}\text{M} \leftarrow \text{M}\underset{\text{OH}}{\overset{\text{OH}}{\quad}}\text{M}$$

$$\text{M}\underset{\text{OH}}{\overset{\text{OH}}{\quad}}\text{M} \rightarrow \text{M}\underset{\text{O—N}}{\overset{\text{N—O}}{\text{——}}}\text{M} \leftarrow 2\,M(H_2O)(N{-}O)$$

or

$$2\,M(H_2O)(N{-}O)$$

Scheme 5.1. Formation of dimeric pyrimidine and α pyridone complexes.

oxidized to a blue is interesting, since it can be oxidized completely to Pt(III).

A related dimer-of-dimers is formed with 1-Me-hydantoin and may be oxidized with difficulty to the mixed-valence blue. Interestingly, the unsubstituted hydantoin gives a blue-green product directly [50]. A further related system is that of α-pyrrolidone which has afforded both a tan-colored dimer of dimers with formal oxidation state of 2.5 [52] and a green nonstoichiometric unit consisting of a mixture of Pt_4^{5+} and Pt_4^{6+} chains [53], one thus corresponding to the blue and one to the tan complex. The major structural features of the complexes discussed are compared in Table 5.II. The principal parameters that change are the tilt angle which describes the dihedral angle between the two platinum coordination planes in the same dimer, the torsion angle which is the twist angle around the Pt—Pt vector, and the Pt—Pt distances themselves. In the α-pyridone system it has been shown that as the Pt—Pt separation decreases the tilt angle decreases while the twist angle increases [37, 50].

One feature of these dimeric complexes is the long intermolecular Pt—Pt distances between dimers in the pyrimidine series; the dimers are essentially unique. The association of two head-to-head dimers appears

TABLE 5.II
Comparison of structural parameters for non-blue complexes of general formula $[Pt(amine)_2L]_2 \cdot X_n$.[a]

am	L	Arrangement[b]	Oxid. State	(Pt—Pt) Å[c]		Angles[d]		Ref.
				outer	inner	τ	ω	
NH_3	α-pyr	htt	2.0	2.898		28.8	13.0	37
NH_3	1-MeT	htt	2.0	2.974		36.1	13.8	46
NH_3	1-MeU	htt	2.0	2.954		35.8	19.1	47
NH_3	1-MeCyt	htt	2.0	2.548				48
NH_3	α-pyr	hth	2.0	2.877	3.129	30.0	20.3	37
en	α-pyr	hth	2.0	2.992	3.236	39.6	24.9	49
NH_3	1-MeHyd	hth	2.0	3.131	3.204	38.6		50
NH_3	1-MeU	hth	2.0	2.937		34.1	25.2	38
NH_3	1-MeT	hth	2.0	2.927		31.4	1.9	51
NH_3	α-pyrl	hth	2.5	2.702	2.709	18.7	4.5	52
NH_3	α-pyrl	hth	2.37	2.764	2.739	24.5	0.8	53
				2.761	2.724	17.4	1.2	
NH_3	α-pyr	hth	3.0	2.547		20.0	23.2	54
NH_3	α-pyr	htt	3.0	2.54		20.3	26.3	54
NH_3	1-MeU	htt	3.0	2.574				55

[a] X omitted for clarity. See Refs. for details.
[b] htt, head-to-tail; hth, head-to-head.
[c] Inner refers to the Pt—Pt bond which bridges the dimer of dimers.
[d] τ is the tilt angle between adjacent platinum coordination planes; ω is the average twist angle about the Pt—Pt vector.

essential for formation of mixed-valent species and Lippert has summarized the possibilities for stacking as of three major types:

Type I. Centrosymmetric with the oxygen atoms bridging the inner platinum atoms.
Type II. Centrosymmetric with the nitrogen atoms bridging the inner platinum atoms.
Type III. Alternated bridging by nitrogen and oxygen atoms.

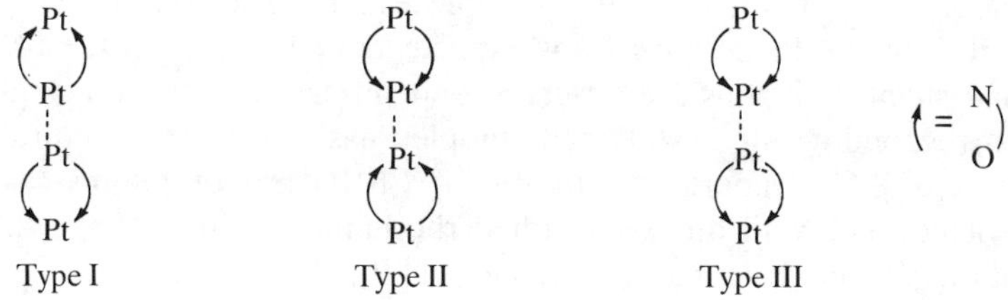

In the examples discussed above the bridging is Type I in all cases, as in all three blue structures (Table 5.I). The hydrogen bonding between the ammine groups and the ligand oxygens assists in giving short Pt—Pt distances. In the 1-MeU dimer the bridging is Type II and, while there are also hydrogen bonding interactions, the repulsions between the non-coordinating oxygens result in a long Pt—Pt distance (4.8 Å). Type III is represented by the 1-MeT dimer where the Pt—Pt separation is even larger (5.66 Å).

It is interesting to note the exigencies of blue formation in these systems; the dimers must first form in a head-to-head arrangement from the monomers and these must then form the dimer of dimers in a Type I manner. Conditions of pH appear critical: thus the 1-MeUracil blue is formed under conditions almost identical to the Pt(II) dimer of dimers (compare Refs. 26 and 38) and by the same chemical reaction. The presence of extra oxygen atoms and the combination of possible attractive and repulsive interactions are clearly much greater for the pyrimidine systems. The presence of the uncoordinated exocyclic oxygen atoms in the pyrimidine dimers has been exploited in a similar fashion to that used with the monomers to obtain a pentanuclear Pt_4,Ag unit whose structure is depicted in Figure 5.4. This is of interest because the stacking is now Type I and indeed, gentle heating affords the blue, along with other products, an example of a chemical redox reaction leading to the mixed-valence species [56]:

$$[Pt_4(NH_3)_8(1\text{-MeU})_4Ag]^{5+} \rightarrow [Pt_4(NH_3)_8(1\text{-MeU})_4]^{5+} + Ag^0$$

5.4.3. COMPLEXES OF Pt(III)

The redox properties of Pt(II) dimers were shown in the last section to afford blue complexes by oxidation either with acid or silver ion. Further oxidation was first exploited by Hollis and Lippard to give Pt(III) dimers with bridging α-pyridonate ligands [54]. The hth complex was obtained, in fact, by further oxidation of the blue and the htt isomer from the Pt(II) precursor with the extra anions occupying the axial positions. The parameters for some of these complexes have also been included in Table 5.II. The Pt—Pt distance is, as expected, smaller than in the Pt(II) dimers and slight variations are observed depending on the axial ligand [57]. The corresponding htt 1-MeUracil complex has also been structurally characterized [55]. Dimeric complexes of Pt(III) are of interest as they are isoelectronic with the well-studied rhodium(II) dimers [58, 59]. This area of Pt(III) complexes has been reviewed [29, 60].

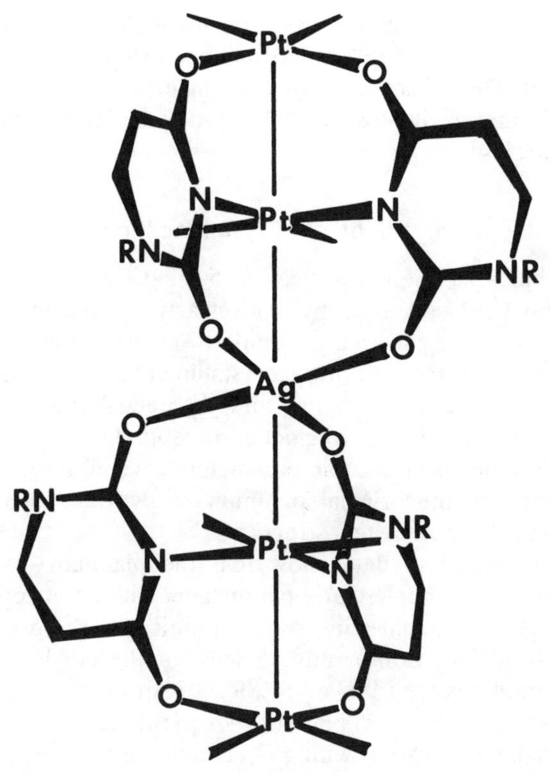

Fig. 5.4. Representation of the structure of a mixed metal Pt, Ag complex which is a platinum—pyrimidine blue precursor, $[Pt_4(NH_3)_8(1\text{-MeU})_4Ag]^+$. From Reference 31.

Electrochemical studies on the Pt(III)—α-pyridone dimers gave an $E_{1/2}$ of +0.63 V for both isomers, the redox process being interpreted as a coupled chemical reaction preceded by an overall two-electron transfer [49]. Concerted two-electron reductions are rare for metal complexes. Some differences were also observed in the electrochemical behavior of the two isomers; combined CV and DPV experiments indicated that whereas in the htt dimer the second electron is easier to remove than the first, the opposite is the case for the hth isomer, although in both species the two processes may be distinguished:

$$\text{Pt(III)Pt(III)} + e^- \rightarrow \text{Pt(II)Pt(III)} + e^- \rightarrow \text{Pt(II)Pt(II)}$$

In accord with the oxidation of the blue, controlled-potential electrolysis of the Pt(III) hth isomer gave the blue color of the mixed-valence species. The electrochemical behavior of the head-to-head 1-MeU dimer has been reported [61]. The reactivity of these species is also of interest, and the $[Pt_2(en)_2(\alpha\text{-pyr})(NO_2)(NO_3)]^{2+}$ cation has been shown to have a reactive NO_2 group [62]. The influence of the axial ligands can lead to asymmetric structures [63] and facile loss of NH_3 has been noted for the Pt(III) 1-MeU complex [64].

5.5. Studies on Crystalline Blues

Many of the blues mentioned in Section 5.2 have been characterized by UV/visible and EPR spectroscopy, and clearly the oligomeric nature of these species warns against any comparisons with the discrete blues discussed above. The isolation of the crystalline blue materials (Table 5.II) has permitted various spectroscopic and theoretical studies to elucidate the origin of the blue color. Magnetic susceptibility, EPR, optical and X-ray photoelectron spectroscopic parameters have all been obtained [65, 66]. The studies on the original platinum—α-pyridone blue have been supplemented by the more recent structures.

Magnetic susceptibility data show that the platinum—pyridone blue complex behaves as a simple Curie paramagnet with an effective magnetic moment of 1.81 μ_B [μ_B(spin-only) = 1.73], indicating the presence of one unpaired electron [24]. Comparative values for the ethylenediamine and 1-MeUracil complexes are 1.934 and 1.89 μ_B, respectively.

The optical spectrum of platinum—α-pyridone blue, as with the pyrimidine analogues, varies with pH, counteranions, temperature and time. Polarized single-crystal spectra in conjunction with a SCF-X Sw calculation have elucidated the major features of the electronic structure [67]. The blue color has been attributed to transitions from the inner Pt—Pt bonding σ orbital to an antibonding one (σ^*) and from the outer Pt—Pt π bonding orbital to outer Pt—Pt σ^*. The effect of alteration of the Pt—Pt distances in the platinum—ethylenediamine—pyridone blue was correlated with the optical spectrum. The delocalization of the unpaired spin density classes the pyridone blue as a Robin—Day Class III-A compound [68].

5.6. Biological Studies on Platinum—Pyrimidine Blue Precursors

The many pyrimidine complexes synthesized and structurally characterized in the foregoing work will clearly be of great use in elucidating the full potential of the biological activity of these complexes. The suggestion has been made that the activity reported could be due to slow release, by hydrolysis, of dimers and monomers [69]. Initial studies on some of these hydrolysis products indicated promising antitumour activity for both the monomeric [$Pt(NH_3)_2$(1-MeU)Cl] and the bridged dimer [$Pt(NH_3)_2$(1-MeU)]$_2(NO_3)_2$, although it was not clear which isomer was used [70].

The binding to DNA of the precursor complexes must also be considered. An early report of selective tumour cell surface staining was of considerable interest [65] but was not confirmed [71] and must be considered erroneous. Initial studies on DNA binding have also been reported [71].

5.7. Summary

The chemical nature of the platinum—pyrimidine blues has been elucidated through extensive studies on the 1-methylpyrimidines and α-pyridone. Crystallographic studies have confirmed that the basic structure of the platinum—pyrimidine blues consists of a dimer of dimers, each dimeric unit containing two *cis*-[$Pt(NH_3)_2$] units bridged through a ring nitrogen and exocyclic oxygen of two pyrimidine or α-pyridone rings. The blue color arises from the formal mixed oxidation state [2.25, 3Pt(II), 1Pt(III)] of the tetramer. The unpaired electron density is delocalized and classes the blues as Robin—Day Class III-A compounds.

Model compounds in the step-by-step formation have been synthesized and structurally characterized. Since there are two bridging pyrimidines or α-pyridones in each dimeric unit, the ligands may be arranged head-to-head (one Pt with two ring nitrogens, the other with two exocyclic oxygens) or head-to-tail (both Pt atoms with one ring nitrogen and one oxygen). Further aggregation of head-to-head dimers and oxidation gives the basic tetramer unit. The discrete Pt(II) dimers may also undergo two-electron oxidation to Pt(III), which species may in some cases be also obtained by reduction of the blues. With α-pyrrolidone a complex with an intermediate oxidation state of 2.5 has also been characterized.

References

1. J. P. Davidson, P. J. Faber, R. G. Fischer, Jr., S. Mansy, H. J. Peresie, B. Rosenberg, and L. VanCamp: *Cancer Chemother Rep.* **59** (Part 1), 287 (1975).
2. W. Bauer, S. L. Gonias, S. K. Kam, K. C. Wu, and S. J. Lippard: *Biochemistry* **17**, 1060 (1978).
3. J. D. Zimbrick, A. Sukrochana, and R. C. Richmond: *Int. J. Radiat. Oncol. Biol. Phys.* **5**, 1351 (1979).
4. B. Lippert: *J. Clin. Hematol. Oncol.* **7**, 26 (1977).
5. A. J. Thomson, I. A. G. Roos, and R. D. Graham: *J. Clin. Hematol. Oncol.* **7**, 242 (1977).

6. K. V. Shooter, R. Howse, R. K. Merrifield, and A. B. Robins: *Chem.-Biol. Interactions* **5**, 289 (1972).
7. R. D. MacFarlane and D. F. Torgerson: *Science* **191**, 920 (1976).
8. J. D. Woollins and B. Rosenberg: *Inorg. Chem.* **21**, 1280 (1982).
9. B.-K. Teo, K. Kijima, and R. Bau: *J. Am. Chem. Soc.* **100**, 621 (1978).
10. K. A. Hoffman and G. Bugge: *Chem. Ber.* **41**, 312 (1908).
11. G. C. Allen and N. S. Hush: *Prog. Inorg. Chem.* **8**, 357 (1967).
12. C. A. Chang, R. B. Marcotte, and H. H. Patterson: *Inorg. Chem.* **20**, 1632 (1981).
13. T. G. Appleton, R. D. Berry, and J. R. Hall: *Inorg. Chim. Acta* **64**, L229 (1982).
14. H. Neubacher, J. Krieger, P. Zaplatynski, and W. Lohmann: *Z. Naturforsch. Part B* **37**, 790 (1982).
15. D. M. L. Goodgame and I. Jeeves: *Z. Naturforsch. Part C* **34**, 1287 (1979).
16. H. Neubacher, P. Zaplatynski, A. Haase, and W. Lohmann: *Z. Naturforsch. Part B* **34**, 1015 (1979).
17. P. Arrizabalaga, P. Castan, and J. P. Laurent: *J. Am. Chem. Soc.* **106**, 1300 (1984).
18. P. Arrizabalaga, P. Castan, and J. P. Laurent: *Inorg. Chim. Acta* **66**, L9 (1982).
19. P. Arrizabalaga, P. Castan, and J. P. Laurent: *Z. Naturforsch. Part B* **35**, 1508 (1980).
20. J. P. Laurent, P. Lepage, P. Castan, and P. Arrizabalaga: *Inorg. Chim. Acta* **67**, 31 (1981).
21. J. H. Burness, M. J. Bandurski, L. J. Passman, and B. Rosenberg: *J. Clin. Hematol. Oncol.* **7**, 508 (1977).
22. P. Arrizabalaga, P. Castan, and J. P. Laurent: *Transition Met. Chem.* **5**, 204 (1980).
23. R. Ettore: *Inorg. Chim. Acta* **46**, L27 (1980).
24. J. K. Barton, D. J. Szalda, H. N. Rabinowitz, J. V. Waszczak, and S. J. Lippard: *J. Am. Chem. Soc.* **101**, 1434 (1979).
25. T. V. O'Halloran, M. M. Roberts, and S. J. Lippard: *J. Am. Chem. Soc.* **106**, 6427 (1984).
26. P. K. Mascharak, I. D. Williams, and S. J. Lippard: *J. Am. Chem. Soc.* **106**, 6428 (1984).
27. J. K. Barton, S. A. Best, S. J. Lippard, and R. A. Walton: *J. Am. Chem. Soc.* **100**, 3785 (1978).
28. J. K. Barton and S. J. Lippard: *Ann. N.Y. Acad. Sci.* **313**, 686 (1978).
29. J. D. Woollins and P. F. Kelly: *Coord. Chem. Rev.* **65**, 115 (1985).
30. B. Lippert: *Inorg. Chem.* **20**, 4326 (1981).
31. B. Lippert: *Platinum(II) Complex Formation with Uracil and Thymine* (Platinum, Gold, and Other Metal Chemotherapeutic Agents, ACS Symposium Series 209, Ed. S. J. Lippard), p. 147. ACS Washington (1983).
32. D. Neuegebauer and B. Lippert: *J. Am. Chem. Soc.* **104**, 6596 (1982).
33. B. Lippert: *Inorg. Chim. Acta* **55**, 5 (1981).
34. H. Schöllhorn, U. Thewalt, and B. Lippert: *Inorg. Chim. Acta* **106**, 177 (1985).
35. R. Faggiani, B. Lippert, and C. J. L. Lock: *Inorg. Chem.* **19**, 295 (1980).
36. L. S. Hollis and S. J. Lippard: *J. Am. Chem. Soc.* **103**, 1230 (1981).
37. L. S. Hollis and S. J. Lippard: *J. Am. Chem. Soc.* **105**, 3494 (1983).
38. B. Lippert, D. Neugebauer, and G. Raudaschl: *Inorg. Chim. Acta* **78**, 161 (1983).
39. J. F. Britten, B. Lippert, C. J. L. Lock, and P. Pilon: *Inorg. Chem.* **21**, 1936 (1982).
40. F. Guay and A. Beauchamp: *J. Am. Chem. Soc.* **101**, 6260 (1979).
41. F. Guay and A. Beauchamp: *Inorg. Chim. Acta* **66**, 57 (1982).
42. D. M. L. Goodgame, R. W. Rollins, and B. Lippert: *Polyhedron* **4**, 829 (1985).
43. D. M. L. Goodgame, M. A. Hitchman, and B. Lippert: *Inorg. Chem.* **25**, 2191 (1986).

44. U. Thewalt, D. Neugebauer, and B. Lippert: *Inorg. Chem.* **23**, 1713 (1984).
45. H. Schöllhorn, U. Thewalt, and B. Lippert: *J. Chem. Soc. Chem. Comm.* 769 (1984).
46. C. J. L. Lock, H. J. Peresie, B. Rosenberg, and G. Turner: *J. Am. Chem. Soc.* **100**, 3371 (1978).
47. R. Faggiani, C. J. L. Lock, R. J. Pollock, B. Rosenberg, and G. Turner: *Inorg. Chem.* **20**, 804 (1981).
48. R. Faggiani, B. Lippert, C. J. L. Lock, and R. A. Speranzini: *J. Am. Chem. Soc.* **103**, 1111 (1981).
49. L. S. Hollis and S. J. Lippard: *Inorg. Chem.* **22**, 2600 (1983).
50. J.-P. Laurent, P. Lepage, and F. Dahan: *J. Am. Chem. Soc.* **104**, 7335 (1982).
51. H. Schöllhorn, U. Thewalt, and B. Lippert: *Inorg. Chim. Acta* **93**, 19 (1984).
52. K. Matsumoto and K. Fuwa: *J. Am. Chem. Soc.* **104**, 897 (1982).
53. K Matsumoto, H. Takahashi, and K. Fuwa: *J. Am. Chem. Soc.* **106**, 2049 (1984).
54. L. S. Hollis and S. J. Lippard: *Inorg. Chem.* **21**, 2116 (1982).
55. H. Schöllhorn, P. Eisenmann, U. Thewalt, and B. Lippert: *Inorg. Chem.* **25**, 3384 (1986).
56. B. Lippert and D. Neugebauer: *Inorg. Chem.* **21**, 451 (1981).
57. L. S. Hollis, M. M. Roberts, and S. J. Lippard: *Inorg. Chem.* **22**, 3637 (1983).
58. E. B. Boyar and S. D. Robinson: *Coord. Chem. Rev.* **50**, 109 (1983).
59. T. R. Felthouse: *Prog. Inorg. Chem.* **29**, 73 (1982).
60. T. V. O'Halloran and S. J. Lippard: *Isr. J. Chem.* **25**, 130 (1985).
61. T. Ramstad, J. D. Woollins, and M. J. Weaver: *Inorg. Chim. Acta* **124**, 187 (1986).
62. T. V. O'Halloran, M. M. Roberts, and S. J. Lippard: *Inorg. Chem.* **25**, 957 (1986).
63. B. Lippert, H. Schöllhorn, and U. Thewalt: *J. Am. Chem. Soc.* **108**, 525 (1986).
64. B. Lippert, H. Schöllhorn, and U. Thewalt: *Inorg. Chem.* **25**, 407 (1986).
65. J. K. Barton, C. Caravana, and S. J. Lippard: *J. Am. Chem. Soc.* **101**, 7269 (1979).
66. J. K. Barton, S. A. Best, S. J. Lippard, and R. A. Walton: *J. Am. Chem. Soc.* **100**, 3785 (1978).
67. A. P. Ginsberg, T. V. O'Halloran, P. E. Fanwick, L. S. Hollis, and S. J. Lippard: *J. Am. Chem. Soc.* **106**, 5430 (1984).
68. M. B. Robin and P. Day: *Adv. Inorg. Chem. Radiochem.* **10**, 247 (1967).
69. C. C. F. Blake, S. J. Oatley, and R. J. P. Williams: *J. Chem. Soc. Chem. Comm.* 1043 (1976).
70. J. D. Woollins and B. Rosenberg: *J. Inorg. Biochem.* **19**, 41 (1983).
71. S. K. Aggarwal, R. W. Wagner, P. K. McAllister, and B. Rosenberg: *Proc. Nat. Acad. Sci. U.S.A.* **72**, 928 (1975).
72. P. K. McAllister, B. Rosenberg, S. K. Aggarwal, and R. W. Wagner: *J. Clin. Hematol. Oncol.* **7**, 717 (1977).

CHAPTER 6

ANTITUMOUR ACTIVITY OF METAL COMPLEXES

The demonstration of the antitumour activity of cisplatin encouraged wider studies on other transition metal complexes. Very few systematic studies on metal complexes in general have been conducted despite the intensive screening of the National Cancer Institute anticancer program. A recent update to 1981 gave a total of nearly 14 000 compounds for all inorganic compounds tested by NCI, of which almost 8000 were transition metal complexes [1]. All but a thousand of these complexes involve metals from iron to mercury, and the percentage of compounds meeting minimal activity standards is approximately 10%.

The early miscellaneous studies have been summarized [2, 3] and cover a wide range of structural types. Systematic studies have included amines, cyclopentadienyls and macrocycles, although much of the recent data has been accumulated for amines, because of the structural similarities to the platinum complexes. As of early 1987 no nonplatinum metal complexes are undergoing full clinical trials, but there are promising leads from animal studies. This chapter, rather than presenting an exhaustive account of all complexes studied, will attempt to demonstrate the great diversity of metal complexes that have been shown to have antitumour activity in standard screens. Some systems have been examined in detail and these will be summarized. Perhaps of more importance than the diversity of structural types is the range of mechanisms by which these complexes are believed to act.

Individual tumour screens may be susceptible or even resistant to a particular structural type, and the activity of initially promising complexes should be confirmed for a range of systems, as indicated in the case of cisplatin. The reasons for this spread of activities is not at all clear. Further, many complexes only show maximal *T/C* values (activity) at maximum tolerated or toxic doses. Much research centers on increasing that difference and, in view of the desirability of expanding our understanding of the biological effects of transition metal complexes, these results are, of course, valid but can also be misleading if compared with the clinically used cisplatin. Claims to 'equivalent activity' to cisplatin are not really meaningful unless the therapeutic index, which relates toxic dose to inhibitory dose, and also medium, are taken into account. The successful application of agents with antitumour activity rests on this selectivity or induction of a selective response between host and tumour cell. Finally, these data on murine tumour systems do not indicate the dose-limiting toxicities which may be expected. On the other hand, there is no reason to expect that an active non-platinum complex will have the same biodistribution and eventually the same range of clinical utility as platinum complexes, and may be active against cisplatin-resistant lines. For these reasons it is of considerable interest to broaden the range of complexes undergoing clinical trials.

6.1. Platinum Group Metal Complexes

A comparison of platinum group metal complexes is illustrative of the relatively narrow window of reactivity and stability which generates complexes with suitable antitumour activity. Table 6.I collects some comparative data of structurally analogous complexes, containing only ammonia and chloride ligands. The palladium complex, like the octahedral iridium analogue, shows little activity, while those of rhodium and ruthenium are somewhat intermediate. Palladium complexes are more labile than those of platinum, reacting approximately 10^5 times faster, and it is reasonable to accept that the palladium complex will be too reactive *in vivo* for any

TABLE 6.I
Comparison of antitumour activity of Group 8—10 ammine complexes.[a]

Complex	Dose Range (mg/kg)	Toxic Level (mg/kg)	Comments
cis-$[PtCl_2(NH_3)_2]$ (S)	0.5—20	9	Active
cis-$[PdCl_2(NH_3)_2]$ (S)	1.25—10	10—50	Inactive
mer-$[RhCl_3(NH_3)_3]$ (S.S)	12—100	70—100	Moderate Activity
mer-$[RhCl_3(NH_3)_3]$ (S)	5—30	20	Inactive
mer-$[IrCl_3(NH_3)_3]$ (W.S.)	50—200	150—200	Inactive
$[RuCl_3(NH_3)_3]$[b]	50	100	Active

[a] Data for Sarcoma 180 under equivalent conditions. Adapted from Ref. 1. Legend in parentheses after complex refers to vehicle of administration, i.e. S = saline; S.S. = saline slurry; W.S. = water slurry.

[b] Data taken from Ref. 26 using P388 leukemia. Probably a mixture of *mer* and *fac* isomers [27].

specificity to be observed. On the other hand, the inactivity and low toxicity of the iridium analogue reflects the inert nature of these Ir(III) species. The complexes of rhodium and ruthenium fall between these extremes and some activity is observed. These two metals probably represent the most promising candidates for development of structurally related platinum analogues.

Interestingly, the rhodium complex, reminiscent of cisplatin, displays some differences of toxicity, depending on the mode of its administration. No *T/C* values are included because no strictly comparable data are available but the general trend is conveyed.

6.1.1. PALLADIUM COMPLEXES

The possibility of decreasing the lability of palladium complexes by chelation has been examined. A number of palladium(II)-diaminocyclohexane complexes have activity but none are exceptional [3]. Modification of the leaving group lability by use of chelating carboxylates also results in some activity, as does the use of oligomeric hydroxo-bridged complexes. There does not appear to be a strict correlation in activities between isostructural complexes of platinum and palladium and in some cases the analogous palladium species seem more active. However, the combination of kinetic and thermodynamic properties required to give suitable pharmacological properties for palladium will require a good deal of basic chemical investigation before fuller studies are warranted.

Some hydrazine and mercaptopurine complexes of palladium were reported to have activity but, where the ligand has also some activity, no significant increases in activity were observed [4]. In this case the rationale for the synthesis was to transport metal ions using biologically active ligands, and it is of interest in this regard to note that sulfur mustard complexes of palladium were suggested as early as 1963, in a monograph on chelation and cancer, as metal transporters [5]. A Pd compound capable of nicking DNA has been reported [6].

6.1.2. RHODIUM COMPLEXES

The complex *mer*-$[RhCl_3(NH_3)_3]$ has been reported to have good activity, the best of the rhodium—amine complexes, but it is very insoluble [2]. Rhodium complexes are of interest because of the demonstration of their antibacterial action [7] (see Chapter 9) in structures of type $[RhCl_2(py)_4]^+$, which show, however, only marginal antitumour activity. Conversely, the most antitumour active rhodium species in general show few antibacterial effects. The relationship of selectivity to structure is, however, of importance with the pyridine complexes representing a defined class with distinct biological properties. The complex *mer*-$[RhCl_3(DMSO)py_2]$ has been reported to be very active in some systems [8] and its structural similarity to the bacteriostatic series is of note. The variation in properties in structurally closely related complexes is easily seen in these examples although these systems have not been as extensively examined as those of platinum.

6.1.2.1. *Studies on Rhodium Carboxylates*

An interesting series of complexes with antitumour activity is that of the rhodium carboxylates, whose structure is a dimer containing bridging carboxylates:

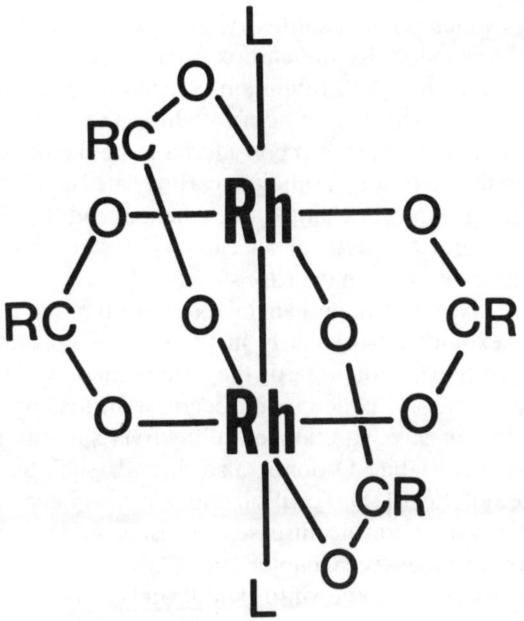

The biological studies on the rhodium carboxylates have been reviewed and discussed [3, 9, 10]. A number of interesting points have emerged, emphasizing the diversity of metal-based complexes with antitumour activity, although for rhodium carboxylates the acute toxicity probably eliminates these complexes from further development. The results are summarized below:

(1) The complexes show best activity against the Ehrlich ascites line and little activity against the standard L1210 and B16 lines. The complexes are also weak radiosensitizers in both mammalian and bacterial systems [11] (see Chapter 8).
(2) The activity increases with lipophilicity of the R group in the bridging carboxylate, and there is no correlation between reduction potential and activity. The $[Rh_2(O_2CR)_4]^+$ cations, with the rhodium atoms in the formal oxidation state of 2.5, have activity equivalent to that of the parent dimers.
(3) The complexes inhibit DNA synthesis over protein and, especially, RNA synthesis both *in vitro* and *in vivo*; the *in vitro* inhibition again correlates with lipophilicity of the different carboxylates [12].
(4) The complexes all inhibit a wide range of enzymes, including DNA polymerase, but there is not a complete correlation between this inhibition and antitumour activity. Of more significance is the fact that those enzymes which contain sulfhydryl groups in or at the active site are inhibited irreversibly by the carboxylates; enzymes without these groups are not affected [13]. Inhibition of cytidine deaminase, which also deactivates analogues of cytidine such as the antitumour agent arabinosylcytosine (ara-C) is also considered to be the possible reason for the synergy shown between rhodium carboxylates and ara-C [14].
(5) The carboxylate groups are claimed to exchange readily [15], although this is not a general property of all carboxylates [15a], and *in vivo* metabolism also breaks down the cage structure [16].
(6) Ligands in the axial positions can increase antitumour efficacy and this has been exploited particularly in the case of polyadenylic acid (poly(A)), which presumably retards the cage breakdown [17]. Further increases in activity, and concomitant decrease in toxicity, have been observed by the *in vivo* addition of sulfhydryls such as glutathione [17]. However, very labile O-donor axial ligands such as cyclophosphamide have no effect [18]. Axial binding has also been exploited to induce selectivity in parasitic diseases by using known antiparasitic drugs as the axial ligands (see Chapter 10).
(7) These complexes are well known to bind ligands in the axial positions [10, 19] and to react with a variety of biomolecules such as singly stranded DNA in this manner. They do not bind to doubly stranded DNA. Their reactions with constituent bases are remarkably specific, binding only occurring for adenine-based nucleosides and nucleotides [20, 21]. The nature of the specificity and the binding sites has been examined, and the specific binding attributed to N_7 binding which would give favorable hydrogen bonding interactions with the oxygen

atoms of the acetate ligand [22, 23], analogous to the case of [Co(acac)$_2$(NO$_2$)(Ado)] [24]. The crystal structures of N$_9$-bound caffeine and theophylline adducts have been reported [25].

In summary, these interesting complexes clearly exert their antitumour activity in a manner different to that understood for platinum—amines, and perhaps by indirect inhibition of DNA synthesis. This inhibition may be related to the affinity displayed for sulfhydryl groups. A number of ways to modify the basic activity appears to be possible, based on a variation in the axial ligands, possibly as carriers for the rhodium cage. The observation of quite a different range of general biological activity, i.e. both *in vitro* and *in vivo*, for the rhodium complexes, in comparison to that of complexes of near neighbor platinum, again underlines the diverse range of metal complexes, but little activity was found *in vivo* in L1210 studies [26].

6.1.3. RUTHENIUM COMPLEXES

The potential of ruthenium complexes has been extensively explored along with fundamental studies on their interactions with molecules of biological interest [27—30]. The most studied series are structurally related to cisplatin and, in principle, ruthenium offers the exploitable property of access to two oxidation states, Ru(II) and (III), which differ in their rates of substitution; the higher oxidation state is more inert and allows the possibility of introduction of a relatively inert and thus inactive complex which could be activated by reduction *in vivo*. This is of particular relevance with respect to the hypoxic or oxygen-deficient areas of tumours which occur at distances from the vascular system as a result of the enhanced respiration of rapidly growing cells (see Chapter 8). The potential for activation of Ru(III) complexes by reduction in this environment has been succintly reviewed [28].

Further, the existence of a number of stable radioisotopes of ruthenium holds promise for development of agents for organ imaging, and it should be remembered that ruthenium will possess some properties similar to those of technetium, whose complexes have been used for some considerable time for this purpose. In this respect, Clarke [28] has pointed out that considerable information on distribution of radioruthenium in animals has been accumulated because it is a major product in nuclear reactor waste.

The systematic studies on ruthenium complexes have involved antitumour and mutagenesis assays, tissue distribution studies and detailed examination of the reactions with nucleic acids and simple bases and nucleosides. The antitumour and mutagenic activity has been tabulated (Table 6.II) [27].

As can be seen, the complexes meet minimal requirements of activity and also, interestingly, they represent a series similar to that discussed for platinum, that is $[Ru(NH_3)_xA_{6-x}]$ (A = anionic ligand). The complex with greatest activity is $[RuCl_3(NH_3)_3]$, as is the case with rhodium(III) amines, and is also notable for activity in the secondary B16 screen. The ruthenium complex can exist as a mixture of *mer* and *fac* isomers [28]. The observation has been made that, in general, ruthenium complexes are approximately an order of magnitude less toxic than platinum derivatives with a corresponding decrease in activity and that, in many cases, the highest *T/C* values are obtained only with doses near the maximum tolerated or toxic dose [27]. However, the complexes clearly represent a class analogous to that of the platinum series.

Some further structurally unique complexes with demonstrated activity are shown in Figure 6.1. Early work on the DMSO complex, *cis*-$[RuCl_2(DMSO)_4]$, has been summarized, [31], and later work has also been published [32, 33]. Although of interest because the complex does not contain amines and is not very toxic (LD_{50} = 1 g/kg) the subsequently large doses of 600 mg/kg needed to achieve effects similar to cisplatin at 0.52 mg/kg [32] seem difficult to translate further beyond animal studies. The DMSO complex is noteworthy in that rhodium and platinum complexes with this ligand do not display any promising activity [31]. The trinuclear cation, $[Ru_3O_2(NH_3)_{14}]^{6+}$ (ruthenium red), has been used for some time as a histological stain [34] and has reasonable antitumour

TABLE 6.II
Antitumour data for some ruthenium complexes.

Complex	Therapeutic Dose (mg/kg)	Toxic Dose (mg/kg)	%*T/C*
K_3RuCl_6	25	100	138
$[RuCl_3(NH_3)_3]$	50	100	189
cis-$[RuCl_2(NH_3)_4]^+$	12.5	50	157
cis-$[Ru(H_2O)_2(NH_3)_4]^{3+}$	100	200	157
$[Ru(O_2CMe)(NH_3)_5]^{2+}$	12.5	25	149
$[Ru(O_2CEt)(NH_3)_5)^{2+}$	12.5	50	163
$[Ru(NH_3)_6]^{3+}$	3.1	25	113

Data from Ref. 27 for P388 leukemia. Counteranions omitted for sake of clarity.

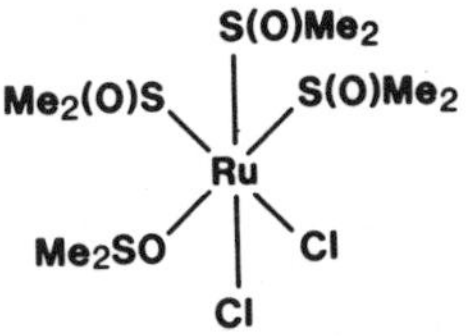

Fig. 6.1. Structures of some novel ruthenium complexes with antitumour activity.

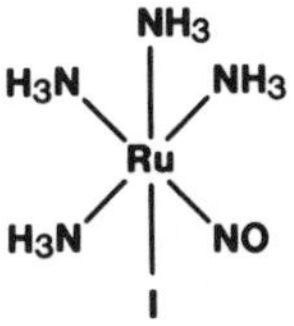

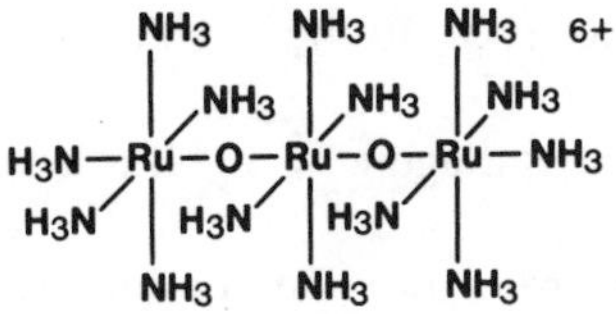

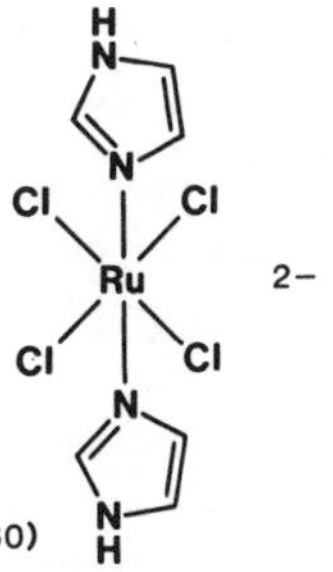

activity [35]. This oxy-bridged cation, whose structure has been elucidated [36], is of considerable interest for its biological properties which include inhibition of mitochondrial Ca^{2+} transport, inhibition of neuromuscular transmission and selective binding of polysaccharides (See Chapter 12). The imidazole groups of the $[RuCl_4(Im)_2]^{2-}$ complex are in the *trans* configuration [37, 120] and the nitrosyl complex listed shows activity at quite low doses. The presence of the NO group is of interest with respect to the presence of this group in nitroprusside, a clinically used vasodilator (Chapter 12).

Ruthenium complexes with bidentate ligands such as those based on 1,10-phenanthroline and bipyridine were some of the earliest complexes studied by Dwyer and coworkers in their perceptive and pioneering work on biological activity of metal complexes [38]. Further work showed that complexes where at least one phenanthroline ligand was substituted by an anionic group, such as acetylacetonate, as in $[Ru(acac)(phen)_2]^+$, were more active *in vitro* than the corresponding trischelates, $[Ru(phen)_3]^{2+}$, probably because of reduced charge [39]. The application of these complexes *in vivo* will be limited by the fact that not only are they eliminated readily but they are also potent neuromuscular poisons [40] (See Chapter 12). The point has also been made [28], that complexes with heterocyclic ligands such as phen will have the Ru(II) oxidation state stabilized with respect to Ru(III), in contrast to the amine complexes, and thus participation in redox cycles which may produce cytotoxicity is then unlikely.

6.1.3.1. *Studies on Ruthenium—Amine Complexes*

Detailed studies on reactions of the $[Ru(NH_3)_5(H_2O)]^{2+}$ cation with nucleic acid and constituent bases have broadened the range of reactions that can take place. The results may be summarized as follows:

(1) The biological target is believed to be DNA, similar to that of platinum complexes [41, 42].
(2) The complexes may induce mutagenesis by frame shift as well as base-pair substitution [27].
(3) The feasibility of *in vivo* reduction by biological reductants in mitochondrial fractions and microsomal redox enzymes has been demonstrated [43].

A number of chemical and biochemical effects of the binding of Ru to DNA have been observed and these include [44]:

(4) Examination of the visible spectra of Ru/DNA complexes reveals that the ruthenium binds primarily at guanine N_7 sites on the DNA with additional binding at the exocyclic nitrogens of adenine and cytosine [45, 46]. Possible further binding sites for a species such as *cis*-$[Ru(H_2O)_2(NH_3)_4]^{2+}$ have been discussed [27].
(5) Strand breakage can occur, confirming the involvement of a Ru(III)/Ru(II) redox reaction in the presence of reductant [44]. Note that not all complexes have equivalent strand-breaking activity.

Work with bases and nucleosides complements that of the platinum systems and in addition has broadened the range of reactions observed. The types of reactions have been discussed in more detail in Chapter 4 and are briefly summarized:

(6) Studies with *trans*-$[Ru(H_2O)(SO_3)(NH_3)_4]$ gave a relative affinity ratio of guanosine to adenosine (K_G/K_A) of 95 [47].
(7) Metallation can catalyze cleavage of the purine—sugar bond [48].
(8) The acid—base properties of the purines, and in particular the pK_a of endocyclic nitrogens, are affected by metallation.
(9) Linkage isomerism has been observed in solution of complexes with 1,3-dimethylxanthine, as has metal ion movement on adenosine [27].
(10) Metallation also labilizes the C_8 proton towards acid hydrolysis [49].

The indications are that the mode of action of Ru complexes is by direct binding to DNA. In contrast to complexes of platinum(II) strand breakage on DNA can occur due to the possibility of the Ru(III)/Ru(II) redox reactions; other reactions, such as metal ion movement, which is pH dependent, can be attributed to the access of two oxidation states.

6.2. Copper Complexes

The role of Cu as an essential trace element has focused attention on possible roles for copper chelation of biologically active ligands, with subsequent interference of normal transport and distribution, as well as the role of the metal in redox reactions due to the accessible oxidation states of (I) and (II). Similarly, the physiological response of copper levels in disease conditions [50] and the overall role of trace metals in health and disease [51, 52] are relevant and of considerable importance. The increase in serum copper content in infections, arthritic diseases, and certain neoplasms is well documented and, in fact, the subsequent decrease in level upon treatment has been used successfully as an indicator of cancer remission [50]. Copper complexes may be effective in therapy due in part to their ability to mimic this physiological response of elevated copper [53] and, clearly, the interplay of introduced copper with pre-existent bound copper and effects on copper–protein mediated processes will affect the ultimate biological fate of the complex. Likewise, while the excess accumulation of free Cu, and indeed Fe and Zn, caused by malfunction or absence of normal metabolic pathways is extremely damaging to the body, the controlled release of such metals may be beneficially cytotoxic. The widespread pharmacological effects of copper complexes have been briefly reviewed [54].

The action of preformed coordination complexes on tumour growth will be discussed in this section. The structures, Figure 6.2, cover a wide variety of classes and may be subdivided into the following principal groups:

(1) Complexes based on thiosemicarbazones and bis(thiosemicarbazones), which represent a mixed N,S donor set of ligands.
(2) Complexes based on chelates formed from 1,10-phenanthroline and macrocyclic ligands containing an all-nitrogen (4N) donor set.
(3) Complexes formed from salicylate-type ligands, and thus having a 4-oxygen (4O) donor set.

6.2.1. BIS(THIOSEMICARBAZONE) COMPLEXES

The carcinostatic properties of these complexes have been summarized

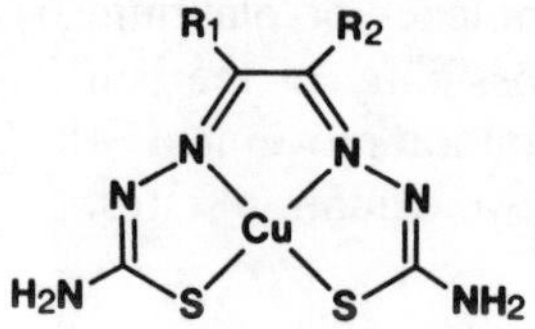

Fig. 6.2. Structures of antitumour copper complexes.

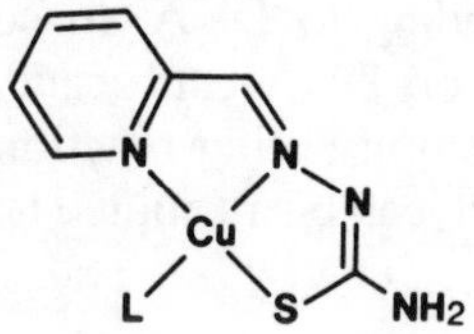

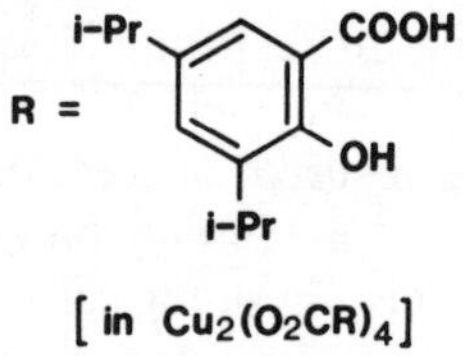

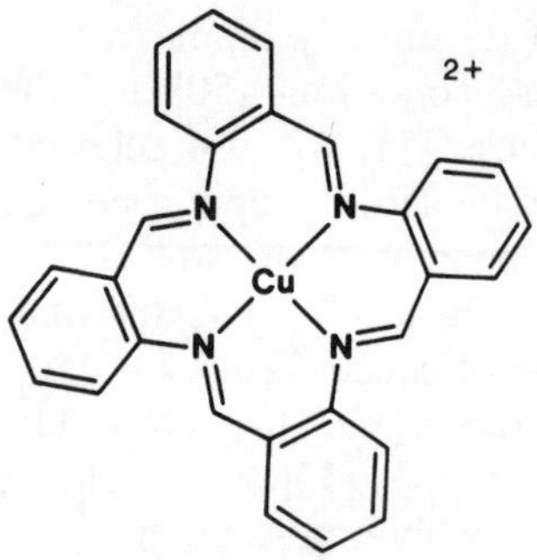

[55]. These studies arose from the original rationale of French and Freelander as early as 1960, who reasoned that good chelating agents such as the bis(thiosemicarbazones) might remove metals from essential sites critical to tumour growth with subsequent inhibition [56, 57]. Further studies showed that the ligands and complexes were indeed active in various systems, with the complex invariably more active than free ligand. The most active ligand was the 3-ethoxy-2-oxobutyraldehyde derivative (H_2KTS with R_1 = ethoxyethyl, R_2 = H in Structure 61, Figure 6.2) and most mechanistic studies were carried out on the CuKTS complex which displays very good activity in Walker 256 carcinoma and Sarcoma 180, although not in leukemias [55].

Consideration of the structure and reactivity of these complexes indicated that intracellular reduction was likely, the reduction potential ($E_0' = -120$ mV at pH 6.6) being easily in the range of biological reducing agents [58]. A general reaction for thiols is:

$$Cu(II)KTS + 2\,RSH \rightarrow Cu(I)SR + \tfrac{1}{2}RSSR + H_2KTS$$

The major features of the biochemistry of these complexes are:

(1) In accordance with the expectation of reduction, the intact complex is rapidly absorbed by Ehrlich cells where reduction occurs with release of the free ligand, H_2KTS [59].
(2) In Sarcoma 180 cells, the ligand was released into the extracellular medium, the copper remaining in the cell [60].
(3) A stimulated rate of oxygen consumption, observed after initial uptake, decreases slowly to values lower than control levels.
(4) Intracellular thiol concentration is concomitantly depleted and the sequestering by formation of Cu(I)SR confirmed. This is nonspecific and copper seems evenly distributed throughout the cell [61]. This copper-catalyzed oxidation accounts for the increased oxygen consumption.
(5) Other biochemical processes affected are DNA synthesis and, to a lesser degree, mitochondrial respiration. The DNA synthesis is very rapid and the concentrations necessary to affect DNA synthesis are compatible with cytotoxic concentrations. There is widespread inhibition of cellular processes involving the nucleus, and this has led to a suggestion of a general poisoning effect on the cell by copper. The exact mechanism is difficult to pinpoint and, besides DNA synthesis in an indirect manner, the involvement of toxic, oxygen-based radicals produced in a Cu(II)/Cu(I) cycle could be the ultimate source of cytotoxicity [55]. Again, DNA synthesis can be inhibited without direct binding.

6.2.2. THIOSEMICARBAZONE COMPLEXES

The thiosemicarbazones (Structure 62, Figure 6.2) have antitumour activity in their own right [62]. Early mechanistic studies led to the postulate that the mode of action is by inhibition of ribonucleotide reductase [63, 64]. This metal-dependent enzyme is a key intermediate in DNA biosynthesis because it converts ribonucleotides to deoxyribonucleotides [65, 66].

The metal dependence of ribonucleotide reductase is cobalt as vitamin B_{12} (prokaryotes) and iron (eukaryotes), with some evidence for manganese as cofactor [66]. The reader is referred to Ref. [66] for a recent discussion on the mechanism of these interesting enzymes. In essence, the reaction can be described:

(ribose ring, HO OH) + HS SH* --> (ribose ring, HO H*) + S—S + H_2O

The role of the metal ion is believed to be to generate and stabilize radicals which mediate the hydrogen transfer from a cysteine-SH to the ribonucleotide substrate.

The carboxaldehyde thiosemicarbazones are the most active inhibitors of the mammalian reductases [66]. Originally, it was proposed that the ligand simply masked the enzyme site, but more recent results suggest that the complex formed from iron and the ligand is the active species, destroying (in an oxygen-dependent reaction) the enzyme's free radical, which is essential to the catalytic process [67]. These results explain the fact that inhibition increases with added iron [68, 69] whereas iron chelators [70, 71] and added dithiol [68, 72] decrease the potency of inhibition.

Both Cu and Fe complexes have significantly greater antitumour activity than the free ligand [55, 73, 74]. Chemical and biochemical studies show behavior similar to that of the bis(thiosemicarbazones). Biochemical studies showed that the metal complexes enter cells to a greater degree than free ligand, and the copper complex is subsequently reduced but, in contrast to the bis(thiosemicarbazone) case there is little dissociation, Cu(I)L being quite stable. EPR studies show intracellular binding of extra ligands, probably thiols, to the intact Cu(I) complex. There is an increase in oxygen utilization with decrease in free thiol concentration. The biochemical effects of decreased thiol can be expected then to be similar in result to those of the bis(thiosemicarbazones) [55].

The free ligands have great affinity for ferrous ion and, in fact, can remove iron from ferritin to form the Fe(III) complex [75]. The *in vivo* existence of the chelate is therefore indicated, and free ligand might then be expected to have the same mechanism of action as the preformed chelate. These results must be placed in context with the newer concepts of mechanism of inhibition of ribonucleotide reductase, where it would appear that thiol depletion could be the determining factor for inhibition. This is mediated by reactive oxygen radicals produced from the redox couple. This reaction and, indeed, the whole system are therefore formally analogous to the bleomycin system to be discussed in Chapter 7, except that the targets (DNA precursor and DNA itself) differ.

The recognition of ribonucleotide reductase as a barrier for cellular proliferation should be of considerable interest for the design of other specific metal complexes as inhibitors and the possibility of selective toxicity (e.g. eukaryotic vs. prokaryotic) that a selective inhibition implies.

6.2.3. NITROGEN CHELATES

In a review of copper, silver and gold complexes [76], complex 64 (Figure 6.2) was the most active species, giving a *T/C* of 170% at a dose range of 6.25—12.5 mg/kg in B16 melanoma, (compare cisplatin with a mean *T/C* of 180% at a dose range of 0.15—20 mg/kg). Complex 64 is inactive in other screens, however [76]. The result is of interest because the macrocycle stabilizes the cuprous oxidation state, possibly because of extensive delocalization inducing a distorted square-planar configuration [76].

This situation is paralleled by the 1,10-phenanthroline complexes which have been studied [38, 39]. Of a series of tetramethyl-1,10-phenanthroline complexes studied against P388 cells *in vitro* the Cu(II) complex was most active, although it was inactive *in vivo* [39], which contrasts with its earlier reported activity against a Landschutz ascites tumour [38]. Despite its good ability to cleave DNA in the presence of oxygen and a reducing agent (see Chapter 1), the unsubstituted 1,10-phenanthroline Cu(II) complex is inactive but the 2,9-dimethyl derivative greatly inhibited growth of P388 *in vivo* [77]. Again, neuromuscular toxicity was a dose-limiting side effect [39, 77]. There is an interesting juxtaposition here in that added Cu overcomes the toxic action of administered free 1,10-phenanthroline, but copper chelation potentiates the action of the 2,9-dimethyl substituted ligand [77]. Further, the cleavage of DNA is not effected by the 2,9-dimethyl-1,10-phenanthroline system.

The Cu(I) oxidation state is also stabilized with respect to Cu(II) in the 2,9-dimethyl-1,10-phenanthroline system [78]. Clearly, the factors which affect *in vitro* and *in vivo* activity are different, but it would be of interest to reconcile these apparently contradictory results that the most active inducers of strand breakage are not antitumour active, while inhibitors of strand breakage are quite active. The cell penetration and lability of the complex once inside the cell will be of importance besides apparent differences in chemistry. A slower redox cycle is actually more selective, being more toxic, than a faster one.

6.2.4. COPPER SALICYLATES

The study of the antitumour effect of thiosemicarbazones evolved simultaneously with, or even preceded, the understanding of the mechanism of action of the proposed target, ribonucleotide reductase. There is little doubt, however, that consideration of the bioinorganic chemistry of non-DNA targets can lead to interesting possibilities for drug design. Another example of this aspect is found in attempts to alter the concentration of 'active oxygen' in tumour cells by mimicking enzyme activity.

An aspect of oxygen utilization in tumour cells that is now well documented is the fact that these cells show diminished superoxide dismutase (SOD) activity, although mitochondrial production of superoxide continues [79]. These enzymes (Cu/Zn, or Mn) catalyze the disproportionation of superoxide, O_2^- [80—83]:

$$2\,O_2^- + 2\,H^+ \rightarrow H_2O_2 + O_2$$

The observation of a distinct biochemical difference between normal and tumour cells (independent of cause/effect relationships) has led to the proposition that superoxide (either produced by exogenous chemicals or inhibition of SOD) could be selectively toxic, if delivered to the tumour cell; and, in parallel, the question arises whether replacement of absent free radical scavengers would allow the neoplastic cells to revert to their original (nontumour) phenotype [83].

Some tumour growth inhibition was observed for Cu(SOD) itself [83]. The demonstrated SOD-mimetic activity (albeit low) of copper complexes such as bis(isopropylsalicylato)copper(II), prompted a study of its antitumour effect, the reasoning being that this lipid-soluble analogue would have greater uptake in cells. The complex (Structure 63) is probably of the cupric acetate type:

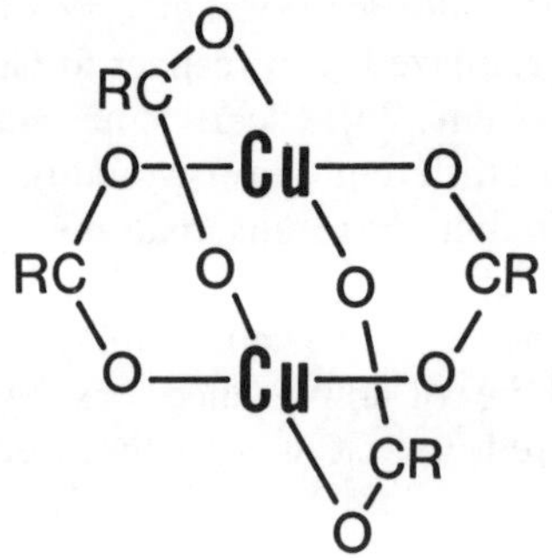

The complex is indeed active in inhibiting Ehrlich carcinoma growth *in vivo*, with a delay in metastasis and increased survival of the host [85]. The exact mechanism is not clear: H_2O_2 is produced and could be responsible for some of the antitumour effect; the Cu complex mimics the activity of SOD but not strongly and is rapidly degraded [86]. Interestingly, these species also inhibit tumour promotion by carcinogens, implying a role for free radicals in this process too [87]. Other SOD-mimetic complexes based on porphyrins have only weak antitumour activity [86]. Reference [83] contains detailed discussion on the relationship of SOD to cancer; its radioprotective properties are presented in Section 8.5. Modification of the inherent cellular balance of a critical metalloenzyme represents a unique bioinorganic approach to the control of cancer.

6.3. Silver and Gold

Of the copper triad, the lightest element has been most extensively studied for antitumour properties [76], although the use of the 'nobler' metals since earliest times is reflected in long-standing medicinal applications, with diverse uses over long periods as bacteriocides (Ag, Chapter 9) and in rheumatoid arthritis (Cu, Au, Chapter 11). The biological chemistry of gold has been reviewed [88, 89]. As a group, these metals and their complexes are not necessarily comparable because of differences in preferred oxidation state and geometry.

Thus, while the most common oxidation states of Cu are (I) and (II) with tetrahedral or square-planar geometry, those of Au are (I) and (III), with Au(I) being predominantly linear. Trigonal planar and tetrahedral geometries are also known for Au(I) but with monodentate ligands these revert easily to the linear form. The Au(III) form is diamagnetic and invariably of square-planar geometry and, as such, is comparable with platinum. Reactions of Au(III) generally occur faster than those of Pt(II) and it is also well to remember that the concepts of *trans*-influence and *trans*-effect widely used in platinum chemistry can be applied to that of gold [90]. The Au(III) oxidation state is expected to be reduced in aqueous solution to Au(I), but the ability of Cu to engage in one-electron redox processes is a fundamental difference in the chemistry of these two elements. No silver complexes are of interest, at this stage, for antitumour properties, and the following material emphasizes recent results with gold.

Early studies on gold complexes have been summarized [76]. The activity of $K[Au(CN)_2]$ is noteworthy as it was originally used in tuberculosis treatment some time ago (see Chapter 9). The use of gold complexes in arthritis treatment and continuing research on new analogues have led to development of auranofin as an oral drug (Chapter 11), whose antitumour properties have also been studied [91]. There is a reciprocal relationship to some extent between drugs active in cancer and arthritis, possibly due to end effects on the immune response system, and it is of interest that cisplatin has antiarthritic activity [92]. A good example of a nonmetal drug with activity in both regimes is methotrexate (Compound 34, Figure 3.3). The underlying reasons for this reciprocity are well worth exploring, and the reciprocal nature of drugs in cancer and trypanosomiasis should also be borne in mind in this respect (Chapter 10).

An evaluation of the antitumour activity of auranofin showed that it was active against P388 leukemia *in vivo* but ineffective in other models such as B16 and M5076 [93]. Auranofin showed no preference for DNA synthesis inhibition in comparison to that of RNA and protein, but the drug appeared to be inactivated by serum [93]. The structure of auranofin (structure 111, Figure 11.1) allows for a considerable number of structure—activity relationships, and this was explored to produce new series of structurally unique gold complexes with antitumour potential.

The structures of some of the complexes are shown in Figure 6.3. Three series may be recognised; linear complexes of monodentate phosphines, $[Au(PR_3)X]$; bis(gold) complexes $[\{AuX\}_2diphosphine]$ with bidentate phosphines, which also preserve the linear coordination; and tetrahedral complexes of formula $[Au(PR_3)_4]^+$ or $[Au(diphosphine)_2]^+$. The basic auranofin analogues (R_2P—Au—X), which have been extensively tabulated [94] are summarized (Table 6.III). These have a somewhat broader range of activity than the parent complex.

The most interesting complexes are the bidentate and tetrahedral complexes. In P388, maximal ILS values of 93% for $[\{AuCl\}_2(dppe)]$ and 86% for [Au(dppe)]Cl were obtained [95]. The nature of the tetrahedral

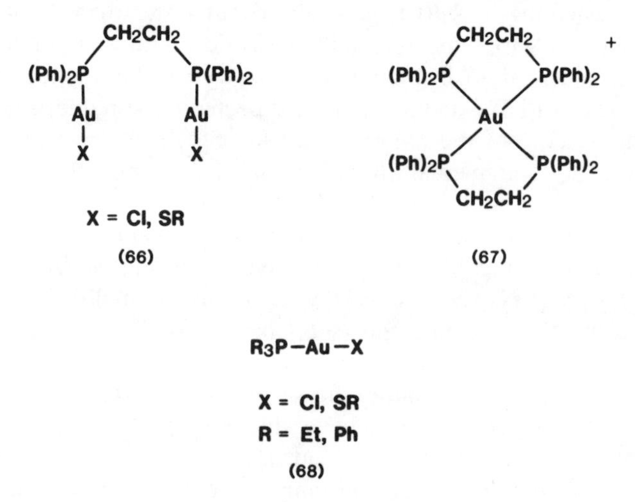

Fig. 6.3. Gold complexes with demonstrated antitumour activity.

TABLE 6.III
Antitumour activity of selected gold complexes.[a]

Complex	MTD[b] (μmol/kg)	IC_{50}[c] (μM)	%ILS[d] (max)
$[Au(PEt_3)(TATG)]$[e]	18	1.5	70
$[Au(PEt_2\textit{i}\text{-}Pr)(TATG)]$	17	2	90
$[Au(PEt_3)CH_3]$	36	1	55
$[Au(PEt_3)CN]$	17	0.4	68
$[Au\{P(CH_2Ph)_3\}Cl]$	19	4	91
$[Au(TATG)]$	110	150	14
$[\{AuCl\}_2(diphos)]$	—	—	93
$[\{AuCl\}_2\textit{cis}\text{-}Ph_2PCH{=}CHPPh_2]$	5	—	91
$[\{AuCl\}_2\textit{trans}\text{-}Ph_2PCH{=}CHPPh_2]$	28	—	34
$[Au(diphos)_2]^+$	3		82

[a] See Reference 94 for details.
[b] Maximum tolerated dose in tumour-bearing mice.
[c] Inhibitory concentration in B16 melanoma cells *in vitro*.
[d] Increase in life span in P388 leukemia *in vivo*, on a 1—5 d schedule, drugs given i.p.
[e] TATG = 2,3,4,5-tetraacetato-thio-β-glucopyranose. See Structure 111, Figure 11.1.

complexes has been confirmed for $[Au(PR_3)_4]^+$ and $[Au(diphosphine)_2]^+$ [96, 97], as has the bridging nature of the bis(gold) complexes [98, 99]. In a study on the $[\{AuX\}_2(diphosphine)]$ series, the most active phosphine was the cyclohexyl derivative and differences in geometry, i.e. *cis* or *trans* ethylene bridge, also gave different activities [95].

As stated, the tetrahedral complexes are the most active, and the conversion of the bis(gold) species to such moieties has in fact been observed in plasma:

$$[\{AuX\}_2(diphos)] \rightarrow [Au(diphos)_2]^+$$

Some dissociation of free phosphine is also observed in solution [97]. A relevant finding in this work is that the free phosphine is also active [95] (originally reported in 1966 [110]), and the implication is that the gold complex serves as a releasing mechanism for the cytotoxic ligand. Since copper salts potentiate the activity of free phosphine [111] there is a possibility of *in vivo* phosphine release followed by copper activation — similar in concept to the possible antitumour mechanisms of 1,10-phenanthroline and the thiosemicarbazones. In pursuit of this line, Cu(II) was shown to displace dppe from $[Au(dppe)_2]Cl$ [112] to give a Cu(I) complex. Further synthesis and testing of copper-diphosphine complexes structurally analogous to the gold [Cu(diphosphine)]Cl and the unusual [{CuCl}$_2$(dppe)$_3$] showed equivalent activity to the gold complexes [113]. These results tend to confirm the mechanism of action of the original gold complexes as that of ligand release. The role of copper in enhancement of cytotoxicity and the biological targets of the phosphines is not at present resolved.

6.4. Organometallic Complexes

The study of cyclopentadienyl complexes is of interest in relation to the carcinostatic properties of the metallocene dihalides MX_2Cp_2, [100—102], as well as the possible tissue imaging properties of the iron and especially ruthenium complexes $RuCp_2$ [28]. The $Fe(Cp)_2^+$ also has some antitumour activity [103] and is of interest as a radiosensitizer (Chapter 8).

The apparent rationale for study of the metallocene dihalides was that the *cis* geometry, Figure 6.4, was reminiscent of that of cisplatin, with a '*cis*-chloride' separation of 3.470 Å for Ti, for example, in contrast to that in cisplatin, 3.349 Å [102]. However, it is clear from studies of the hydrolysis of the cyclopentadienyl complexes that the aqueous chemistry is very different from that of platinum and no structural analogies should be drawn. In general, the complexes inhibit Ehrlich ascites tumours and solid growths of this tumour, are active in Lewis Lung, but are only very

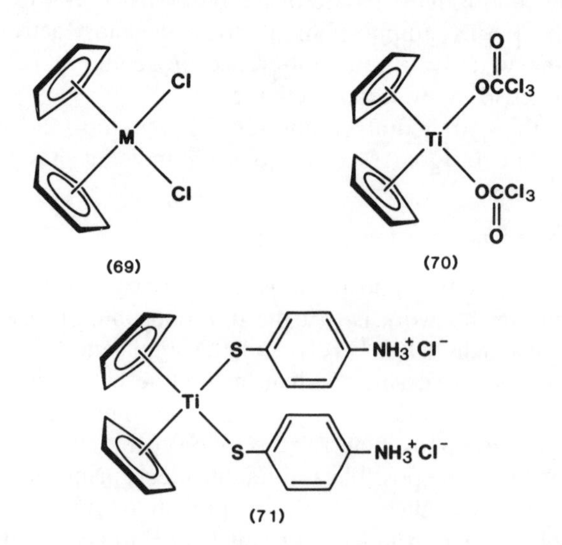

Fig. 6.4. Cyclopentadienyl complexes that display antitumour activity.

marginally active in L1210 and P388. The data in Table 6.IV show that V, Ti and Nb complexes are most active and that, for a series of X-substituted Ti complexes, little difference is found. The activity of the ferricenium cation, although less than that of titanium salts [103], confirms the relative unimportance of the halide groups. The high LD_{50} is noteworthy, however. Substitution in the Cp ligands and use of bridging Cp groups do not enhance activity, and so the series is best represented by MCl_2Cp_2, where M = Ti, Nb or V. Recent modifications include carboxylato and thiolato groups instead of halides [104]. A point to note, however, is that for the halide complexes activities are always obtained at maximum tolerated doses, and thus the therapeutic indices are not high. The data have been extensively summarized [114].

A number of chemical, biological and morphological studies have been carried out on these complexes. The toxicity has been compared with that of cisplatin [105]. Morphological effects of metal accumulation in nucleic acid-rich areas such as chromatin and production of giant cells are also reminiscent of those induced by cisplatin. In structure—activity terms, the tetrahedral geometry of the complexes results in the central metal influencing such factors as the Cl—Cl distance, and it has been pointed out that the inactive complexes of Zr and Hf have a bite angle larger than that of *cis*-$[PtCl_2(NH_3)_2]$ [102].

However, other factors such as relative lability may be more determinant of activity, especially as alteration of halide (leaving group?) in the

TABLE 6.IV
Antitumour activity of metal-cyclopentadienyl complexes.[a]

Complex	Optimum Dose (mg/kg)	LD_{50} (mg/kg)	TI[b]
$[VCl_2Cp_2]$	80—90	110	1.4
$[TiCl_2Cp_2]$	30—60	100	3.3
$[TiF_2Cp_2]$	40	60	2.0
$[TiBr_2Cp_2]$	40—80	135	4.5
$[NbCl_2Cp_2]$	20—25	35	3.5
$[MoCl_2Cp_2]$	75—100	175	2.9
$[FeCp_2]^{+}$[c]	220—300	400	

[a] In Ehrlich ascites tumour-bearing mice. Consult References 100 and 105 for details.

[b] TI defined as LD_{50}/ED_{75}, where ED is effective dose.

[c] From Reference 103.

Ti series does not produce dramatic effects, unlike the case of the platinum amines. The metallocenes are expected to be more labile than Group 8—10 complexes, and this is borne out by hydrolysis studies where polymeric products from both halide and cyclopentadienyl loss have been isolated [102—104]. Interestingly, differences between the complexes have been observed — the Ti and V complexes lose Cp much more readily than Mo, for example [115, 116]. The relative lack of toxic side effects (nonnephrotoxic) may well be a reflection of the fact that, especially in aqueous medium, decomposition products may be unreactive, unlike those of the platinum systems.

The mechanism of action is intriguing because of the organometallic nature of these species. Biochemical precursor studies with $[MCl_2Cp_2]$, M = Ti, V, show selective inhibition of DNA synthesis [106, 107]. Binding *in vitro* has been demonstrated. If a direct binding mechanism is involved then we would expect substantially different binding sites from the platinum case, because of the relative hard/soft properties of the metals involved, and this is shown by the crystal structure of a theophylline adduct with an N_7—O_6 chelate [108], which has eluded all searches with platinum. The high oxygen affinity of Ti and V species is also shown by formation of polynuclear uracil-bridged complexes [109]. These structures have been observed in nonaqueous solution but evidence is accruing in aqueous solution to indicate a totally different reactivity pattern to cisplatin and the later transition metals. As expected, strong phosphate interaction is found in nucleotide complees [116, 117], while a unique 4-membered chelate formed by N_1—$6NH_2$ binding has been structurally characterized in $[Mo(Cp)_2 9\text{-MeAde}]^+$ [117].

The different behavior of $[V(O)(acac)_2]$ in comparison to $[Co(acac)_2(NO_2)_2]^-$, and especially acac displacement by nucleic acid bases with NH groups also emphasizes the different chemistry involved [118]. The reactivity of β-diketonate complexes is indeed relevant due to the demonstrated antitumour activity of the benzoylacetonato (complexes $[TiX_2(Bzacac)_2]$ (X = halide, OR) (119). Indeed, one complex from this new series of Ti complexes, budotitane {diethoxybis(1-phenylbutane-1,3-dionato)titanium(IV), X = OEt} is receiving limited clinical trials in Europe.

There is no necessity to invoke a Pt-type adduct in order for Ti complexes to inhibit DNA synthesis and, if direct binding is involved, then the lesions will almost certainly be different than in the Pt case. Can they, however, result in similar conformational changes? The differences, therefore, in the binding of the two species should be stressed.

6.5. Summary

Non-platinum metal complexes with antitumour activity in murine screens are known for a wide diversity of structural types and the transition metals most studied include rhodium, ruthenium, gold, copper and titanium. The spectrum of activity is generally different to cisplatin (Ti, Au) and in some cases the complexes show lack of cross-resistance with cisplatin-resistant lines.

The diversity of structures is matched by a wide range of mechanisms. While the ruthenium—amines may act by a DNA-binding mechanism the rhodium carboxylates inhibit DNA synthesis but do not bind directly to the doubly stranded polynucleotide. In the latter case a possible mechanism is inhibition of precursor enzymes. Copper thiosemicarbazones and gold diphosphines appear to act by release of the toxic ligand. For thiosemicarbazones the ultimate target is ribonucleotide reductase. Other copper complexes have been developed to mimic the action of superoxide dismutase.

Organometallic complexes based on cyclopentadienyl and exemplified by $[TiCl_2Cp_2]$ and also $[FeCp_2]^+$ also present a different spectrum of activity compared to cisplatin. Again, DNA synthesis is selectively inhibited but the exact mechanism is unclear. The titanium β-diketonate, budotitane, is another Ti complex with antitumour activity.

References

1. M. J. Cleare: *Coord. Chem. Rev.* **12**, 349 (1974).
2. M. J. Cleare and P. C. Hydes: *Metal Ions in Biol. Syst.* **11**, 1 (1980).
3. D. S. Gill: *Structure Activity Relationship of Antitumour Palladium Complexes* (Platinum Coordination Complexes in Cancer Chemotherapy, Eds. M. P. Hacker, E. B. Douple, and I. H. Krakoff), p. 267. Martinus Nijhoff, Boston (1984).
4. S. Kirschner, Y. K. Wei, D. Francis, and J. G. Bergman: *J. Med. Chem.* **9**, 369 (1966).
5. A. Furst: *The Chemistry of Chelation in Cancer*, C. C. Thomas, Springfield (1963).
6. G. R. Newkome, M. Onishi, W. E. Puckett, and W. A. Deutsch: *J. Am. Chem. Soc.* **102**, 4551 (1980).
7. R. J. Broomfield, R. H. Dainty, R. D. Gillard, and B. T. Heaton: *Nature* (*London*) **223**, 735 (1969).
8. P. Colamarino and P. Orioli: *J. Chem. Soc.* (*Dalton*) 845 (1979).
9. J. L. Bear: *Rhodium Compounds for Antitumour Use* (Proceedings of the Ninth International Precious Metals Conference Eds. E. E. Zysk and J. A. Bonucci) pp. 337—344. Int. Precious Metals Inst. Allentown (1986).
10. E. B. Boyar and S. D. Robinson: *Coord. Chem. Rev.* **50**, 109 (1983).

11. R. Chibber, I. J. Stratford, P. O'Neill, P. W. Sheldon, I. Ahmed, and B. Lee: *Int. J. Radiat. Biol.* **48**, 513 (1985).
12. P. N. Rao, M. L. Smith, S. Pathak, R. A. Howard, and J. L. Bear: *J. Natl. Cancer Inst.* **64**, 905 (1980).
13. R. A. Howard, T. G. Spring, and J. L. Bear: *Cancer Res.* **36**, 4402 (1976).
14. R. G. Hughes J. L. Bear, and A. P. Kimball: *Proc. Am. Assoc. Cancer Res.* **13**, 120 (1972).
15. J. L. Bear, H. B. Gray, Jr., L. Rainen, I. M. Chang, R. Howard, and A. J. Kimball: *Cancer Chemother. Rep.* **59**, 611 (1975).
15a. F. Pruchnik, B. R. James, and P. Kvintovics: *Can. J. Chem.* **64**, 936 (1986).
16. A. Erck, E. Sherwood, J. L. Bear, and A. P. Kimball: *J. Natl. Cancer Inst.* **36**, 2204 (1978).
17. J. L. Bear, R. A. Howard, and A. M. Dennis: *Curr. Chemother.* 1321 (1978).
18. M. D. Joesten: *Metal Ions in Biol. Syst.* **11**, 285 (1980).
19. T. R. Felthouse: *Prog. Inorg. Chem.* **29**, 73 (1982).
20. K. Das and J. L. Bear: *Inorg. Chem.* **15**, 2093 (1976).
21. K. Das, E. L. Simmons, and J. L. Bear: *Inorg. Chem.* **16**, 1268 (1977).
22. N. Farrell: *J. Inorg. Biochem.* **14**, 261 (1981).
23. N. Alberding, D. E. Crozier, and N. Farrell: *J. Am. Chem. Soc.* **107**, 384 (1985).
24. L. G. Marzilli and T. J. Kistenmacher: *Acc. Chem. Res.* **10**, 146 (1977).
25. K. Aoki and H. Yamazaki: *J. Chem. Soc. Chem. Comm.* 186 (1980).
26. L. M. Hall, R. J. Speer, and H. J. Ridgway: *J. Clin. Hematol. Oncol.* **7**, 391 (1980).
27. M. J. Clarke: *Metal Ions in Biol. Syst.* **11**, 231 (1980).
28. M. J. Clarke: *The Potential of Ruthenium in Anticancer Pharmaceuticals* (Inorganic Chemistry in Biology and Medicine, ACS Symposium Series v. 140, Ed. A. E. Martell) pp. 157—180. ACS, Washington (1980).
29. M. J. Clarke: *Ruthenium Anticancer Agents and Relevant Reactions of Ruthenium-Purine Complexes* (Chemistry and Biochemistry of Platinum, Gold and Other Metal Chemotherapeutic Agents, ACS Symposium Series v. 209, Ed. S. J. Lippard) pp. 335—354. ACS, Washington (1983).
30. M. J. Clarke: *The Chemistry of Ruthenium Compounds in Potential Anticancer Agents* (Proceedings of the Ninth International Precious Metals Conference, Eds. E. E. Zysk and J. A. Bonucci) pp. 369—386 (1986).
31. N. Farrell: *Chemical and Biological Activity of Metal Complexes Containing Dimethyl Sulfoxide* (Platinum, Gold and Other Metal Chemotherapeutic Agents, ACS Symposium Series v. 209, Ed. S. J. Lippard) p. 279. ACS, Washington (1983).
32. G. Sava, T. Giraldi, G. Mestroni, and G. Zassinovich: *Chem.-Biol. Inter.* **45**, 1 (1983).
33. G. Sava, S. Zorzet, T. Giraldi, G. Mestroni, and G. Zassinovich: *Eur. J. Cancer Clin. Oncol.* **20**, 841 (1984).
34. J. H. Luft: *Anat. Rec.* **171**, 369 (1971).
35. T. Tsuruo, H. Iida, S. Tsukagoshi, and Y. Sakurai: *Jap. J. Cancer Res.* **71**, 151 (1980).
36. M. A. A. F. de C. T. Carrondo, W. P. Griffith, J. P. Hall, and A. C. Skapski: *Biochim. Biophys. Acta* **627**, 332 (1980).
37. B. K. Keppler and W. Rupp: *J. Cancer Res. Clin. Oncol.* **111**, 166 (1986).
38. F. P. Dwyer, E. Mayhew, E. M. F. Roe, and A. Shulman: *Brit. J. Cancer* **19**, 195 (1965).
39. A. Schulman, G. M. Laycock, and T. R. Bradley: *Chem.-Biol. Inter.* **16**, 89 (1977).
40. F. P. Dwyer, E. C. Gyarfas, W. P. Rogers, and J. H. Koch: *Nature (London)* **170**, 490 (1952).

41. A. D. Kelman, M. J. Clarke, S. D. Edmonds, and H. J. Peresie: *J. Clin. Hematol. Oncol.* **7**, 274 (1978).
42. R. E. Yasbin, C. R. Matthews, and M. J. Clarke: *Chem.-Biol. Inter.* **31**, 355 (1980).
43. M. J. Clarke, S. Bitler, D. Rennert, M. Buchbinder, and A. D. Kelman: *J. Inorg. Biochem.* **12**, 279 (1980).
44. M. J. Clarke, B. Jansen, K. A. Marx, and R. Kruger: *Inorg. Chim. Acta* **124**, 13 (1986).
45. M. J. Clarke, M. Buchbinder, and A. D. Kelman: *Inorg. Chim. Acta* **27**, L87 (1978).
46. M. E. Kastner, K. F. Coffey, M. J. Clarke, S. E. Edmonds, and K. Eriks: *J. Am. Chem. Soc.* **103**, 5747 (1981).
47. G. M. Brown, J. E. Sutton, and H. Taube: *J. Am. Chem. Soc.* **100**, 2767 (1978).
48. M. J. Clarke and P. E. Morrissey: *Anal. Biochem.* **122**, 404 (1982).
49. M. J. Clarke and P. E. Morrissey: *Inorg. Chim. Acta* **80**, L69 (1984).
50. Z. A. Karcioglu, G. Karcioglu, R. M. Sarper, and M. Hrgovcic: *Zinc and Copper in Neoplastic Diseases* (Zinc and Copper in Medicine, Eds. Z. A. Karcioglu and R. M. Sarper) pp. 464—534. C. C. Thomas, Springfield (1980).
51. E. D. Harris: *Copper in Human and Animal Health* (Trace Elements in Health, Ed. J. Rose) pp. 44—73. Butterworth, London (1983).
52. *Trace Metals in Health and Disease* (Ed. N. Kharasch). Raven Press (1979).
53. J. R. J. Sorenson: *Prog. Med. Chem.* **15**, 211 (1978).
54. J. R. J. Sorenson: *Chem. in Brit.* 1110 (1984).
55. D. H. Petering: *Metal Ions in Biol. Syst.* **11**, 197 (1980).
56. F. A. French and B. L. Freelander: *Cancer Res.* **18**, 1298 (1958).
57. F. A. French and B. L. Freelander: *Cancer Res.* **20**, 505 (1960).
58. D. H. Petering: *Bioinorg. Chem.* **1**, 255 (1972).
59. D. T. Minkel and D. H. Petering: *Cancer Res.* **38**, 117 (1978).
60. B. A. Booth and A. C. Sartorelli: *Mol. Pharmacol.* **3**, 290 (1967).
61. D. T. Minkel, L. A. Saryan, and D. H. Petering: *Cancer Res.* **38**, 124 (1978).
62. K. C. Agrawal and A. C. Sartorelli: *Handbook of Experimental Pharmacology* **38**(2), 793 (1975).
63. F. A. French, E. J. Blanz Jr., J. R. doAmaral, and D. A. French: *J. Med. Chem.* **13**, 1117 (1970).
64. A. C. Sartorelli, K. C. Agrawal, and E. C. Moore: *Biochem. Pharmacol.* **20**, 3119 (1971).
65. L. Thelander and P. Reichard: *Ann. Rev. Biochem.* **48**, 133 (1979).
66. M. Lammers and H. Follman: *Structure and Bonding* **54**, 27 (1983).
67. L. Thelander and A. Graslund: *J. Biol. Chem.* **258**, 4063 (1983).
68. E. C. Moore, M. S. Zedeck, K. C. Agrawal, and A. C. Sartorelli: *Biochememistry* **9**, 4492 (1970).
69. A. C. Sartorelli: *Adv. Enz. Regul.* **15**, 117 (1977).
70. J. G. Cory, L. Lasater, and A. Sato: *Biochem. Pharmacol.* **30**, 979 (1981).
71. J. G. Cory, A. Sato, and L. Lasater: *Adv. Enz. Regul.* **19**, 139 (1981).
72. P. J. Preidecker: *Mol. Pharmacol.* **18**, 507 (1980).
73. L. A. Saryan, E. Ankel, C. Krishnamurti, D. H. Petering, and H. Elford: *J. Med. Chem.* **22**, 1218 (1979).
74. W. Antholine, J. Knight, H. Whelan, and D. H. Petering: *Mol. Pharmacol.* **13**, 89 (1977).
75. W. E. Antholine, J. Knight, H. Whelan, and D. H. Petering: *Mol. Pharmacol.* **16**, 569 (1977).
76. P. J. Sadler, M. Nasr, and V. L. Narayanan: *The Design of Metal Complexes as

Anticancer Agents (Platinum Coordination Complexes in Cancer Chemotherapy, Eds. M. P. Hacker, E. B. Douple, and I. H. Krakoff) p. 290. Martinus Nijhoff, Boston (1984).
77. A. Mohindru, J. M. Fisher, and M. Rabinowitz: *Biochem. Pharmacol.* **32**, 3627 (1983).
78. B. R. James and R. J. P. Williams: *J. Chem. Soc.* 2007 (1961).
79. L. W. Oberley and G. R. Buettner: *Cancer Res.* **39**, 1141 (1979).
80. *Superoxide Dismutase* vols. 1 and 2, Ed. L. W. Oberley, CRC Press, Boca Raton, Florida (1982).
81. J. A. Fee: *Superoxide, Superoxide Dismutases, and Oxygen Toxicity* (Metal Ion Activation of Dioxygen, Ed. T. G. Spiro), p. 209. Wiley, New York (1980).
82. J. S. Valentine: *J. Chem. Ed.* 925 (1985).
83. L. W. Oberley: *Superoxide Dismutase and Cancer* (Ref. 80), pp. 127—165.
84. U. Weser, C. Richter, A. Wendel, and M. Younes: *Bioinorg. Chem.* **8**, 201 (1978).
85. S. W. C. Leuthauser, L. W. Oberley, T. D. Oberley, J. R. J. Sorenson, and K. Ramakrishna: *J. Natl. Cancer Inst.* **66**, 1077 (1981).
86. L. W. Oberley, S. W. C. Leuthauser, T. D. Oberley, J. R. J. Sorenson, and R. F. Pasternack: *Antitumour Activities of Compounds with Superoxide Dismutase Activity* (Inflammatory Diseases and Copper, Ed. J. R. J. Sorenson) pp. 423—433. Humana (1982).
87. T. W. Kensler, D. M. Bush, and W. J. Kozumbo: *Science* **221**, 75 (1983).
88. P. J. Sadler: *Structure and Bonding* **29**, 171 (1976).
89. C. F. Shaw III: *Inorg. Perspect. Biol. Med.* **2**, 287 (1979).
90. R. J. Puddephat: *The Chemistry of Gold*, Elsevier, Amsterdam (1978).
91. T. M. Simon, D. H. Kunishima, G. J. Vilbert, and A. Lorbert: *Cancer Res.* **41**, 94 (1981).
92. J. R. Bowen, G. R. Gale, W. A. Gardner, and W. A. Bonner: *Agents and Actions* **4**, 108 (1974).
93. C. K.Mirabelli, R. K. Johnson, C. M. Song, L. Faucette, K. Muirhead, and S. T. Crooke: *Cancer Res.* **45**, 32 (1985).
94. C. K. Mirabelli, R. K. Johnson, D. T. Hill, L. F. Faucette, G. R. Girard, G. Y. Kuo, C. M. Sung, and S. T. Crooke: *J. Med. Chem.* **29**, 218 (1986).
95. C. K. Mirabelli, D. T. Hill, L. F. Faucette, F. L. McCabe, G. R. Girard, D. B. Bryan, B. M. Sutton, J. O'L. Bartus, S. T. Crooke, and R. K. Johnson: *J. Med. Chem.* **30**, 2181 (1987).
96. P. A. Bates and J. M. Waters: *Inorg. Chim. Acta* **81**, 151 (1984).
97. S. J. Berners-Price, M. A. Mazid, and P. J. Sadler: *J. Chem. Soc. Dalton* 969 (1984).
98. C. A. McAuliffe, R. V. Parish, and P. D. Randall: *J. Chem. Soc. Dalton* 1730 (1979).
99. D. S. Eggleston, D. F. Chodosh, G. R. Girard, and D. T. Hill: *Inorg. Chim. Acta* **108**, 221 (1985).
100. H. Kopf and P. Kopf-Maier: *Tumour Inhibition by Metallocene Dihalides of Early Transition Metals: Chemical and Biological Aspects* (Platinum, Gold and Other Metal Chemotherapeutic Agents, ACS Symposium Series v. 209, Ed. S. J. Lippard), p. 315. ACS, Washington (1983).
101. P. Kopf-Maier and H. Kopf: *The Metallocene Dihalides — A Class of Organometallic Early Transition Metal Complexes as Antitumour Agents* (Platinum Coordination Complexes in Cancer Chemotherapy, Eds. M. P. Hacker, E. B. Douple, and I. H. Krakoff) p. 279. Martinus Nijhoff, Boston (1984).
102. P. Kopf-Maier and H. Kopf: *Dev. Oncol.* **17**, 279 (1984).

103. P. Kopf-Maier, H. Kopf, and E. W. Neuse: *Angew. Chem. Int. Ed. Engl.* **23**, 456 (1984).
104. P. Kopf-Maier, S. Grabowski, J. Liegener, and H. Kopf: *Inorg. Chim. Acta* **108**, 99 (1985).
105. P. Kopf-Maier and H. Kopf: *Anticancer Res.* **6**, 227 (1986).
106. P. Kopf-Maier and H. Kopf: *Naturwiss.* **67**, 415 (1980).
107. P. Kopf-Maier, W. Wagner, and H. Kopf: *Naturwiss.* **68**, 272 (1981).
108. D. Cozak, A. Mardhy, M. J. Olivier, and A. L. Beauchamp: *Inorg. Chem.* **25**, 2600 (1986).
109. B. F. Fieselman, D. N. Hendrikson, and G. D. Stucky: *Inorg. Chem.* **17**, 1841 (1980).
110. R. F. Struck and Y. F. Shealy: *J. Med. Chem.* **9**, 414 (1966).
111. R. M. Synder, C. K. Mirabelli, R. K. Johnson, C. M. Sung, L. F. Faucette, F. L. McCabe, J. P. Zimmerman, M. Whitman, J. C. Hempel, and S. T. Crooke: *Cancer Res.* **46**, 5054 (1986).
112. S. J. Berners-Price, C. K. Mirabelli, R. K. Johnson, M. R. Mattern, F. L. McCabe, L. F. Faucette, C.-H. Sung, S.-M. Mong, P. J. Sadler, and S. T. Crooke: *Cancer Res.* **46**, 5486 (1986).
113. S. J. Berners-Price, R. K. Johnson, C. K. Mirabelli, L. F. Faucette, F. L. McCabe, and P. J. Sadler: *Inorg. Chem.* **26**, 3383 (1987).
114. P. Kopf-Maier and H. Kopf: *Chem. Rev.* **87**, 1137 (1987).
115. J. H. Toney and T. J. Marks: *J. Am. Chem. Soc.* **107**, 947 (1985).
116. J. H. Toney, C. P. Brock, and T. J. Marks: *J. Am. Chem. Soc.* **108**, 7263 (1986).
117. L. Y. Kuo, M. G. Kanatzidis, and T. J. Marks: *J. Am. Chem. Soc.* **109**, 7207 (1987).
118. N. Farrell and S. B. Barczewski: *J. Inorg. Biochem.* **30**, 315 (1987).
119. J. H. Keller, B. Keppler, and D. Schmahl: *J. Cancer Res. Clin. Oncol.* **105**, 109 (1983).
120. B. K. Keppler, W. Rupp, U. M. Juhl, H. Endres, R. Niebl, and W. Balzer: *Inorg. Chem.* **26**, 4366 (1987).

CHAPTER 7

METAL-MEDIATED ANTIBIOTIC ACTION

The chemical production of hydroxyl radical, either by metal-mediated redox reactions or as a result of radiolysis products, can induce strand breaks in DNA. An example of a chemotherapeutic agent whose chemical production of activated oxygen is believed to be the critical step in its action is bleomycin, a naturally occurring antibiotic. The binding of metals to this and other antibiotics has been summarized in detail [1—3] and the material presented here is based on these reviews.

7.1. Discovery and Pharmacology of Bleomycin

The bleomycins are a group of antibiotics derived from a strain of the fungus *Streptomyces verticillus* and were discovered by Umezawa and co-workers in 1966 [4]. The subsequent biological studies indicated promising antitumour activity and currently the drug is clinically used in the treatment of testicular tumours, squamous cell carcinoma and various lymphomas. The clinical spectrum of bleomycin and its successors has been extensively reported in a number of monographs and review volumes [5—7]. The bleomycins fall within the scope of this monograph because of their purported mechanism of action. The clinical and toxicological aspects will not be dealt with at length, since the antibiotic cannot really be classified as a 'robust' synthetic coordination complex, although it is isolated as a blue copper(II) chelate.

The precursor of bleomycin, phleomycin, was too nephrotoxic for development and the search for less toxic analogues resulted in the discovery of bleomycin [4, 8]. The major dose-limiting toxicity is pulmonary fibrosis and other serious toxicities noted have been hyperpyrexia, acute hypotensive responses and mucositis. The pulmonary toxicity appears to be dose-related and has led to fatalities. Bleomycin is administered as the metal-free chelate and is absorbed when administered orally; excretion is mainly by renal clearance.

7.2. Structure of Bleomycin and Analogues

The structure of bleomycin is shown in Figure 7.1. There are essentially three critical parts to the molecule. First, the sugar groups which may be considered to confer water solubility on the overall structure, then the bleomycin nucleus which consists of a metal-binding set of ligands such as the imidazole and pyrimidine groups, and then the connection of these via peptide groups to a bis(thiazole) grouping containing a terminal amine. In fact the isolated bleomycins are a mixture of 12—13 antibiotics differing in their terminal amine.

The nature of the terminal amine may be varied chemically and slight modification of the bleomycinic acid nucleus gives second-generation analogues such as isobleomycin and zorbamycin; third-generation analogues are considered to be those isolated which differ significantly in the nucleus, one of interest being tallysomycin [9—11].

As stated, the antibiotics are isolated as their blue copper(II) chelates. The copper can be easily removed as sulfide with H_2S and the blue color regenerated by addition of excess copper(II). The metal is believed to be

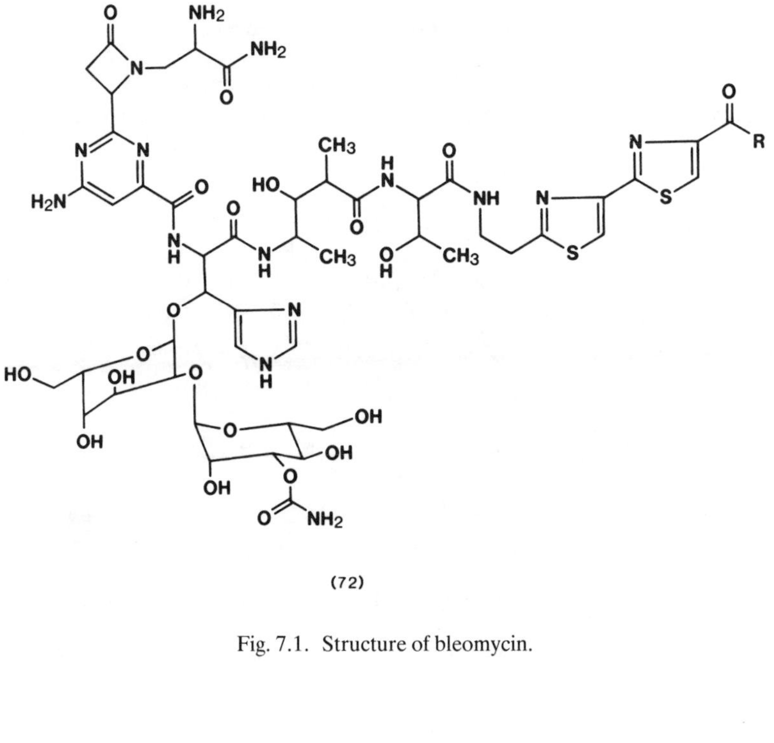

(72)

Fig. 7.1. Structure of bleomycin.

essential for the biosynthesis of the antibiotic, and Cu-mediated hydrolysis reactions are critical in the construction of the bleomycin nucleus [1]. The complete stereochemical structure was resolved by Umezawa and co-workers [12] and a total synthesis has also been achieved [13].

The crystal structure of a Cu(II) complex of P-3A, a biosynthetic intermediate of bleomycin, resolved the metal binding site (Figure 7.2) [14]. The metal is bound by the nitrogen atoms of the α-amino group, a secondary amine, a pyrimidine ring, an imidazole nitrogen and a deprotonated amido group from the histidine residue. The Cu is in a distorted square-pyramidal structure and is displaced 0.2 Å from the coordination plane toward the α-amino nitrogen. In the Cu(II)-bleomycin molecule itself, an oxygen from the sugar amide completes the coordination sphere. The metal-ligating atoms need not necessarily be the same for all metals, but are considered to be the same for the biologically important iron [15—17].

In bioinorganic terms, the chemistry of bleomycin can be divided into that of its metal complexes, their reaction with and activation of oxygen, and the interaction and inactivation of DNA.

7.3. Metal Complexes of Bleomycin

The metal binding ability of bleomycin was apparent from its isolation as the copper chelate, and derivatives of interest besides Cu and Fe (see

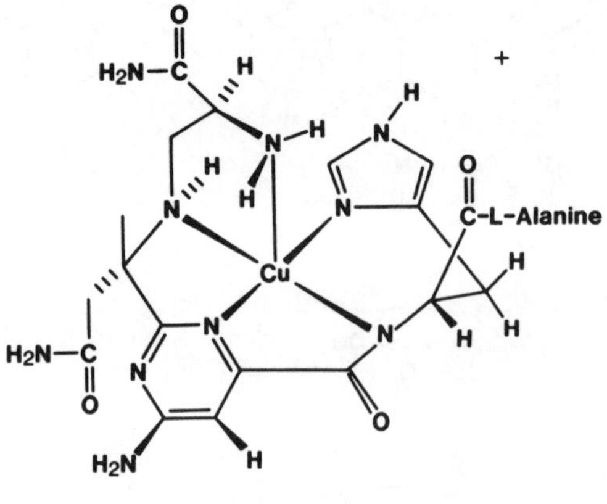

(73)

Fig. 7.2. Structure of the copper complex of P-3A, a biosynthetic intermediate of bleomycin showing the probable metal binding sites of bleomycin.

Section 7.4), include Co(II) and Co(III) [18–22], Mn(II) [23, 24], Zn(II) [25, 26], Ni(II) [27], Ru(III) [28], Tc [29], and V(IV) [30]. Earlier results have been summarized [3]. Many of these examples may be considered to be attempts at producing tumour-localizing agents, combining the known affinity of bleomycin for tumour cells with a suitably stable metal radiolabel. The metallation does appear to affect tissue distribution, which may imply an alteration in functional structure of the drug. A study of the cytotoxic and antitumour properties of some metal complexes gave an order of activity in culture and antitumour potency of Cu(II)BLM $>$ BLM $>$ Zn(II)BLM $>$ Fe(III)BLM $\gg$ Co(II)BLM [31].

7.3.1. COPPER BLEOMYCIN

The formation of Cu(II)—BLM species is pH dependent with at least two complexes identified [32, 33]. A reduction potential of −319 mV (vs. NHE) has been assigned to the Cu(II)/Cu(I) couple [34]. The Cu(II) complex is kinetically and thermodynamically stable to ligand substitution, being only slowly reduced by sulfhydryls [35]. Metal exchange is fast for Cu(I) and the binding sites may not be identical to those for Cu(II), as indicated by potentiometric and ^{1}H NMR data [34]. Further, reduction in the presence of the Cu(II) species of Fe(II) results in metal displacement [36].

Compared to free BLM, the affinity of Cu(II)BLM for DNA is greater, and shows a two-step binding process with a lifetime for the slow step of 0.1 s [37]. The ability of CuBLM to cleave DNA has been a matter of contention — the long held opinion that the copper complex is inactive was altered following a report on cleavage [38], but these data have been challenged and the cleavage has been attributed to trace iron contamination [39]. These results show again that, as with simple copper macrocycles (Section 6.3), complexes which appear most stable to redox activity have most antitumour activity.

7.3.2. COBALT BLEOMYCIN

The Co(II) adduct has ESR features indicative of a low-spin configuration with the unpaired electron in the d_{z^2} ground state [18, 19]. In the presence of oxygen, the ESR features are characteristic of a monooxygenated low-spin Co(II)—O_2 adduct [19]. In the presence of DNA, the ESR parameters again change, and this was attributed to a different orientation of the bound O_2 molecule. The presence of more than one complex, however, was indicated by other studies [40, 41]. The brown Co(II)—O_2 complex slowly decomposes, releasing superoxide radical [3], while the formation of a green dinuclear μ-peroxo Co(III) complex by dimerization and loss of 1 molecule of O_2 has also been confirmed [21, 22]:

$$2\,Co(II)O_2 \rightarrow Co(III){-}O_2{-}Co(III) + O_2$$

The dimerization has a second-order rate constant of $200 \pm 50\ M^{-1}s^{-1}$ at 25°C, which is lowered by a factor of 2×10^3 in the presence of DNA [21]. An apparent equilibrium binding constant of $8.4 \times 10^4\ M^{-1}$ was found for the Co(III)—O_2—Co(III)/DNA adduct, with binding of one μ-peroxo complex approximately every three base pairs [21]. Of interest is the fact that the variously colored Co complexes have the ability to nick DNA in the presence of light [42, 43], reminiscent of the reaction of simple Co(III) complexes (Section 1.3.1). As the ^{57}Co-substituted bleomycin has been extensively investigated for tumour imaging and found to accumulate selectively in certain cancer cells, the possibility of selective attack by a later light-induced reaction is attractive.

7.3.3. MANGANESE BLEOMYCIN

Both Mn(II) and Mn(III) degrade DNA in the presence of oxidizing agents such as H_2O_2 or oxygen donors such as iodosobenzene [44, 45]. In the case of Mn(III)BLM, olefin oxidation may also be effected, reinforcing the analogy between bleomycin and oxygen-activating enzymes such as cytochrome P450 [44] (see also Section 7.3.4). The DNA degradation is at least an order of magnitude less efficient than with Fe(II/III)BLM and the products are not fully defined, but the results are of interest for comparison with heme proteins and Mn—porphyrin models. Further comparisons have been made between the Fe, Co and Mn compounds and the sequence cleavage was noticeably similar for Mn and Fe [46]. The design of nontoxic Mn agents, in view of the oxidation states available, the known oxygen chemistry from model studies of hemes and chlorophyll, and the analogies which may be drawn for Cu/ZnSOD and MnSOD, could be a fruitful area to pursue.

7.4. Mechanism of Action of Bleomycin

The Fe—bleomycin system has been described in detail and the mechanism of action with DNA *in vitro* substantially clarified [1, 2].

7.4.1. DNA–BLEOMYCIN INTERACTIONS

The terminal 'tripeptide S' portion of bleomycin, consisting of the terminal amine group, the bis(thiazole) derivative and the threonine residues is essential for binding of the drug to DNA, which occurs with an apparent binding constant of 10^5 M^{-1} and a saturation ratio of 1 drug per 5 nucleotides [47]. The intercalation of the bis(thiazole) unit was inferred from studies on the fluorescence quenching of the moiety and hydrodynamic changes on closed circular DNA–drug complexes [48]. Electrostatic interactions are attributed to the terminal amine, and are reinforced by the fact that some of the amines found in the antibiotic mixture include spermine and spermidine [11, 49]. Both Cu(II) and Fe(III)–BLM complexes lengthen DNA, by 4.6 and 3.2 Å/bound molecule, respectively [50]. The binding of the drug to DNA is necessary, but not uniquely sufficient, for strand breakage. A parallel observation is that small, structurally similar analogues to the purported DNA-activating site of the molecule cannot independently cleave DNA, but can transfer oxygen to stilbene [51].

Both single and double strand breaks are found in BLM-degraded DNA and the breakage is sequence-specific, being preferential for —GC— and —GT— (5′—3′) sequences [52, 53]. The analogous 3′—5′ sequences are not recognized. BLM analogues, modified either on the DNA-binding site or on the metal-binding site, have substantially similar but not identical sequence specificity and modes of cleavage [1, 54, 55]. There is also a guanine dependence on DNA binding as evidenced by experiments with homopolynucleotides [49], which may explain the sequence specificity of cleavage. DNA-binding agents alter the sequence specificity and, with actinomycin D and ethidium bromide, —GA— and —GT— sites are cleaved [56]. Binding of cisplatin produces new G-rich cleavage sites [57] but, interestingly, other guanine-reacting agents such as mitomycin C do not substantially alter the mode of cleavage [58].

7.4.2. IRON AND IRON–OXYGEN ADDUCTS

The requirement for reducing agents and oxygen in the DNA cleavage *in vitro*, and the inhibition of the cleavage by chelating agents such as EDTA, led to the suggestion that trace metal ions were responsible for the mechanism of action of the drug [59—62] and, indeed, 0.02% Fe was found by AA in clinical grade BLM [63]. A scheme for binding, activation and DNA-reaction is presented below. Many similarities with schemes for oxygen activation by hemes will be apparent.

$$
\begin{array}{ccccc}
Fe(II) + BLM & \longrightarrow & Fe(II)BLM & \xrightarrow{O_2} & Fe(II)BLM(O_2) \\
\uparrow & & \downarrow {\scriptstyle +O_2 \atop +DNA} & & \downarrow {\scriptstyle +DNA} \\
Fe(III) + BLM & \longleftarrow & Fe(III)BLM + \text{damaged DNA} & \longleftarrow & Fe(II)BLM(O_2)/DNA
\end{array}
$$

The pink Fe(II) complex of BLM is high spin ($S = 2$) and can form adducts with CO and EtNC; all these Fe(II) species are ESR silent, having either even or zero spin [1, 2, 64—66]. The properties of many complexes have been summarized [1]. The NO adduct displays an ESR spectrum which is altered in the presence of DNA [67, 68], and the metal environment of the Fe(II)—CO adduct is also changed upon binding to polynucleotides [1]. The low spin Fe(III)—drug complex shows ESR signals similar to those of Fe(III)-containing oxygenases [64, 69], and such characteristics suggest that similar chemistry might also be expected from the two systems.

Addition of O_2 to the Fe(II)—BLM complex produces an ESR-silent adduct which rapidly breaks down but is stabilized in the presence of DNA; the Fe(II)BLM—O_2 reaction thus results in considerable drug decomposition, unless DNA is present [64]. Spectral and kinetic studies distinguished three events in the Fe(II)—O_2 reaction; (a) formation of the short-lived, ESR silent species with $t_{1/2} = 6$ s at 2°C; (b) the decay of this species to an 'activated' complex, capable of cleaving DNA and (c) a slower decay of this second species to Fe(III)BLM, $t_{1/2} = 60$ s at 2°C [70]. Upon addition of Fe(II) to a bleomycin—DNA mixture in air, a long-lived ($t_{1/2} = 45$ min), ESR silent species is formed [71]. The formation of this species consumes one mole of oxygen and the complex eventually decomposes to Fe(III)BLM [71]. The initial Fe(II)O_2BLM species was shown initially to yield a 1 : 1 mixture of activated BLM and Fe(III)BLM, the reaction perhaps occurring by reduction of Fe(II)O_2—BLM by Fe(II)—BLM [72]:

$$FeO_2\text{—BLM} + \text{Fe(II)BLM} \rightarrow \text{Activated BLM} + \text{Fe(III)BLM}$$

The reduction potential of Fe(III)BLM is +129 ± 12 mV (vs. NHE) and thus the Fe(II) form can be easily regenerated under biological conditions by reducing agents such as GSH or O_2^- [73]. The chemistry of activated bleomycin has been recently reviewed [111].

The Fe(II)O_2—BLM adduct formation is first order with respect to both [O_2] and [Fe(II)BLM] [70]. Mössbauer studies confirm that the complex is best described as low spin ferric iron bound to superoxide [74]. Activated bleomycin, the kinetically competent form that cleaves DNA, is formed by one-electron reduction of Fe(II)O_2—BLM [75]. The activated form has an ESR spectrum (g = 2.26, 2.17, 1.94) different from that of Fe(III)BLM (g = 2.45, 2.18, 1.89), although the optical spectra are very similar [75]. Mössbauer spectroscopy also indicates the iron to be low spin ferric [74]. Further ESR studies on the activated form show it to contain at least one atom of O_2 and chemical studies indicate that it has two more oxidizing equivalents than Fe(III)BLM [76], e.g. activated BLM reacts with a 1 : 1 stoichiometry with two-electron reductants such as NADH and thio-NADH [76]. The oxidation level is the same as compounds of horseradish peroxidase and chloroperoxidase and also the proposed activated form of cytochrome P450:

$$\mathrm{Fe(III)O_2^-} \rightarrow \mathrm{Fe(III)O_2^{2-}} \rightarrow \mathrm{Fe(V)O_2} \rightarrow \text{`OFe(V)'} + \mathrm{H_2O}$$

The final species in the above scheme contains the active oxygen (independent of exact electronic description). This analogy is reinforced by the similar chemical reactivity of the activated form of bleomycin and chloroperoxidase [77].

The products of DNA cleavage by bleomycin are various [2] and will not be dealt with at length here. The most specific proposals for initiation of DNA cleavage involve hydrogen abstraction from the C_4' deoxyribose carbon to form a reactive free radical [2, 78—80]. The point is that the redox chemistry on the metal ion produces active oxygen which initiates the strand breakage of DNA, eventually leading to cell death. An excellent review covers bleomycin-induced DNA degradation along with mechanistic aspects [112].

7.5. Other Metal Binding Antibiotics

Many antitumour antibiotics have structures which allow for metal binding. Their mechanism of action in some cases is also implied to be production of activated oxygen and DNA inactivation, although the metal co-factor is not essential because of the presence of quinone groups, which can function as active oxygen reagents through the quinone/semiquinone couple. Two examples of such antibiotics which will be discussed are anthracyclines such as daunorubicin (and doxorubicin) and streptonigrin (Figure 7.3). Their metal binding has been reviewed [3] and updated recently [81, 82]. The chemistry of these antibiotics can be considered from the points of view of metal complex formation, and possible relevance of these complexes to their biological activity.

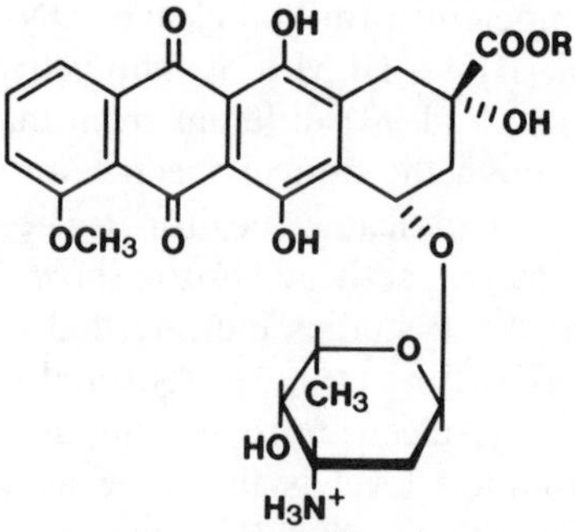

Fig. 7.3. Structures of the metal binding antibiotics adriamycin (doxorubicin), R = CH_3 (74), daunorubicin, R = H (74) and streptonigrin (75).

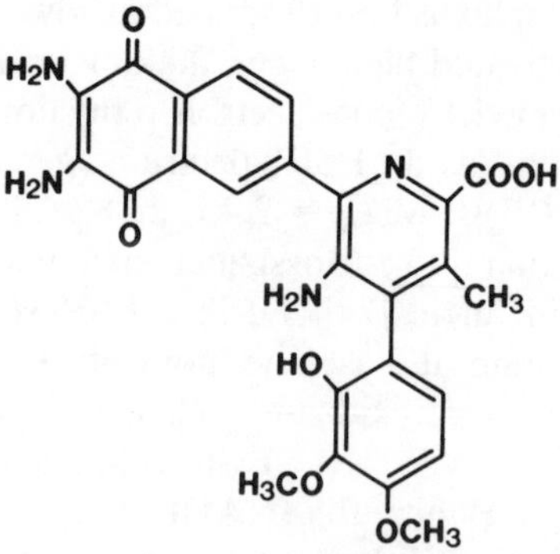

7.5.1. ANTHRACYCLINES

The structures of daunorubicin (daunomycin) and doxorubicin (adriamycin) differ in only one substituent in the acyl side chain. There are three acidic protons in these molecules, the two phenolic groups and the sugar ammonium group. There is discrepancy in the literature on the assignment of the pK_a values [83–85]. One phenolic group deprotonates with $pK =$ 10.0 but the pK at 8.6 [83] (8.94 [84]) has been assigned both to the ammonium group [83] and the other phenolic group [84]. The arguments against the latter assignment have been summarized and appear valid [81].

Metallation of anthracyclines releases protons, and using a pK_a of 10.0 for the first phenolic group, Martin calculated log K_m values (for $M + H_2L \rightarrow M(HL) + H^+$, where H_2L refers to the neutral ligand) of 11.0 and 7.3 for Fe^{3+} and Cu^{2+}, respectively [81]. Metal ion binding must take into account possible formation of hydroxo and polymeric metal complexes at basic pH, and such events make analysis difficult, e.g. a polymeric 1 : 1 complex CuL forms at high pH [84]. With variation of pH and molar ratio, various complexes are formed between the ligand and these metals. The 2 : 1 $Cu(HL)_2$ adduct predominates at $5 < pH < 8$ [84, 86], with a reported log $\beta = 16.66$ [84], which seems high in comparison to the K_m value of 7.3 reported [81]. Resonance Raman spectroscopy has been particularly useful in analysis of these systems and studies on the Cu(II)–adriamycin–DNA adduct indicated that an intercalated adduct could be formed [87].

Oxidation of Fe^{2+} to Fe^{3+} occurs in the presence of daunorubicin [88] and, from potentiometric and spectroscopic measurements, the species $Fe(HL)_3$ has been identified at physiological pH [89]. The binding sites for both Cu and Fe almost certainly form the chelate from one keto oxygen and the deprotonated phenolic group (Compound 74, Figure 7.4). CD measurements indicate that the Fe complex can adopt different conformations and that intercalation into DNA can occur [89].

Palladium(II) and platinum(II) complexes of adriamycin have also been reported recently [90, 91].

7.5.1.1. *Biochemical Aspects of Metal—Anthracyclines*

Severe cardiotoxicity is a major dose-limiting factor for anthracyclines and an early report indicated that an iron adduct of adriamycin, called quelamycin, alleviated this toxicity [92]. This finding, however, has not been substantiated [93] and chemical studies indicated quelamycin to be a mixture of $Fe(HL)_3$ and ferric hydroxide [89].

The iron(III)—adriamycin complex can catalyze the transfer of electrons from reduced GSH to O_2, giving superoxide and hydrogen peroxide and the process requires no enzymatic activation [88]. Since the iron chelate may also bind to cell membranes, the possibility exists that the system can function as a source of damage to target molecules, independent of interaction with DNA [88]. In a similar vein, a ternary complex of Fe(II)—ADR—ADP can be reduced by microsomal P450 with resulting lipid peroxidation [94, 95]. Other workers have indicated an iron requirement in the ternary complex [96, 97]. The oxidative destruction of erythrocyte ghost membranes is also catalyzed by the iron complex [88]. The Fe(III)ADR complex does not catalyze electron transfer from NADH to O_2 by NADH dehydrogenase [89], although NADH oxidation is reported to inhibit lipid peroxidation [89].

The proposed mechanisms of cardiac toxicity by adriamycin include free radical damage, membrane effects and inhibition of ubiquinone-requiring enzymes, and in each case superoxide anion production is implicated [99]. The results summarized above imply that attribution of superoxide production with subsequent toxicity to the quinone/semiquinone couple is not straightforward. These effects may be produced by reactions catalyzed by metal binding; the difficulty is in separating out the competing pathways. The trace levels of iron in many laboratory reagents is a further complicating factor for correlation of *in vivo* and *in vitro* data. The results do underline the interesting point that (organic) drug metabolism can be affected by metal binding, which may serve to either activate or deactivate the system, independent of metal ion involvement in binding with the target. Elucidation of the role of metal ions in toxicity can point toward methods of avoidance.

7.5.2. STREPTONIGRIN

Streptonigrin is a potent antitumour antibiotic whose clinical usefulness is very limited because of toxic side effects, especially severe bone marrow depression. The mechanism of action is believed to be interference with cell respiration and interruption of cell replication [100].

The antibiotic produces single strand breaks in DNA in the presence of O_2 and a reducing agent [101, 102]. Breakage is dependent on O_2, metal ions and a reducing agent, and is inhibited by catalase and superoxide dismutase [103]; production of hydroxyl radical and superoxide during DNA scission was confirmed [104]. The role of metal ions was proposed by Lown to be catalysis of the oxidation of the reduced hydroquinone to semiquinone with subsequent electron transfer to O_2, leading to superoxide and thus production of an activated oxygen [105]:

$$\begin{aligned} SN + NADH + H^+ &\longrightarrow SNH_2 + NAD^+ \\ SNH_2 + O_2 &\xrightarrow{M} SNH^{\cdot} + HO_2^{\cdot} \\ HO_2^{\cdot} &\longrightarrow O_2^{-} + H^+ \text{ or} \\ HO_2^{\cdot} &\longrightarrow \tfrac{1}{2} H_2O_2 + \tfrac{1}{2} O_2 \end{aligned}$$

The protective effect of SOD and catalase can be explained by either decomposition of superoxide or decomposition of hydrogen peroxide, respectively. This protection has been observed in the presence of DNA *in vitro* [102]. An alternative mechanism to that of Lown has been forwarded by Bachur and co-workers who favor a 'site-specific free radical' with the semiquinone directly bound to DNA [106, 107].

The exact role of the metal must await more details on the binding. Complexation involves loss of one proton and 1 : 1 adducts are apparently formed [108, 109]. The structure of streptonigrin shows the quinoline ring to be essentially coplanar with the substituted pyridine ring, the aryl ring being almost perpendicular to these [110]. The presence of the acid group (see Figure 7.3) may suggest a binding site, but this would probably be too far away from the redox-active group to effect catalysis. Metal-binding in aqueous media is also difficult to study because of the poor water solubility of the antibiotic. The possibility of ternary complexation (M, DNA, drug) has also been invoked as a role for the metal ion and these aspects have been reviewed [82].

The proposed metal involvement in production of semiquinone production is another example of a mechanism whereby endogenous metal ion may influence drug activation.

7.6. Summary

The bleomycins are a family of naturally-occurring antibiotics clinically used in the treatment of cancer. The antitumour activity is manifested by DNA binding followed by DNA strand breakage. The cleavage requires O_2 and a metal ion, with Fe(II) being the most effective *in vivo* and *in vitro*. The mechanism involves binding of O_2 to an iron—bleomycin complex, reduction to an 'activated' species with binding to DNA of this species, cleavage and regeneration of free bleomycin and Fe(III).

Antibiotics such as streptonigrin and the anthracyclines also bind metals and cleave DNA and metal ion involvement through similar mechanisms to that of bleomycin is possible. In the case of the clinically-used adriamycin, the acute cardiotoxicity may have as a chemical cause free radical damage produced by an iron—drug interaction.

References

1. Y. Sugiura, T. Takita, and H. Umezawa: *Metal Ions in Biol. Syst.* **19**, 81 (1986).
2. R. M. Burger, J. Peisach, and S. B. Horwitz: *Life Sciences* **28**, 715 (1981).
3. J. C. Dabrowiak: *Metal Ions in Biol. Syst.* **11**, 305 (1980).
4. H. Umezawa, K. Maeda, T. Takeuchi, and Y. Okami: *J. Antibiot.* (Tokyo) **19A**, 200 (1966).
5. *Bleomycin: Current Status and New Developments* (Eds. S. K. Carter, S. T. Crooke, and H. Umezawa) Academic Press, London (1978).
6. *Bleomycin: Chemical, Biochemical and Biological Aspects* (Ed. S. M. Hecht) Springer-Verlag, Berlin (1979).
7. S. T. Crooke: *Cancer and Chemotherapy v. III* pp. 97—109 and pp. 343—352. Academic Press, London (1981).
8. H. Umezawa, Y. Suhara, T. Takita, and K. Maeda: *J. Antibiot.* **19A**, 210 (1966).
9. S. T. Crooke and W. T. Bradner: *J. Med.* (*Basel*) **7**, 333 (1977).
10. T. Takita, Y. Muraoka, T. Nakatani, A. Fujii, Y. Umezawa, H. Naganawa, an H. Umezawa: *J. Antibiot.* **31**, 801 (1978).
11. H. Umezawa: *Bleomycin. Origin, Chemistry, and Artificial Bleomycins* (Antibiotics III. Mechanism of Action of Antimicrobial and Antitumour Agents, Eds. J. W. Corcoran and F. E. Hahn) pp. 21—33. Springer (1975).
12. T. Takita, Y. Muraoka, T. Nakatani, A. Fujii, Y. Umezawa, H. Naganawa, and H. Umezawa: *J. Antibiot.* **31**, 801 (1978).
13. S. Saito, Y. Umezawa, T. Yoshioka, T. Takita, H. Umezawa, and Y. Muraoka: *J. Antibiot.* **36**, 92 (1983).
14. Y. Iitaka, H. Nakamura, T. Nakatani, Y. Murata, A. Fujii, T. Takita, and H. Umezawa: *J. Antibiot.* **31**, 1070 (1978).

15. Y. Muraoka, H. Kobayashi, A. Fujii, M. Kunishima, T. Fujii, Y. Nakayama, T. Takita, and H. Umezawa: *J. Antibiot.* **29**, 853 (1976).
16. J. C. Dabrowiak, F. T. Greenway, W. E. Longo, M. van Husen, and S. T. Crooke: *Biochim. Biophys. Acta* **517**, 517 (1978).
17. T. Takita, Y. Muraoka, T. Nakatani, A. Fujii, Y. Iitaka, and H. Umezawa: *J. Antibiot.* **31**, 1073 (1978).
18. Y. Sugiura: *J. Antibiot.* **31**, 1206 (1978).
19. Y. Sugiura: *J. Am. Chem. Soc.* **102**, 5216 (1980).
20. A. Kono, Y. Matsushima, M. Kojima, and T. Maeda: *Chem. Pharm. Bull. Jap.* **25**, 1725 (1977).
21. J. P. Albertini and A. Garnier-Suillerot: *Biochemistry* **21**, 6777 (1982).
22. A. Garnier-Suillerot, J. P. Albertini, and L. Tosi: *Biochem. Biophys. Res. Comm.* **102**, 499 (1981).
23. R. M. Burger, J. H. Freedman, S. B. Horwitz, and J. Peisach: *Inorg. Chem.* **23**, 2217 (1984).
24. G. M. Ehrenfeld, N. Murugesan, and S. M. Hecht: *Inorg. Chem.* **23**, 1498 (1984).
25. J. C. Dabrowiak, F. T. Greenway, and R. Grulich: *Biochemistry* **17**, 4090 (1978).
26. N. J. Oppenheimer, L. O. Rodriguez, and S. M. Hecht: *Biochemistry* **18**, 3439 (1979).
27. Y. Sugiura, K. Ishizu, and K. Myoshi: *J. Antibiot.* **32**, 453 (1979).
28. P. H. Stern, S. E. Halpern, P. L. Hagan, S. B. Howell, J. E. Dabbs, and R. M. Gordon: *J. Natl. Cancer Inst.* **66**, 807 (1981).
29. T. Mori, K. Hamamoto, and K. Torizuka: *J. Nucl. Med.* **14**, 431 (1973).
30. L. Banci, A. Dei, and D. Gatteschi: *Inorg. Chim. Acta* **67**, L53 (1982).
31. E. A. Rao, L. A. Saryan, W. E. Antholine, and D. H. Petering: *J. Med. Chem.* **23**, 1310 (1980).
32. J. P. Albertini and A. Garnier-Suillerot: *J. Inorg. Biochem.* **25**, 15 (1985).
33. D. Solamain, E. A. Rao, W. E. Antholine, and D. H. Petering: *J. Inorg. Biochem.* **12**, 201 (1980).
34. N. J. Oppenheimer, C. Chang, L. O. Rodriguez, and S. M. Hecht: *J. Biol. Chem.* **256**, 1514 (1981).
35. W. E. Antholine, D. Solaiman, L. A. Saryan, and D. H. Petering: *J. Inorg. Biochem.* **17**, 75 (1982).
36. J. H. Freedman, S. B. Horwitz, and J. Peisach: *Biochemistry* **21**, 2203 (1982).
37. L. F. Povirk, M. Hogan, N. Dattagupta, and M. Buechner: *Biochemistry* **20**, 665 (1981).
38. G. M. Ehrenfeld, L. O. Rodriguez, S. M. Hecht, C. Chang, V. J. Basus, and N. J. Oppenheimer: *Biochemistry* **24**, 81 (1985).
39. T. Suzuki, J. Kuwahara, and Y. Sugiura: *Biochemistry* **24**, 4719 (1985).
40. C. M. Vos, G. Westera, and D. Schimmer: *J. Inorg. Biochem.* **13**, 165 (1980).
41. L. H. DeReimer, C. F. Meares, D. A. Goodwein, and C. I. Diamanti: *J. Med. Chem.* **22**, 1019 (1979).
42. C. H. Chang and C. F. Meares: *Biochemistry* **21**, 6332 (1982).
43. C. H. Chang and C. F. Meares: *Biochemistry* **23**, 2268 (1984).
44. G. M. Ehrenfeld, N. Murugesan, and S. M. Hecht: *Inorg. Chem.* **23**, 1498 (1984).
45. R. M. Burger, J. H. Freedman, S. B. Horwitz, and J. Peisach: *Inorg. Chem.* **23**, 2217 (1984).
46. T. Suzuki, J. Kuwahara, M. Goto, and Y. Sugiura: *Biochim. Biophys. Acta* **824**, 330 (1985).
47. M. Chien, A. P. Grollman, and S. B. Horwitz: *Biochemistry* **16**, 3641 (1977).

48. L. F. Povirk, M. Hogan, and N. Dattagupta: *Biochemistry* **18**, 96 (1979).
49. H. Kasai, H. Naganawa, T. Takita, and H. Umezawa: *J. Antibiot.* **31**, 1316 (1978).
50. L. F. Povirk, M. Hogan, N. Dattagupta, and M. Buechner: *Biochemistry* **20**, 665 (1981).
51. R. E. Kilkuskie, H. Suguna, B. Yellin, N. Murguesan, and S. M. Hescht: *J. Am. Chem. Soc.* **107**, 260 (1985).
52. A. D. D'Andrea and W. A. Haseltine: *Proc. Natl. Acad. Sci. USA* **75**, 3608 (1978).
53. M. Takeshita, A. Grollman, E. Otshubo, and H. Otshubo: *Proc. Natl. Acad. Sci. USA* **75**, 5983 (1978).
54. Y. Sugiura, T. Suzuki, M. Otsuka, S. Kobayashi, M. Ohno, T. Takita, and H. Umezawa: *J. Biol. Chem.* **258**, 1328 (1983).
55. C. H. Chang, C. K. Mirabelli, Y. Jan, and S. T. Crooke: *Biochemistry* **20**, 233 (1981).
56. Y. Sugiura and T. Suzuki: *J. Biol. Chem.* **257**, 10544 (1982).
57. P. K. Mascharak, Y. Sugiura, J. Kuwahara, T. Suzuki, and S. J. Lippard: *Proc. Natl. Acad. Sci. USA* **80**, 6795 (1983).
58. T. Suzuki, J. Kuwahara, and Y. Sugiura: *Biochem. Biophys. Res. Comm.* **117**, 916 (1983).
59. E. A. Sausville, J. Peisach, and S. B. Horwitz: *Biochem. Biophys. Res. Comm.* **73**, 814 (1976).
60. R. Ishida and T. Takahashi: *Biochem. Biophys. Res. Comm.* **66**, 1432 (1975).
61. E. A. Sausville, J. Peisach, and S. B. Horwitz: *Biochemistry* **17**, 2740 (1978).
62. E. A. Sausville, R. W. Stein, J. Peisach, and S. B. Horwitz: *Biochemistry* **17**, 2746 (1978).
63. J. W. Lown and S. K. Sim: *Biochem. Biophys. Res. Comm.* **77**, 1150 (1977).
64. R. M. Burger, J. Peisach, W. E. Blumberg, and S. B. Horwitz: *J. Biol. Chem.* **254**, 10906 (1979).
65. M. Otsuka, M. Yoshida, S. Kobayashi, M. Ohno, Y. Sugiura, T. Takita, and H. Umezawa: *J. Am. Chem. Soc.* **103**, 6986 (1981).
66. Y. Sugiura: *J. Am. Chem. Soc.* **102**, 5208 (1980).
67. Y. Sugiura, T. Takita, and H. Umezawa: *J. Antibiot.* **34**, 249 (1981).
68. Y. Sugiura and K. Ishizu: *J. Inorg. Biochem.* **11**, 171 (1979).
69. Y. Sugiura and T. Kikuchi: *J. Antibiot.* **31**, 1310 (1978).
70. R. M. Burger, S. B. Horwitz, J. Peisach, and J. B. Wittenberg: *J. Biol. Chem.* **254**, 12299 (1979).
71. J. P. Albertini, A. Garnier-Suillerot, and L. Tosi: *Biochem. Biophys. Res. Comm.* **104**, 557 (1982).
72. H. Kuramochi, T. Takahashi, T. Takita, and H. Umezawa: *J. Antibiot.* **34**, 576 (1981).
73. D. L. Melnyk, S. B. Horwitz, and J. Peisach: *Biochemistry* **20**, 5327 (1981).
74. R. M. Burger, T. A. Kent, S. B. Horwitz, E. Munck, and J. Peisach: *J. Biol. Chem.* **258**, 1559 (1983).
75. R. M. Burger, J. Peisach, and S. B. Horwitz: *J. Biol. Chem.* **256**, 11636 (1981).
76. R. M. Burger, J. S. Blanchard, S. B. Horwitz, and J. Peisach: *J. Biol. Chem.* **260**, 15406 (1985).
77. G. Padbury and S. S. Sligar: *J. Biol. Chem.* **260**, 7820 (1985).
78. J. C. Wu, J. W. Kozarich, and J. A. Stubbe: *J. Biol. Chem.* **258**, 4694 (1983).
79. A. P. Grollman, M. Takeshita, K. M. R. Pillai, and F. Johnson: *Cancer Res.* **45**, 1127 (1985).
80. R. M. Burger, J. Peisach, and S. B. Horwitz: *J. Biol. Chem.* **257**, 2401 (1982).

81. R. B. Martin: *Metal Ions in Biol. Syst.* **19**, 19 (1985).
82. J. Hajdu: *Metal Ions in Biol. Syst.* **19**, 53 (1985).
83. R. Kiraly and R. B. Martin: *Inorg. Chim. Acta* **67**, 13 (1982).
84. H. Beraldo, A. Garnier-Suillerot, and L. Tosi: *Inorg. Chem.* **22**, 4117 (1983).
85. P. M. May, G. K. Williams, and D. R. Williams: *Inorg. Chim. Acta* **46**, 221 (1980).
86. F. T. Greenway and J. C. Dabrowiak: *J. Inorg. Biochem.* **16**, 91 (1982).
87. P. K. Dutta and J. A. Hutt: *Biochemistry* **25**, 691 (1986).
88. C. E. Myers, L. Gianni, C. B. Simone, R. Klecker, and R. Greene: *Biochemistry* **21**, 1707 (1982).
89. H. Beraldo, A. Garnier-Suillerot, and L. Tosi: *Biochemistry* **24**, 284 (1985).
90. M. M. L. Fiallo and A. Garnier-Suillerot: *Biochemistry* **25**, 924 (1986).
91. A. Pasini, G. Pratesi, G. Savi and F. Zunino: *Inorg. Chim. Acta* **137**, 123 (1987).
92. M. Gosvalez, M. F. Blanco, C. Vivero, and F. Valles: *Eur. J. Cancer* **14**, 1185 (1978).
93. M. J. Egorin, R. E. Clawson, L. A. Ross, R. D. Friedman, S. D. Reich, A. Pollak, and N. R. Bachur: *Cancer Res.* **43**, 3253 (1983).
94. K. Sugioka and M. Nakano: *Biochim. Biophys. Acta* **713**, 333 (1982).
95. K. Sugioka, H. Nakano, T. Noguchi, J. Tschiya, and M. Nakano: *Biochem. Biophys. Res. Comm.* **100**, 1251 (1981).
96. E. G. Mimnaugh, M. A. Trush, and T. E. Gram: *Biochem. Pharmacol.* **30**, 2797 (1981).
97. H. Kappus, H. Muirawan, and M. E. Schuelen: *In Vivo Studies on Adriamycin Induced Lipid Peroxidation and Effects of Ferrous Ions* (Mechanism of Toxicity and Hazard Evaluation, B. Holmstedt, R. Lauwerys, M. Mericer, and M. Roberfroid), p. 635. Elsevier, New York (1980).
98. E. J. F. Demant: *Eur. J. Biochem.* **137**, 113 (1983).
99. L. Gianni, B. J. Corden, and C. E. Myers: *The Biochemical Basis of Anthracycline Toxicity and Antitumour Activity* (Reviews in Biochemical Toxicology, Eds. Hodgson, Bead, and Philpot) p. 1. Elsevier, New York (1983).
100. N. S. Mizuno: *Antibiotics V(2)* Ed. F. E. Hahn, pp. 372–384. Springer-Verlag (1979).
101. H. L. White and J. R. White: *Biochim. Biophys. Acta* **123**, 648 (1966).
102. R. Cone, S. K. Hasan, J. W. Lown, and A. R. Morgan: *Can. J. Biochem.* **54**, 219 (1976).
103. J. W. Lown, S. K. Sim, K. D. Majundar, and R. Y. Chang: *Biochem. Biophys. Res. Comm.* **76**, 705 (1977).
104. J. W. Lown and A. V. Joshua: *Can. J. Chem.* **59**, 390 (1981).
105. J. W. Lown: *Acc. Chem. Res.* **15**, 381 (1982).
106. N. R. Bachur, S. L. Gordon, and M. V. Gee: *Cancer Res.* **38**, 1745 (1978).
107. N. R. Bachur, S. L. Gordon, M. V. Gee, and H. Kon: *Proc. Natl. Acad. Sci. USA* **76**, 914 (1979).
108. J. Hadju and E. C. Armstrong: *J. Am. Chem. Soc.* **103**, 232 (1981).
109. K. V. Rao: *J. Pharm. Sci.* **68**, 853 (1979).
110. Y. H. Chiu and W. N. Lipscomb: *J. Am. Chem. Soc.* **97**, 2525 (1975).
111. S. M. Hecht: *Acc. Chem. Res.* **19**, 383 (1986).
112. J. Stubbe and J. W. Kozarich: *Chem. Rev.* **87**, 1107 (1987).

CHAPTER 8

METALS, METAL COMPLEXES, AND RADIATION

An important modality in cancer treatment is that of radiotherapy, either alone or in combination with chemotherapy. Indeed, the first treatment of cancer by X-rays followed within a year of their discovery by Röntgen [1] and it is estimated that today approximately half of all cancer patients receive radiation in some form. Conventional sources such as X-rays and ^{60}Co-γ rays are now supplemented with particles such as protons, neutrons and π^- mesons [2] and also by internal delivery (brachytherapy) using various implanted isotopes [3].

8.1. Interaction of Radiation and Biological Tissue

The radiolysis reaction of oxygenated aqueous solutions may be written, with the yields (G-values) for γ-rays in parentheses [4]:

$$\begin{array}{llllllll} H_2O \rightarrow & e^-_{aq} & + \; H^{\cdot} & + \; H_2 & + HO^{\cdot} & + H_3O^+ & + H_2O_2 \\ & (2.7) & (0.55) & (0.45) & (2.7) & (2.7) & (0.7) \end{array}$$

The formation of both $HO^{\cdot}$ and e^-_{aq} is complete in 10^{-11} s with molecule formation from recombination or chemical reaction complete in 10^{-8} s. Both e^- and $H^{\cdot}$ react at diffustion controlled rates with oxygen giving superoxide:

$$e^- + O_2 \rightarrow O_2^-$$

$$H^{\cdot} + O_2 \rightarrow O_2^- + H^+$$

The hydroxyl radical is by far the most reactive, and thus damaging, of the water radiolysis products. Indirect damage from this species can then occur either by hydrogen abstraction or $^{\cdot}OH$ addition, subsequent chemical rearrangement 'fixing' the lesion. Direct damage of the target molecule implies that the target itself suffers a localized polarization which, through electron donation by chemical reaction, also succeeds in inflicting permanent damage. The relative contributions in oxygenated solutions have been calculated as 60% indirect and 40% direct [5]. The time scale (in seconds) for such radiobiological events is [6]:

Energy deposition and ionization	10^{-18} to 10^{-14}
Formation of primary radicals	10^{-14} to 10^{-10}
Radical reaction leading to target lesions	10^{-10} to 10^{-1}
Enzymatic repair of lesions	10^{-1} to 10^{4}
Expression of cellular functions	10^{2} to 10^{6}

The vast majority of evidence implicates DNA as the primary radiation target and so the situation above can be summarized:

$$T_i \underset{\text{repair}}{\rightleftharpoons} T_i\cdot \xrightarrow{\text{fixation}} T_f$$

T_i may be considered to be the undamaged target, $T_i\cdot$ the target upon initial radiation damage, and T_f the target 'fixed' through electron transfer to the accepting molecule. Chapter 7 described the chemical production of hydroxyl radical as a cytotoxic species, and the radical may also be produced by physical, i.e. radiation, means. Chemical agents may modify the biological effects of radiation by protection of normal tissue or sensitization of tumour cells.

The effectiveness of cell kill by radiation, either *in vitro* or *in vivo*, is measured by the cell survival or dose survival curve [7]. In Figure 8.1 are represented the typical curves for cells exposed to X-radiation under N_2 and under O_2. A number of mathematical models have been used to describe these curves and these have been compared [8, 9]. Most models converge at high doses but they give disparate results at low doses. This is of importance because, whereas doses of 300—3000 rads are routinely used to construct these curves for X-rays, the clinically relevant doses are up to 200 rads, with multiple doses over the period of treatment. Accurate description of the survival curve at low doses is essential for prediction of clinical behavior. Two of the most common models used are the 'multi-target' model where $S = 1 - (1 - e^{-D/D_0})^n$ [10], and the '$\alpha\beta$' model, where $S = e^{-\alpha D - \beta D^2}$ [11], with S being the surviving fraction, D the dose, and n, α and β are independent parameters.

An important point to note for the survival curve is that the dose required to obtain a given effect (i.e. same cell kill) is approximately three times greater in deoxygenated, or hypoxic, cells than in normal aerobic cells. This ratio, which varies with the type of radiation but which, for any given radiation, is generally independent of survival level is called the oxygen enhancement ratio (OER), and is the ratio of hypoxic to aerated doses needed to produce a given effect. The concentration of oxygen

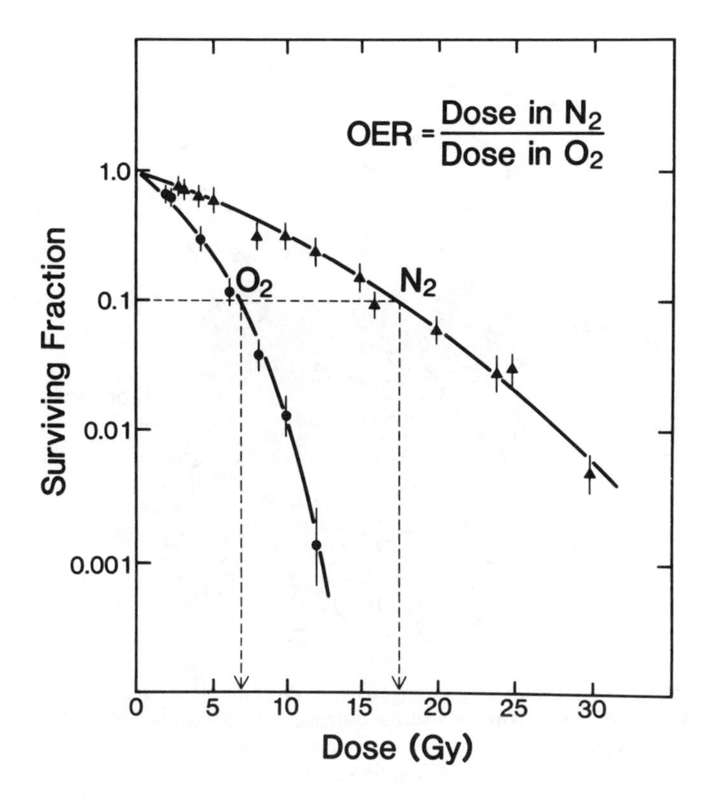

Fig. 8.1. A typical survival curve for CHO cells in O_2 and N_2 fitted using the '$\alpha\beta$' model and showing the oxygen enhancement ratio. From Reference 8.

required for full sensitization is ~2% (30 μM) so most normal tissues ($[O_2]$ between 2—5%) have sufficient oxygen for the full radiation effects to be observed. Again, the importance of the OER at low doses has been discussed [12]. For agents other than oxygen, ER or dose-modifying factor (DMF) is used to describe the effect of the agent vs. N_2.

The fact that tumour cells showed diminished radiation sensitivity in the absence of oxygen had been known for some time and the full relevance of this physicochemical effect in therapy was demonstrated in the classic studies by Gray and co-workers [13, 14] who showed that many tumours contain necrotic areas usually separated by some distance (150—200 μm) from the vascular system. This situation is depicted in Figure 8.2 and is explained by the fact that the disordered nature of tumour cell replication and the subsequent requirements of oxygen metabolism produce an oxygen gradient rendering the cells farthest from

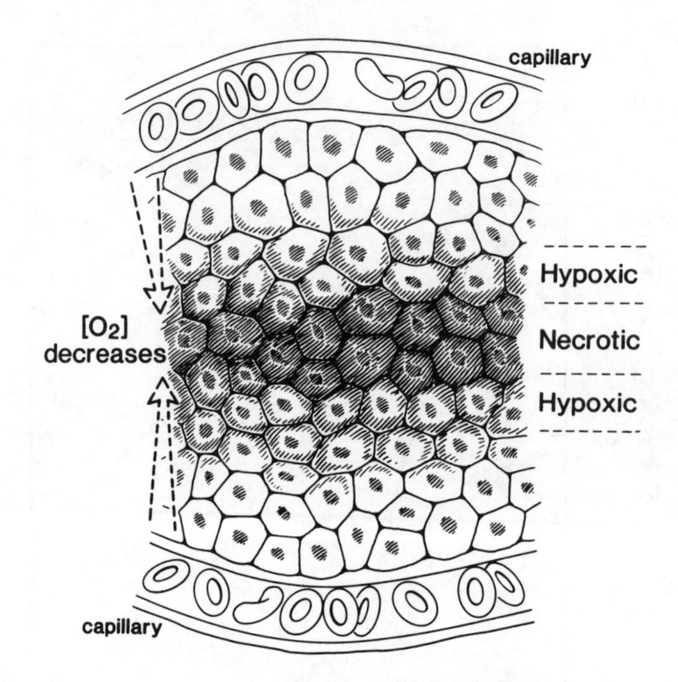

Fig. 8.2. Schematic diagram illustrating oxygen diffusion from the vascular capillary through a tumour mass. With increasing distance, oxygen depletion renders the cells hypoxic and eventually necrotic. The intermediate hypoxic layer may be viable but radioresistant.

the capillaries hypoxic. The clinical relevance is that because of the oxygen effect these cells, which may represent up to 30% of the total tumour mass, will be radioresistant and upon termination of treatment may become aerobic and tumour growth will recommence.

Oxygen may therefore be considered to be the classic dose-modifying agent or sensitizer and must be present during delivery of radiation, at least for all practical purposes, to exert its sensitizing effect. The mechanism is considered to be electron acceptance from target radicals generated by the radiation:

$$R^{\cdot} + O_2 \rightarrow RO_2^{\cdot} \rightarrow R^+ + O_2^-$$

These $R^{\cdot}$ radicals have a very short lifetime and may undergo chemical changes; reaction of a product with oxygen to give an organic peroxide is nonreversible and results in biological damage by fixation of a lethal lesion on the target, i.e. DNA, as stated. In the absence of oxygen, lower amounts of peroxide are formed and thus more are repaired. The radioprotecting ability of thiols such as cysteamine have been known for some time. Thiols are believed to exert their protective effect by hydrogen atom donation:

$$R^{\cdot} + RSH \rightarrow RH + RS^{\cdot}$$

which would compete with the oxygen fixation at the damaged site. With respect to radiation-induced repair processes, the radiobiologist categorizes radiation-induced damage as: (a) lethal and thus irreversible, (b) sublethal, which is reversed or repaired in a short time span i.e. hours, and (c) potentially lethal, which is that component of damage whose repair can be influenced by environmental conditions pertaining after irradiation [5]. The molecular lesions are classified as: (a) double strand breaks, (b) single strand breaks, (c) base damage and (d) cross linking [6]. These indeed correspond to the physical interactions discussed in Chapter 1. Hydroxyl radicals cause strand breaks, of which double strand breaks are considered most difficult to repair and may be related to cell kill [88].

Attempts to overcome this hypoxic problem include use of hyperbaric oxygen and use of radiation such as π-mesons with a high linear energy transfer (LET). This parameter describes energy loss through a medium and the magnitude of the oxygen effect decreases the higher the LET of the radiation [2].

8.2. Chemical Modification of Radiation Damage

Approaches to chemical modification include use of 'electron-affinic' sensitizers which act as oxygen mimics; thiol-binding agents which increase the effective concentration of oxygen by competitive reactions with endogenous reducing thiols, and DNA-binding agents which may alter template activity either by incorporation or inhibition of repair processes. These may be considered to be the major mechanisms of radiosensitization. Both bacterial and mammalian cell lines have been used for *in vitro* studies.

Design of compounds that are selectively active in hypoxia and also the development of nontoxic radioprotectants are active areas of current research. The concepts briefly outlined above can be found in more detail in some relatively recent reviews [15—19]. Many metals and metal complexes have been shown to modify radiation damage and what is of interest here is how these metal-based actions can be understood through, and incorporated into, the general understanding of radiation sensitization and protection.

8.2.1. OXYGEN MIMICS AND ELECTRON AFFINITY

The explanation of the oxygen effect implied that other compounds which react with free radicals might also sensitize, and indeed the sensitizing properties of nitric oxide were reported in 1957 [20]. Interestingly, hydroxyl radical can be produced from irradiation of N_2O in a clean reaction [4]. The systematic chemical approach using electron affinity dates from 1963 when Adams and Dewey [21] reasoned that in a biological system damage at a particular site would be dependent upon competing nondamaging processes, and that, for the hydrated electron, the magnitude of the biological response would therefore depend on its lifetime or diffusion limit and the distribution of electrophilic centers throughout the system. Moderation of this response by chemical agents could occur through electron acceptance with stabilization of the transient radical ion, thereby enhancing the probability of reaction at a relevant (i.e. damaging) site. Organic molecules containing conjugated systems, and therefore capable of delocalization, were considered suitable candidates and initial studies confirmed the utility of this hypothesis, sensitization being observed for such compounds as benzophenone and diacetyl.

Subsequent development of the electron affinity theory produced many molecules with sensitizing activity. The introduction of the nitro group in compounds such as *para*-nitroacetophenone [22, 23] led to studies on nitrofurans [24] and nitroimidazoles [25, 26]. The nitrofurans, although antibacterial, proved to be too toxic at the high doses required for sensitization and currently the species of clinical interest are the nitroimidazoles substituted in either the 2- or 5-position. The structures and nomenclature of representative examples, along with other compounds discussed here, are shown in Figure 8.3. Of major importance are misonidazole and metronidazole but again clinical trials have shown unacceptable toxic side effects for misonidazole, especially neurotoxicity, at clinically useful doses for radiosensitization. Recently attention has turned to nitroimidazoles containing on the side-chain aziridine moieties, capable of alkylating behavior (i.e. RSU 1069) [27], a compound with a piperidinyl side chain (Ro 03-8799) [28], and a peptide side linkage with structure similar to misonidazole (SR 2508) [29]. These are so far unproven in clinical trials; ongoing studies on RSU 1069 indicate unacceptable toxicity. More recent results may be found in conference proceedings [30]. The need, then, for efficient nontoxic radiosensitizers is as acute as ever.

A clear requirement for an efficient radiosensitizer is good electron

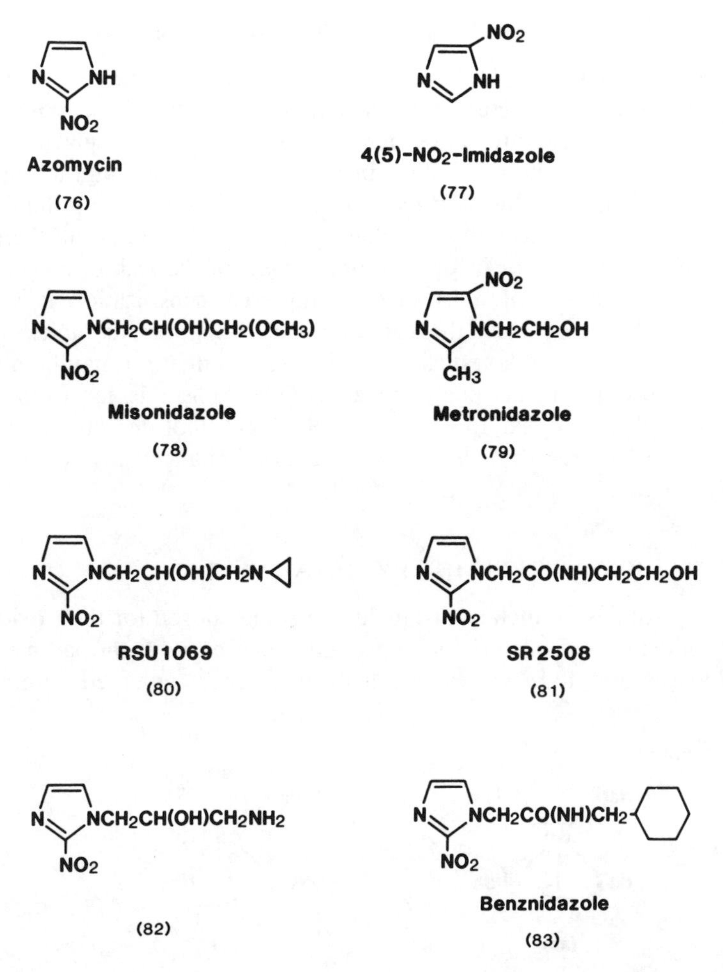

Fig. 8.3. Structures of some nitroimidazoles with radiosensitizing properties of current interest.

affinity and the ability to undergo a one electron reduction reaction. An ideal radiosensitizer would be nontoxic to aerobic cells and thus be selective to hypoxic regions; pharmacokinetic parameters such as rate of metabolism and ability to reach sufficiently high concentrations eventually dictate the overall efficacy. Metal complexes would appear to represent ideal candidates for systematic studies but few such studies have been carried out.

8.3. Metal Complexes as Radiation Sensitizers

In a manner similar to the periodic table of metals with some chemotherapeutic activity, Figure 8.4 shows those metals whose complexes, independent of structure, have demonstrated radiosensitizing ability. A point to be clarified here is the distinction between sensitization, implying an oxygen-mimetic pathway or mechanism for the effect, and potentiation, which implies modification of radiation damage by other mechanisms, such as DNA binding or incorporation. In view of the lack of mechanistic detail on the action of most metal complexes, sensitization is a valid concept and will be used throughout for non-platinum complexes. For platinum—amine complexes, as we shall see, potentiation may be a more accurate description, although the actual mechanism is far from clear. Some complexes appear to exert their effect by thiol depletion, whereas for other complexes the concept of reduction-enhanced cytotoxicity, e.g. Cu(II) → Cu(I) (see below), has been utilized.

8.3.1. RADIOSENSITIZATION BY METAL SALTS

The early studies on metal salts, including some chosen for their oxidizing ability, mostly used bacterial systems and, since no uniform parameter of activity was adopted, no real mechanistic pointers emerged from these

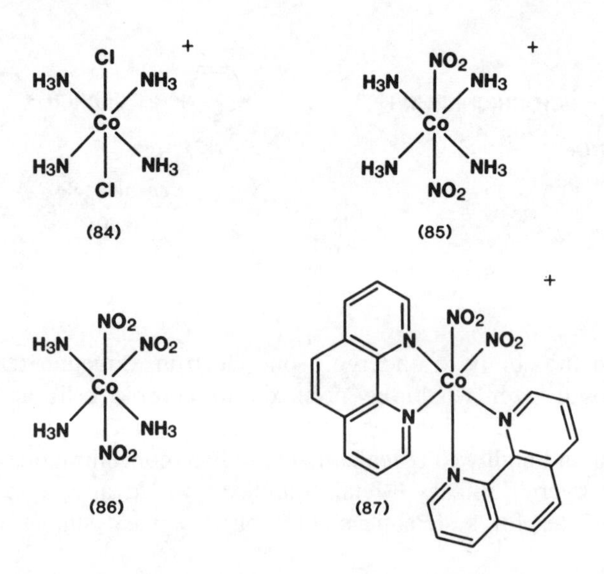

Fig. 8.4. Structures of cobalt—amine complexes with radiosensitizing properties.

results. They do, however, demonstrate the potential for study and the need for systematic evaluation of closely related complexes. The results, along with some comments, are summarized in Table 8.I.

The radiosensitizing effects of copper salts in both bacterial [31] and mammalian [32] cells have been demonstrated. Whereas Cu(II) salts had little or no toxicity in normal oxygenated cells, incubation under hypoxic conditions and thus reduction to Cu(I), or the introduction directly of cuprous solutions in the oxygen-free medium, resulted in enhanced sensitivity to radiation, a DMF of 1.54 being obtained in the mammalian system. The interaction of the hydroxyl radical with Cu(II) to produce Cu(III) has also been suggested as a mechanism of sensitization [33]. The effect of copper on radiation damage of T7 bacteriophage has also recently been investigated and at low concentrations enhanced the damage [34].

TABLE 8.I
Metal salts with radiosensitization ability.

Metal	System	Comments/Examples	Ref.
Cu	Bacterial, Mammalian	Reduction? Sensitizes/Protects	31, 32
Co	Bacterial	$[Co(NH_3)_5Cl]^{2+} > [Co(NH_3)_6]^{3+}$	36
Co	Mammalian	$[Co(NH_3)_4X_2]^+$	40
	Mammalian	Vit. B_{12}, hypoxic	41
Ag	Bacterial	Hypoxic; inhib. by *t*-BuOH	37
Mn	Bacterial Spores	MnO_4^-; color change *in situ* Hypoxic	38
Zn	Bacterial	Sensitizes/Protects	37, 39
Fe	Mammalian	$[Fe(Cp)_2]^+$	40
	Mammalian	$[Fe(CN)_5NO]^{2-}$	41
	Bacterial	$[Fe(CN)^6]^{3-}$, inactive in mammalian cells	42

The above example of reduction-enhanced cytotoxicity is of particular relevance because of the role of copper as an essential trace element and its function in many intracellular redox processes. The arguments for an overall mechanism of radiobiological action invoking copper participation have, in fact, been presented [35] and will be summarized later.

The fact that the ferricenium ion ($FeCp_2^+$) is an effective radiosensitizer, DMF = 2.0 at 100 μM, is of interest because of its good oxidizing power and stability. The analogous Co complex is inactive, no difference between effects on hypoxic and oxygenated cells was found in this study [40]. The nitroprusside salt was an effective sensitizer in V-79 cells at 10 μM [41]. This effect was attributed to toxic ligand (CN^-) release [41] but other factors should be considered, such as oxidizing power and possible chemistry of the NO group. The need for systematic study of metal complexes under identical conditions is emphasized by the fact that while ferricyanide, $[Fe(CN)_6]^{3-}$, is a sensitizer of bacterial cells, the activity being attributed to its thiol binding capacity (see Section 8.3.5) [42], physicochemical studies of hydroxyl damage on DNA showed that the reduced ferrocyanide salt, $[Fe(CN)_6]^{4-}$, protected against strand breaks [43]. The relationships between isostructural metal complexes differing only in their oxidation state and number need to be explored. Initial studies show that, in CHO cells, nitroprusside, being weakly active with an ER = 1.1 at 100 μM is a more effective sensitizer than ferricyanide [44]. No detailed mechanistic results are at present available for oxidizing metal complexes.

Another metal complex which may be mentioned here is platinum—thymine blue (see Chapter 4), which, because of its mixed-valent nature, may well involve reduction in its as yet undefined mechanism of sensitizing action [45].

8.3.2. COBALT–AMINE COMPLEXES

A further apparent example of reduction-enhanced cytotoxicity with metal complexes is that of the hexamminecobalt ion, $[Co(NH_3)_6]^{3+}$, where reduction to Co(II) also resulted in enhanced sensitivity to radiation [36]. Cobalt chelates such as $[Co(bipy)_3]^{3+}$, on the other hand, did not show any sensitization. Study of the Co—amine system in the EMT6 cell line has given a series of complexes with radiosensitizing activity and some correlations are emerging [40, 47]. The structures of some of these are shown in Figure 8.4.

The more recent results on the interaction of these substitution inert complexes with DNA (see Chapter 1), and especially the photosensitized cleavage of DNA by Co(III) complexes [46], may well be relevant to these biological results, especially as a 1,10-phenanthroline chelate, *cis*-$[Co(phen)_2(NO_2)_2]^+$, is also claimed to be an efficient sensitizer. Differing activities have been observed in isomeric forms of $[Co(NH_3)_4X_2]^+$, X = Cl, NO_2. Abrams *et al.* [40] have observed that all active complexes show a reduction potential of <260 mV, which represents a cut-off point as there is no strict correspondence between reduction potential and OER. No selectivity is apparent for these complexes in hypoxic cells. The hexammine and pentammine complexes originally reported [36] are essentially inactive in this cell line (EMT6), again pointing out the difference in bacterial and mammalian systems.

8.3.3. RADIOSENSITIZATION BY ELECTRON AFFINITY

8.3.3.1. *Metal—Radiosensitizer Complexes*

The rationale for electron affinity has been summarized Section 8.2.1. The oxidizing properties of the complexes discussed above are metal-centered. Ligands which contain redox-active sites may have their chemical and biological properties modified by complexation, for example the nitro-aromatics with ligating properties.

The identification of DNA as the ultimate target of radiation damage by 'oxygen mimics' prompts the question as to how potential drugs with both DNA-binding and radiosensitizing properties may be designed. The interaction between radiation and cisplatin (Section 8.3.4), and the presence of potential donor atoms in nitroimidazoles, have prompted the study of the the modification of the properties of these molecules by metallation. The strong binding of platinum and other platinum metals, in their complexes, to purines and pyrimidines suggested that the metal atom could carry the reductive moiety, in this case a nitroimidazole, to the purported site of action.

The structure of a typical complex, *trans*-[$PtCl_2$(misonidazole)$_2$], is shown in Figure 8.5 and the factors affecting the synthesis (basicity of nitroimidazoles) and products formed such as *cis/trans* isomers have been discussed [48]. The deactivating nature of the nitro group means that the heterocyclic nitrogen is much less basic than that of, say, 1-MeIm, and the extra lability is reflected in an unusually facile isomerization for these complexes [49]; indeed, so far, only the *trans*-isomer has been observed for 2-NO_2-imidazoles in [$PtCl_2L_2$]. Structurally, the complexes are straight-forward and the crystal structures of both isomers of the metronidazole complex and the *trans*-isomer of the misonidazole complex have been reported [48, 49]. One feature worth noting is that the NO_2 group is out of the plane of the imidazole ring for the misonidazole complex (2-NO_2Im, NO_2 group on carbon adjacent to platinum-bonded nitrogen), in contrast to the metronidazole complexes (5-NO_2Im, NO_2 group two carbons away) and the free ligands where the nitro group is coplanar. This could reflect a steric effect.

Initial results on *cis*-[$PtCl_2$(metronidazole)$_2$] (abbreviated FLAP from Flagyl-Platinum) were very promising with a reported ER of 2.4 and little toxicity [50] but later studies [48, 51, 52] failed to corroborate this finding

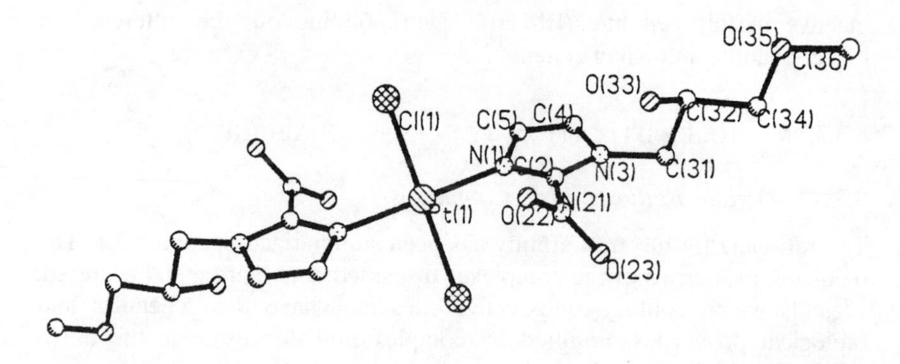

Fig. 8.5. The structure of a typical Pt—nitroimidazole complex, *trans*-$[PtCl_2(misonidazole)_2]$. From Reference 48.

and the complex is at best a weak to moderate radiosensitizer. Table 8.II summarizes the biological results, and relevant chemical data, for some nitroimidazole complexes. In sensitization, the *trans*-isomers were equally as effective as the *cis*-counterparts, which is of interest mechanistically as both isomers of $[PtCl_2(NH_3)_2]$ are also effective sensitizers (Section 8.3.4). Thus, the structural constraints for chemotherapeutic activity are not present in this case.

TABLE 8.II
Chemical and biological data for some metal—nitroimidazole complexes

Complex	Enhancement Ratio[a] (Conc.)	ΔE (mV)[b]	DNA binding[c]
cis-$[PtCl_2(NH_3)(Miso)]$	1.25(100 μM)	+110	+
cis-$[PtCl_2(NH_3)(Metro)]$	1.07(100 μM)	—	+
cis-$[PtCl_2(NH_3)(4\text{-}NO_2Im)]$	1.16(100 μM)	—	+
trans-$[PtCl_2(Miso)_2]$	1.2(100 μM)	+150	−
cis-$[PtCl_2(Metro)_2]$	1.1(100 μM)	+200	−
cis-$[PtCl_2(NH_3)_2]$	1.3(10 μM)	—	+
$[RuCl_2(DMSO)_2(4\text{-}NO_2Im)_2]$[d]	1.6(200 μM)	+70	+

[a] In hypoxic CHO cells. Values for cisplatin for comparison.
[b] Refers to difference between complex and free ligand under identical conditions.
[c] Refers to inhibition of restriction endonuclease activity.
[d] Probably *trans,trans,trans*-isomer. See Refs. 55 and 56 for details.

Extensive structure—activity relationship studies on free nitroimidazoles have established a correlation between the *in vitro* radiosensitizing efficiency and the one-electron reduction potential of the nitro group [53]. Metal binding renders this potential more positive because of the inductive effect resulting from bond formation of the lone-pair nitrogen of the imidazole ring to the metal. Theoretically this should allow for 'fine tuning' of the redox potential of the nitro group and more efficient sensitization. Studies on the intracellular concentrations attained by FLAP, versus the radiosensitization achieved at that dose, indicates strongly that a major problem is uptake, the intracellular concentrations simply not being high enough to show a reasonable effect [54]. A combination of chemical (stability, solubility) and pharmacological factors (partition coefficients, distribution) needs to be considered before the biological activities of these systems can be optimized.

The general problem with the bis(nitroimidazole) complexes is their lack of water solubility and general chemical instability. Methods to circumvent this lability include synthesis of complexes with only one nitroimidazole and also containing amine, such as $[PtCl_2(NH_3)(NO_2Im)]$, which may help absorption and solubility [55]. Use of more kinetically inert metals such as ruthenium has also been invoked to produce more robust species and in at least one case, that of $[RuCl_2(DMSO)_2(4\text{-}NO_2Im)_2]$ [56], the complex is a better sensitizer than the free ligand (Table 8.II). These results are encouraging since they confirm that, with suitable choice of complex, metallation can produce less toxic but more active nitroimidazoles and their study can help elucidate the requirements for metal-based sensitizers.

The original rationale for use of metal complexes of this type was that they may target the nitroimidazole to DNA. An assessment of DNA binding may be obtained by studying inhibition of restriction enzyme activity on plasmid DNA (Chapter 1.3.3.3). Interestingly, the complexes containing one nitroimidazole ligand inhibit restriction endonuclease activity, whereas the bis complexes do not. The relative inhibitions compared to *cis*-$[PtCl_2(NH_3)_2]$ are also shown in Table 8.II. The mono-(nitroimidazole) species are much more effective than the bis(nitroimidazole) complexes, emphasizing the requirement for a primary amine in DNA binding. The Ru complex also inhibits strongly.

A set of complexes has been reported where the metallation is through a secondary amine on the aliphatic side chain, rather than the imidazole ring nitrogens, of the substituted nitroimidazole (Figure 8.3). This strategy to avoid the instability of the nitro-substituted ring gives active complexes [51, 57] but since the platinum is now some distance away from the nitro group it is not clear what, if any, electronic effects are involved.

Enhanced *in vitro* activity for complexes containing nitroimidazoles axially bound to the rhodium carboxylates (see Chapter 2.3) has also been reported [58, 59], although *in vivo* data were not as promising. The sensitization by the rhodium carboxylates themselves is discussed in Section 8.3.5. The crystal structure of $[Rh_2(O_2CCH_3)_4\{$1-(2-hydroxy-3-methoxypropyl)-2-methyl-5-nitroimidazole$\}_2]$ again shows the loss of coplanarity between the NO_2 group and the imidazole ring [59].

The complexation of the substituted thiazole, 2-amino-5-nitrothiazole (ANT) has also been examined:

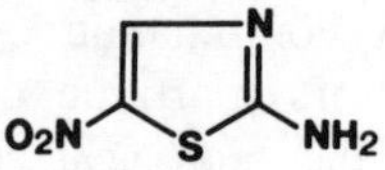

(89)

Studies on palladium complexes showed that linkage isomerization could in fact take place between the ring nitrogen and exocyclic amine, depending on solvent used in the reaction [60]. Similar linkage isomerism occurs for Pt [61], making the system more complicated than was initially understood or perceived [62]. The different linkage isomers of *trans*-$[PtCl_2(ANT)_2]$ (ring and amine bound) show the interesting feature of different DNA-binding and radiosensitizing properties, the DNA-binding form being the better sensitizer [61].

A further set of complexes which may be mentioned are the platinum nitroxyls, derived from piperidine and pyrollidine *N*-oxides [63, 64]:

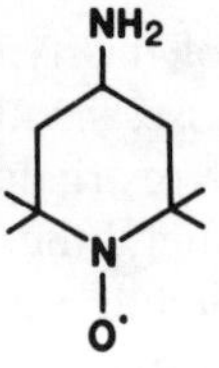

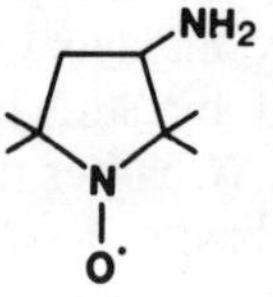

(90) (91)

These species, because of the nitroxyl group, are ESR-traceable analogues of *cis*-$[PtCl_2(NH_3)_2]$ and thus useful as probes [65]. Again, however, their instability and lack of aqueous solubility render them of little potential value for their dose-modifying behavior.

8.3.3.2. *Metal Chelates*

In essence, metal—radiosensitizer complexes contain a redox-active ligand combined to the metal. Enhanced radiosensitization could occur through better binding to the target or modification of the redox properties of the ligands. The requirement for a redox-active ligand may be met by many systems and one such complex is that of a nickel chelate of lapachol, 2-OH-3-isoprenyl-1,4-naphthoquinone, a natural product with carcinostatic activity [66]:

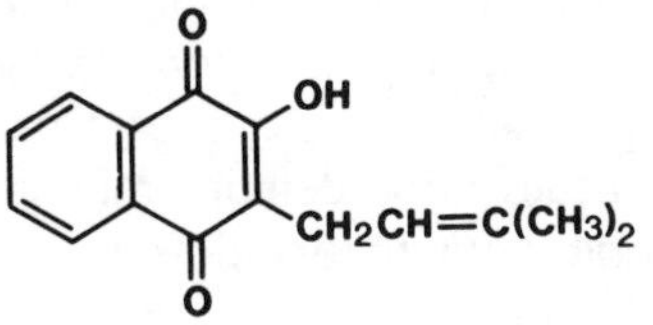

O
OH
$CH_2CH{=}C(CH_3)_2$
O

(92)

The chelation is through the 2-OH and adjacent keto group. This complex and its Cu and Zn analogues were first studied because, in an independent study, they were shown to be catalysts for the chemiluminescence of luminol [67], as are the chelates of the simple analogues lawsone and phthiocol [68]. Since the mechanism of chemiluminescence implies hydroxyl radical production [69], the complexes were tested to examine their interaction with radiation-produced radicals. Despite this somewhat tenuous rationale (but see Section 8.5) the interesting result was found that the Ni chelate was a moderately good sensitizer (CHO cells, ER = 1.3 at 100 μM); neither the Cu or Zn complex showing any activity. The Ni complex also is the best chemiluminescence catalyst. This is a good example of the variable properties of metal complexes, where isostructural species differing by only one electron (Ni d^8 to Zn d^{10}) show different biological activity. The nickel chelate does not deplete thiols, does not bind to DNA (as assessed by inhibition of restriction endonuclease activity), but does enhance radiation induced breaks. These results imply an electron-affinic pathway similar to that of O_2 and the complex may be the first example of a metal chelate acting in this manner. The fact that this chelate is nontoxic at the levels used indicates wide scope for this approach.

8.3.4. POTENTIATION OF RADIATION DAMAGE BY PLATINUM–AMINES. DIFFERENTIAL REPAIR PROCESSES

The clinical utility of the platinum amines makes their interaction with radiation of particular interest, from the point of view of their inherent ability to modify radiation damage, their interaction with sensitizers such as the nitroimidazoles and their potential clinical use in combined chemotherapy/radiation treatment regimens [70]. Indeed, highly promising results, in terms of survival of patients, are emerging from clinical trials with cisplatin/radiation combinations [71].

The first report of the interaction of cisplatin with radiation was in 1971 when the effect of cisplatin on the post-irradiation lethality of mice after irradiation with X-rays was reported [72]. A recent review [52] complements the earlier one of Douple and Richmond [73]. The general points to be made are that the ability to modify radiation damage does not parallel the antitumour activity of platinum complexes and that the most active antitumour agents do not necessarily demonstrate potentiation. Indeed, *trans* isomers are also good potentiators despite their lack of antitumour effect.

The survival curves show that at 100 μM the *trans*-isomer is as effective as 10 μM of the *cis*, with a DMF of 1.3—1.4; the higher dose for the *trans*-isomer possibly results because of the reduced toxicity [74]. So, from purely *in vitro* considerations, the complexes are only moderately active. As stated, it is desirable that the enhancement selectively affects hypoxic regions.

The results seem to be general for the other platinum complexes studied. The results obtained with CHIP, the Pt(IV) complex (see Figure 2.4) are somewhat dubious because of the problem of its constitution (see Chapter 2.4). The solvate form with H_2O_2 may affect interpretation of the biological effects observed and early results [75, 76] apparently are complex and not easily repeatable [52]. The interactions of CBDCA, because of its clinical promise, with radiation are of interest. Initial results indicate promising activity in both bacterial and mammalian systems [77].

8.3.4.1. *Mechanism of Action of Platinum—Amine Complexes*

A true sensitizing action for platinum complexes is likely to require a reduction to Pt(I). The rate constants for reaction of *cis*-$[PtCl_2(NH_3)_2]$ with e^-_{aq} and $^{\cdot}OH$ have been calculated as 1.8×10^{10} and 2.0×10^9 $M^{-1}s^{-1}$, respectively [75]. Other workers calculate the reaction rates with the hydrated electron as 1.2×10^{10} and 1.3×10^{10} $M^{-1}s^{-1}$ for the *cis*- and *trans*-isomers respectively [72]. Upon reduction of the Pt(II) by the hydrated electron a number of disproportionation reactions occur [78, 79]. Interestingly, chloride ion is lost almost 40 times faster from the *cis*- as from the *trans*-isomer [79]. At concentrations used for the pulse radiolysis experiments, platinum metal is deposited indicating disproportionation:

$$2\,Pt(I) \rightarrow Pt(0) + Pt(II)$$

Reoxidation will involve Pt(II) aqua and hydroxo complexes because of loss of chloride ligand, and the expected high reactivity of Pt(I) species, especially in aqueous solutions, predicts that the radiation chemistry is in all likelihood deactivating (protecting) rather than activating (sensitizing). The reduction potentials of both *cis*- and *trans*-$[PtCl_2(NH_3)_2]$ are < -1.0 V and therefore the complexes are extremely poor electron scavengers (compare E^1_7 $O_2/O_2^- = -155$ mV) [79]. Therefore any mechanism involving electron acceptance from target radicals (such as DNA) must also be eliminated from consideration.

These points also explain the semantic differences between a modifier or potentiator of radiation damage (by whatever mechanism) and a sensitizer (implying oxygen-mimetic or electron-accepting mechanisms); although the literature is not always consistent on this point. The radiation potentiation of platinum amine complexes *in vitro* is remarkably, and somewhat frustratingly, dependent on experimental conditions such as the time of incubation before irradiation and the phase of the cell cycle used; occasional protection has even been seen by most workers. Consideration of these factors requires careful analysis of the shape and slope of the survival curves and is beyond the scope of this summary, but the factors need to be known for a complete description of the potentiating effect.

The mechanism of action would appear to be most strongly connected with the ability of the platinum complexes to inhibit DNA repair processes, although thiol depletion has also been proposed [80, 81]. However, other results [82] do not show intracellular thiol depletion. Studies of $\varkappa$-irradiation of platinated DNA showed that an additive conformational distortion of the polymer, over and above what might be expected by independent modalities, is obtained [83]. No strand breaks are induced by platinum complexes alone, yet the binding to DNA inhibits repair of such breaks. The conformational changes occurring on Pt—DNA binding may also affect the radiation damage.

Radiation-induced thymine release from DNA is also inhibited by cisplatin [78]. The possibility exists that platination of DNA (presumably via the purine bases at biologically relevant concentrations) may affect the inherent radiosensitivity of the bases, which for aerated solutions, follows the order Thy > Cyt > Ade > Gua. An analysis of this type of damage should take into consideration profiles of base release and decomposition with and without platination, as well as the effects the conformational and electronic changes in platinated DNA will have on the radiation interaction. In this respect the studies on radiosensitivity of Ag and Hg—DNA complexes [85], as well as the effects on the radiation chemistry of simple bases of Cu(II) [86] and Fe(III) [87], are relevant.

Attempts have been made to separate out the components of damage caused by platinum complexes [52, 84]. The extra burden placed on the cells repair processes by the two types of damage (Pt—base, HO) may be additive or synergistic. The ability of the *trans*-isomer (albeit at higher concentrations) to achieve potentiation equivalent to that of *cis*-$[PtCl_2(NH_3)_2]$ implies that the specific DNA-binding mode required for the manifestation of the chemotherapeutic effect is not essential for enhancement of radiation damage and perhaps inhibition of repair. A recent study indicates that sensitization by the *trans*-isomer involves hydroxyl radical, which is not the case for the *cis*-isomer [89].

8.3.5. RADIOSENSITIZATION BY THIOL DEPLETION

The radioprotecting properties of thiols and aminothiols have been known for some time and this is generally attributed to their quenching of radiation-induced radicals such as O_2^-. Compounds which remove endogenous thiols thus increase radical lifetime and can have sensitizing properties. This mechanism contributes to the action of, for instance, *N*-ethylmaleimide [90] and any thiol depletion by artificial means should indeed produce enhancement of sensitization [91].

The great affinity of mercuric salts for thiols led to their examination as sensitizers and these results have been summarized [92]. In bacteria, phenylmercuric acetate (PMA), was found to sensitize under both aerobic and anaerobic conditions, the dose modifying factor being twice as great under nitrogen. Studies with *p*-hydroxymercuribenzoate (HMB) have also been reported and activity in various strains, including radioresistant ones, correlated with thiol binding. Treatment of bacteria with PMA or HMB after irradiation did not result in sensitization. A brief report on *p*-chloromercuribenzoate in mice claimed some sensitizing action [93].

The known use of ferricyanide as a thiol labelling group also prompted its examination, and excellent sensitization in bacteria was found [42] and attributed to the reaction:

$$2\,FeCN_6^{3-} + 2\,RSH \rightarrow RS{-}RS + 2\,FeCN_6^{4-}$$

As stated earlier, the number of possibilities for reaction of these species with radiation-induced radicals is large and the exact mechanism may be different from that suggested by the above reaction. In recent results the *in vitro* sensitizing action of rhodium carboxylates has been attributed to their thiol binding capacity [58], and this is of interest in view of their known affinity for sulfur-containing cellular constituents (see Chapter 6). The *in vitro* sensitizing efficiency of the carboxylates follows the order butyrate > propionate > acetate > methoxyacetate, which parallels the antitumour effect and is related to the intracellular uptake. The sensitization, except for the butyrate, is greater under hypoxic conditions. Unfortunately, essentially no sensitization was seen *in vivo*.

8.4. Metal Complexes and Radioprotection

A further aspect to be considered is the interaction of endogenous metals and metalloenzymes with both radiolysis products and extraneous chemicals which may result in radioprotection. Indeed, in any class of compounds there is a reciprocal nature of sensitization and protection that demands that examples exist illustrating both effects. The need for good radioprotectors today, of course, is something which transcends the uses discussed here, given the ever-present threat of radiation damage from nuclear accidents.

As stated, thiols and aminothiols are efficient protectors. The compound WR2721 {*S*-2-(3-aminopropylaminoethylphosphorothioic acid} (see Figure 1.3) is of current clinical interest [94]. Attempts have been made to relate the protecting ability of thiols to their chelating ability, with the chelate formed presumably being responsible for radical reactions [35, 95]. This is certainly attractive for the aminothiols with two chelating groups in the appropriate positions for metal binding but no clear-cut results are apparent.

Hydroxyl radical scavenging by species such as DMSO is also a recognized mechanism of protection [96].

8.4.1. METAL COMPLEXES AS RADIOPROTECTORS

The production of superoxide from secondary reactions of radiation-produced radicals can contribute to the oxygen enhancement of radiation damage, because superoxide can further react to give reactive oxidizing agents. The enzyme superoxide dismutase (Cu/Zn SOD) catalyzes the disproportionation of superoxide [97]:

$$2\,O_2^- + 2\,H^+ \rightarrow H_2O_2 + O_2$$

The radioprotectant properties of SOD have been examined; the enzyme is indeed effective *in vivo*. These properties have recently been reviewed [98], as has the general area of SOD and cancer [99].

The SOD-like activity of some simple copper salts such as bis(3,5-diisopropylsalicylato)copper(II) (see Figure 6.3) prompted its comparison with SOD and radioprotection was also found, with a survival rate of 58% upon 24 h pretreatment before a dose of 12 Gy [100]. The protection must be related to other factors besides superoxide dismutation, because the complex is much less efficient at disproportionation than the enzyme but is considered to be a better protector. Factors such as lipophilicity and ease of membrane permeability may in fact be more important in dictating overall efficiency. This particular copper system is quite complicated *in vivo* (see Chapter 6.4). The fundamental point is, clearly, that a biological effect such as radioprotection may be obtained from consideration of the bioinorganic chemistry of small molecules, in this case superoxide disproportionation, and known radiation chemistry. A further copper complex that has been reported to have protecting properties is the complex with the 3-mercapto-2-hydroxypropyl ether of the polysaccharide dextran [101]. Further, the sensitization afforded by zinc salts was dependent on conditions and protection was also observed in certain instances [39].

The protection of isolated DNA from radiation damage by metal salts such as $[Fe(CN)_6]^{4-}$ has been mentioned earlier [43]. Other studies on metal salts have been reported [102].

8.5. Activation of Metal Complexes

We have seen that there are examples of metal complexes in all widely accepted mechanisms of radiation modification, and metal complexes in general must be considered as excellent potential for further *in vitro* and *in vivo* studies. A review paralleling this chapter has appeared [103]. Many of the effects of ionizing radiation on DNA with respect to modification have also been reviewed in a comprehensive volume [104]. Systematic *in vitro* studies to correlate activity with chemical parameters e.g. reaction with the hydrated electron, hydroxyl radical etc. are clearly required. The fact that some metal complexes show a fine line between sensitization and protection is indicative of competing processes. In this respect, the point has been made that these processes are the 'two sides of a coin'. The basic mechanism for both reactions is closely related and we can consider protection as a reductive process or repair and implying donation to a particular molecule, while sensitization can be thought of as abstracting electrons, an oxidative process [105].

The rationale of radiation sensitization by electron-affinic compounds involves the chemical interaction with e^-_{aq}. An alternative statement is that sequestering of e^-_{aq} can increase concentrations of the damaging $^\cdot OH$ [106]. Suffice it to say that the reactions of metal cations with the hydrated electron are well characterized and, indeed, cover a wide range of rate constants depending on the electronic structure of the complex [107].

Similarly, reactions with superoxide and peroxide are well categorized. The protection by copper salts has, as we have seen, been discussed in terms of their chemical reactivity toward superoxide [100]; the radiation enhancement in T7 bacteriophage has also been rationalized in terms of Fenton-type chemistry [34], whereby reactive intermediates are produced:

$$M^{n+} + H_2O_2 \xrightarrow{H^+} \text{`}{}^{\cdot}OH\text{'} + M^{(n+1)+} + H_2O$$

In the presence of a substrate such as DNA this can give:

$$\text{`}{}^{\cdot}OH\text{'} + RH \rightarrow \text{`}R\text{'} + H_2O$$

Clearly these events, requiring as they do the presence of peroxide produced from secondary reactions of initial radiolysis products, may affect aerobic rather than hypoxic cells. The production of hydroxyl radical from H_2O_2 was shown not to result in sensitization [108]; however if peroxide or hydroxyl radical is produced randomly the radicals will react in a nonspecific manner, whereas localization near the target, e.g. by metal complex binding, could enhance damage. The fact that SOD-mimetic compounds protect, whereas complexes which activate or have some xanthine oxidase-mimetic activity, such as [Ni(Lapachol)$_2$], sensitize, is intriguing and is worthy of confirmation for a wider range of complexes. An increase in strand breaks observed upon irradiation of Cu^{2+}/DNA mixtures is a demonstration that activation chemistry can indeed occur [109] and further emphasizes the duality of metal complexes — depending on their structure, complexes may either sensitize or protect. The factors relating to the chemistry of metal complexes with oxygen substrates which may induce a biological event (sensitization or protection) are worth exploring. Similarly, the conformational changes upon metal binding to DNA could also result in sensitization or protection [104], and systematic differentiation may be possible, depending on the structure of the complex.

8.6. Summary

Radiosensitization and radipotentiation refer to the enhancement of cellular radiation damage by chemical agents. The locus of radiation damage is DNA and oxygen is necessary for damage to occur. Thus, the clinical effectiveness of radiation therapy in cancer treatment may be offset by the presence of hypoxic (oxygen-deficient) cells. Chemical mechanisms for radiopotentiation can be recognized as 'oxygen-mimetic', thiol depletion (thiols protect against radiation damage, thus depletion potentiates) and DNA modification. Radiosensitization strictly refers to the 'oxygen-mimetic' or electron-affinic mechanism and the most studied examples are the nitroimidazoles.

Metal complexes acting by each of these mechanisms are known. Cisplatin potentiates *in vitro* and is giving encouraging results in the clinic in combination with radiation. Platinum and ruthenium have been used to target organic sensitizers such as the nitroimidazoles to DNA. Mercury salts act by thiol binding in their sensitization of bacterial cells. A number of metal chelates of Co and Ni and oxidizing agents such as the ferricenium ion may act by mechanisms similar to the electron-affinic approach. The ability to design chelates of known reduction potential and which bind to DNA gives considerable scope for design of metal-based radiosensitizers.

Radioprotection has also been observed for metal chelates, particularly those of copper.

References

1. J. T. Case: *Progress in Radiation Therapy* (Ed. F. Buschke) pp. 1—18. Grune and Stratton, New York (1958).
2. E. J. Hall: *Cancer, a Comprehensive Treatise* (v. 6, Ed. F. Becker) pp. 281—312. Plenum, New York (1977).
3. R. J. Shalek: *ibid.*, pp. 39—50.
4. G. Scholes: *Effects of Ionizing Radiation on DNA Physical, Chemical and Biological Aspects* (Molecular Biology Biochemistry and Biophysics v. 27, Eds. A. Kleinzeller, G. F. Springer, and H. G. Wittman) pp. 153—170. Springer-Verlag (1978).
5. J. D. Chapman, A. P. Reuvers, J. Borsa, and C. L. Greenstock: *Radiat. Res.* **56**, 291 (1973).
6. M. M. Elkind and J. L. Redpath: Reference 2, pp. 51—99.
7. E. J. Hall: *Radiobiology for the Radiologist* Harper and Row, New York, Ch. 3, 29 (1978).
8. B. Palcic, J. W. Brosing, G. Y. K. Lam, and L. D. Skasgard: *Proceedings of the Berkeley Conference in Honor of Jerzy Neyman and Jack Kiefer* (v. 1, Eds. L. M. Le Cam and R. A. Olshen), pp. 331—342. Wadsworth (1985).
9. B. C. Millar, E. M. Fielden, and J. L. Millar: *Int. J. Radiat. Biol.* **33**, 599 (1978).
10. M. M. Elkind and G. F. Whitmore: *The Radiobiology of Cultured Mammalian Cells* Gordon and Breach (1967).
11. B. C. Millar, E. M. Fielden, and J. L. Millar: *Int. J. Radiat. Biol.* **33**, 599 (1978).
12. B. Palcic, J. W. Brosing, and L. D. Skasgard: *Br. J. Cancer* **46**, 980 (1982).
13. L. H. Gray, A. D. Conger, M. Ebert, S. Hornsey, and O. C. A. Scott: *Brit. J. Radiol.* **26**, 638 (1953).
14. R. H. Thomlinson and L. H. Gray: *Br. J. Cancer* **9**, 539 (1955).
15. K. A. Kennedy, B. A. Teicher, S. A. Rockwell, and A. C. Sartorelli: *Biochem. Pharmacol.* **29**, 1 (1980).
16. C. L. Greenstock: *J. Chem. Ed.* **58**, 157 (1981).
17. D. Nori, J. H. Kim, B. S. Hilaris, and F. Chu: *Cancer Investigation* **2**(4), 321 (1984).
18. M. A. Shenoy and B. B. Singh: *Int. J. Radiat. Biol.* **48**, 315 (1985).
19. *Oxygen and Oxy-Radicals in Chemistry and Biology* (Eds. M. A. J. Rodgers and E. L. Powers), Academic Press, New York (1981).
20. P. Howard-Flanders: *Nature* **180**, 1191 (1957).

21. G. E. Adams and D. L. Dewey: *Biochem. Biophys. Res. Comm.* **12**, 473 (1963).
22. J. D. Chapman, R. G. Webb, and J. Borsa: *Int. J. Radiat. Biol.* **19**, 561 (1971).
23. G. E. Adams, J. C. Asquith, D. L. Dewey, J. L. Foster, B. D. Michael, and R. L. Willson: *Int. J. Radiat. Biol.* **19**, 575 (1971).
24. J. D. Chapman, A. P. Reuvers, and J. Borsa: *Br. J. Radiol.* **46**, 623 (1973).
25. J. L. Foster and R. L. Willson: *Br. J. Radiol.* **6**, 234 (1973).
26. J. D. Chapman, A. P. Reuvers, J. Borsa, J. S. Henderson, and R. D. Migliore: *Cancer Chemother. Rep. Part 1* **58**, 559 (1974).
27. G. E. Adams, I. Ahmed, P. W. Sheldon, and I. J. Stratford: *Br. J. Cancer* **49**, 571 (1984).
28. J. T. Roberts, N. M. Bleehen, P. Workman, and M. I. Walton: *Int. J. Radiat. Biol. Oncol. Phys.* **10**, 1755 (1984).
29. J. M. Brown, N. Y. Yu, D. M. Brown, and W. W. Lee: *Int. J. Radiat. Oncol. Biol. Phys.* **7**, 695 (1984).
30. See for instance *Int. J. Radiat. Oncol. Biol. Phys.* **12**(7), 1019—1267 (1986).
31. W. A. Cramp: *Nature* (*London*). **206**, 636 (1965).
32. I. P. Hesselwood, W. A. Cramp, D. C. H. McBrien, P. Williamson, and K. A. K. Lott: *Br. J. Cancer* **37** (Supp. III), 95 (1978).
33. I. Kirschner, N. Citri, A. Levitzki, and M. Anbar: *Int. J. Rad. Biol.* **17**, 81 (1970).
34. A. Samuni, M. Chevion, and G. Czapski: *Radiat. Res.* **99**, 562 (1984).
35. *Copper and Peroxides in Radiobiology and Medicine* J. Schubert C. C. Thomas, Springfield, U.S.A. 1964.
36. R. C. Richmond, M. Simic, and E. L. Powers: *Radiat. Res.* **63**, 140 (1975).
37. R. C. Richmond and E. L. Powers: *Radiat. Res.* **58**, 470 (1974).
38. A. Tallentire and A. B. Jones: *Int. J. Radiat. Biol.* **24**, 345 (1973).
39. M. Kiortsis: *Int. J. Radiat. Biol. Relat. Stud. Phys. Chem. Med.* **32**, 583 (1977).
40. B. A. Teicher, J. L. Jacobs, K. N. S. Cathcart, M. J. Abrams, J. F. Vollano, and D. H. Picker: *Radiat. Res.* **109**, 36 (1987).
41. E. B. Douple, C. J. Green, and M. G. Simic: *Int. J. Radiat. Oncol. Biol. Phys.* **6**, 1545 (1980).
42. H. Moroson and D. Tenney: *Experientia* **24**, 1041 (1968).
43. P. Achey and H. Duryea: *Int. J. Radiat. Biol.* **25**, 595 (1974).
44. N. Farrell and K. A. Skov: unpublished results.
45. J. D. Zimbrick, A. Sukrochana, and R. C. Richmond; *Int. J. Radiat. Oncol. Biol. Phys.* **5**, 1351 (1979).
46. J. K. Barton and A. L. Raphael: *J. Am. Chem. Soc.* **106**, 2466 (1984).
47. C. H. Chang and C. F. Meares: *Biochemistry* **23**, 2268 (1984).
48. N. Farrell, T. M. Gomes Carneiro, F. W. B. Einstein, T. Jones, and K. A. Skov: *Inorg. Chim. Acta* **92**, 61 (1984).
49. J. R. Bales, C. G. Coulson, D. W. Gilmour, M. A. Mazid, S. Neidle, R. Turoda, B. J. Peart, C. A. Ramsden, and P. J. Sadler: *J. Chem. Soc.* (*Chem. Comm.*) 432 (1983).
50. J. R. Bales, P. J. Sadler, C. J. Coulson, M. Laverick, and A. H. W. Nias: *Br. J. Cancer* **46**, 701 (1982).
51. R. Chibber, I. J. Stratford, I. Ahmed, A. B. Robbins, D. M. L. Goodgame, and B. Lee: *Int. J. Radiat. Onc. Biol. Phys.* **10**, 1213 (1984).
52. A. H. W. Nias: *Int. J. Radiat. Biol.* **48**, 297 (1985).
53. G. E. Adams, I. R. Flockhart, C. E. Smithen, I. J. Stratford, P. Wardman, and M. E. Watts: *Radiat. Res.* **67**, 9 (1976).
54. I. J. Stratford: personal communication.

55. K. A. Skov, H. Adomat, D. C. Konway, and N. P. Farrell: *Chem.-Biol. Inter.* **62**, 117 (1987).
56. P. K. L. Chan, K. A. Skov, B. R. James, and N. P. Farrell: *Int. J. Radiat. Oncol. Biol. Phys.* **12**, 1059 (1986).
57. G. E. Adams and I. J. Stratford: *UK Patent Appl.*, GB2, 131, 020, (1984); *C.A.* 102, 6489n (1985).
58. R. Chibber, I. J. Stratford, P. O'Neill, P. W. Sheldon, I. Ahmed, and B. Lee: *Int. J. Radiat. Biol.* **48**, 513 (1985).
59. D. M. L. Goodgame, A. S. Lawrence, A. M. Z. Slawin, D. J. Williams, and I. J. Stratford: *Inorg. Chim. Acta* **125**, 143 (1986).
60. N. Farrell and T. M. Gomes de Carneiro: *Inorg. Chim. Acta* **126**, 137 (1987).
61. K. A. Skov and N. Farrell: *Int. J. Radiat. Biol.* **52**, 49 (1987).
62. B. A. Teicher, S. Rockwell, and J. B. Lee: *Int. J. Radiat. Oncol. Biol. Phys.* **11**, 937 (1985).
63. H. R. Claycamp, E. I. Shaw, and J. D. Zimbrick: *Radiat. Res.* **106**, 141 (1986).
64. A. Mathew, B. Bergquist, and J. D. Zimbrick: *J. Chem. Soc.* (*Chem. Comm.*) 222 (1979).
65. H. G. Claycamp and J. D. Zimbrick: *J. Inorg. Biochem.* **26**, 257 (1986).
66. N. Farrell and K. A. Skov: in press.
67. N. Farrell: unpublished results.
68. B. P. Geyer and G. McP. Smith: *J. Am. Chem. Soc.* **63**, 3071 (1941).
69. E. H. White, O. Zafiriou, H. H. Kagi, and J. H. M. Hill: *J. Am. Chem. Soc.* **86**, 940 (1964).
70. L. DeWit: *Int. J. Radiat. Oncol. Biol. Phys.* **13**, 403 (1987).
71. E. B. Douple: *Plat. Met. Rev.* **29**, 118 (1985).
72. M. Zak and J. Drobnik: *Strahlentherapie* **142**, 112 (1971).
73. E. B. Douple and R. C. Richmond: *Cisplatin, Current Status and Clinical Developments* (Eds. A. W. Prestayko, S. T. Crooke, and S. K. Carter) pp. 125. (Academic Press) 1980.
74. E. B. Douple and R. C. Richmond: *Br. J. Cancer* **37** (Supp. III), 98 (1978).
75. M. Laverick and A. H. W. Nias: *Br. J. Radiology* **54**, 529 (1981).
76. A. H. W. Nias and I. Szumiel: *J. Clin. Hematol. Oncol.* **7**, 562 (1977).
77. E. B. Douple, R. C. Richmond, J. A. O'Hara, and C. T. Coughlin: *Cancer Treat. Rev.* **12** (Supp. A), 111 (1985).
78. R. C. Richmond and M. G. Simic: *Br. J. Cancer* **37** (Supp. III), 20 (1978).
79. J. Butler, B. M. Hoey, and A. J. Swallow: *Radiat. Res.* **102**, 1 (1985).
80. V. Alvarez, G. Cobreros, A. Heras, and C. Lopez Zumel: *Br. J. Cancer* **37** (Supp. III), 68 (1978).
81. A. K. Murthy, A. H. Rossof, K. M. Anderson, and F. R. Hendrickson: *Int. J. Radiat. Oncol. Biol. Phys.* **5**, 1411 (1979).
82. K. A. Skov: personal communication.
83. O. Vrána and V. Babec: *Int. J. Radiat. Biol.* **50**, 995 (1986).
84. E. B. Douple: *Pharmac. Ther.* **25**, 297 (1984).
85. P. C. Beaumont and E. L. Powers: *Int. J. Radiat. Biol.* **43**, 485 (1983).
86. J. Holian and W. M. Garrison: *Nature* **212**, 394 (1966).
87. S. N. Bhattacharyya and P. C. Mandal: *Int. J. Radiat. Biol.* **43**, 141 (1983).
88. H. P. Leenhouts and K. H. Chadwick: *Adv. Radiat. Biol.* **7**, 56 (1978).
89. E. L. Powers and M. A. Centilli: *Int. J. Radiat. Biol.* **50**, 31 (1986).

90. B. A. Bridges: *Nature (London)*. **188**, 415 (1960).
91. J. E. Biaglow, M. E. Varnes, E. P. Clark, and E. R. Epp: *Radiat. Res.* **95**, 437 (1983).
92. B. A. Bridges: *Adv. Radiat. Biol.* **3**, 159 (1974).
93. H. Moroson and H. A. Spielman: *Int. J. Radiat. Biol.* **11**, 87 (1966).
94. J. M. Yuhas and J. B. Storer: *J. Natl. Cancer Inst.* **42**, 331 (1969).
95. S. S. Block, D. D. Mulligan, J. P. Weidner, and D. G. Doherty: *Radiation Protection and Sensitization* (Eds. H. Moroson and M. Quintillani) p. 163, Taylor and Francis (London), 1970.
96. J. D. Chapman, A. P. Reuvers, J. Borsa, and C. L. Greenstock: *Radiat. Res.* **56**, 291 (1973).
97. I. Fridovich: *Science* **201**, 875 (1978).
98. A. Petkau and C. A. Chuaqui: *Radiat. Phys. Chem.* **24**, 307 (1984).
99. L. W. Oberley: *Superoxide Dismutase and Cancer* (Superoxide Dismutase v. 2, Ed. L. W. Oberley), pp. 127—165. CRC Press (1982).
100. J. R. J. Sorensen: *J. Med. Chem.* **27**, 1749 (1984).
101. Z. Wieczorek, J. Gieldanowski, M. Zimecki, J. Z. Mioduszewski, S. Szymaniec, and R. Daczynska: *Arch. Immunol. Ther. Exp.* **31**, 715 (1983).
102. E. Rotlevi, H. M. Moss, S. Kominami, and P. Riesz: *Ann. N.Y. Acad. Sci.* **222**, 387 (1973).
103. K. A. Skov: *Radiat. Res.* **112**, 217 (1987).
104. *Effects of Ionizing Radiation on DNA (Physical, Chemical and Biological Aspects)* (Molecular Biology Biochemistry and Biophysics v. 27, Eds. J. Hutterman, W. Kohler, and R. Teoule), Springer-Verlag (1978).
105. G. Hotz: *Modification of Radiation Damage* (Ref. 104, pp. 304—311).
106. E. L. Powers: *Israel J. Chem.* **10**, 1199 (1972).
107. E. J. Hart and M. Anbar: *The Hydrated Electron* p. 170, Wiley (New York) 1970.
108. J. F. Ward, W. F. Blakeley, and E. I. Joner: *Radiat. Res.* **103**, 383 (1985).
109. P. M. Cullis, J. D. McClymont, M. N. O. Bartlett, and M. C. R. Symons: *J. Chem. Soc. Chem. Comm.* 1859 (1987).

CHAPTER 9

ANTIBACTERIAL EFFECTS OF METAL COMPLEXES

The almost complete elimination of debilitating and fatal bacterial diseases such as pneumonia, tuberculosis and infections arising from surgery and wounds, not to mention the reduction in the drastic consequences of bacterial epidemics, must undoubtedly be ranked as one of the major achievements of applied chemotherapy; this has been brought about by the discovery and application of the sulfonamides, penicillins and tetracyclines. The history of these discoveries is well documented and, for the sulfonamides, can be traced back to the original postulates of Paul Ehrlich at the turn of this century [1, 2]. Antimetabolites such as the sulfonamides and antibiotics, as exemplified by penicillin, have been subjected to extensive study of their structure–activity relationships and are further important in the development of systematic chemotherapy because of the recognition of clear targets associated with design of chemotherapeutic agents (for penicillin the bacterial cell wall and for sulfonamides tetrahydrofolate synthesis).

The introduction of Salvarsan, an organoarsenical whose antitrypanosomal activity had been studied by Ehrlich (see Chapter 10), for the treatment of syphilis in 1912, is considered to be the first synthetic therapeutic agent to achieve widespread use. During the latter half of the nineteenth century the isolation and recognition of the bacterial causes of various diseases also laid the ground for the systematic study of antibacterial chemotherapy. Indeed, this dates to the experiments of Koch in 1881 when he tested mercuric chloride for activity on anthrax spores, and this was one of the first laboratory antibacterial experiments [3]. This preceded Koch's discovery of the tubercle bacillus in the early 1890s and indeed some of the agents he tested against this latest bacterium were also inorganic, notably gold salts. The activity of gold complexes, such as the cyanide and later the thiosulfate, led to their extensive use in tuberculosis treatment, especially in the period of the so-called 'Gold Decade' of 1925–1935, before once again being essentially discarded and supplanted by the newer antibiotics and antimetabolites [4]. Despite the fact that at the time no better remedies were available, the use and development of both gold and mercury antibacterial agents suffered considerably from inadequate perception and resolution of pharmacokinetic and toxicity problems, and the lack of good collaboration between laboratories on both the biochemical and chemical properties of these substances. In some senses, it is unfortunate for bioinorganic aspects of chemotherapy that the emphasis in antibacterial chemotherapy naturally switched to antibiotics and the antimetabolites at the same time that inorganic chemistry was attaining its present day sophistication as well as the realization of the involvement of trace metals in so many enzymatic processes. These historical coincidences certainly contributed to the lack of attention in postwar times to inorganic pharmacology. The common bacteria discussed here are described in the glossary.

9.1. Mercury and Silver Salts as Antibacterial Agents

9.1.1. MERCURY

The use of mercury in bacterial infections, of course, dates from some considerable time ago, and references may be found to the use of mercury as early as the Arab Rhazes in the tenth century. The physician Theodoric popularized the use of mercuric ointment for various skin diseases in the thirteenth century, and it was this already established use which may have been responsible for its almost immediate application to syphilis when this bacterial disease reached epidemic proportions in Europe in the early sixteenth century [5]. The fact that mercury was of undoubted chemotherapeutic utility was well documented and quite specific instructions as to its application can be found [6]. Mercury was in fact the first 'specific' to be used for syphilis and, despite the introduction of organic natural products such as guaiac, its utility remained unchallenged over the centuries, with some ups and downs, up to the introduction of Salvarsan. Unprincipled use and bad prescription earned for mercuric treatment an undeserved reputation but, despite this, its utility remained widespread. At one time, early in the period of the use of mercury in syphilis treatment, the drug was prescribed by charlatans and surgeons in unnecessarily large quantities and at exalted prices while they restricted the supply to limited numbers of people; the abuse was such that the popular malicious opinion of the time was that the surgeons had discovered the secret of alchemy and had transformed mercury into gold. A brief and interesting summary of the history of the uses of mercury is to be found in reference [7].

The mercury used over the centuries took many forms, including chloride, iodide, cyanide, oxide and sulfide and indeed it is of interest that renal toxicity and the concept of biotransformation were both noted, it being recognized that the administered substance was probably bound in 'some combination with albumen' before entering circulation [6].

The introduction of mercurochrome in 1919 was a consequence of the search for improved mercurials, initiated after Koch's observation of the activity of mercuric chloride in anthrax infections. The promise of mercuric chloride at that time may be gauged by the fact that it was the most active of almost 70 chemicals tested by Koch. Many other organomercurials were introduced around the same time, and Figure 9.1 shows representative structures. The fortuitous observation of the diuretic effects on syphilitic patients treated with organomercurials also stems from this time (see Chapter 12). The status of the antiseptic properties of

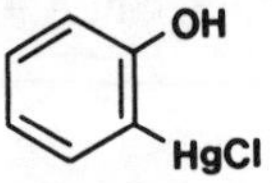

OH
HgCl
2-phenolmercuric chloride
(93)

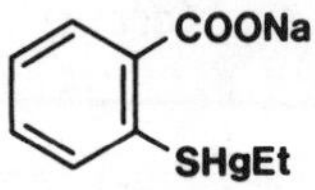

COONa
SHgEt
Merthiolate
(94)

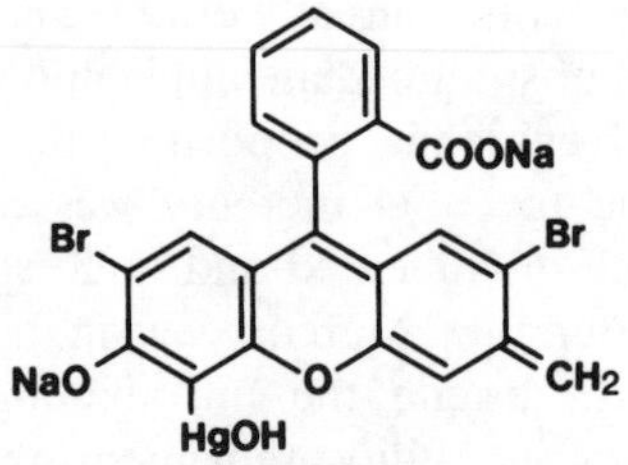

COONa
Br
Br
NaO
O
CH_2
HgOH
Mercurochrome
(95)

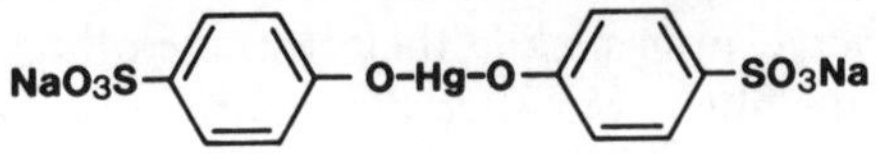

NaO_3S—O-Hg-O—SO_3Na
Hermophenyl
(96)

Fig. 9.1. Structures of some common mercuric compounds used in skin and wound infections.

mercury and silver complexes, and related mechanistic studies, have been reviewed and recently updated [8, 9]. The spectrum of activity has also been tabulated; mercury salts are active against strains of *Salmonella, Staphylococcus* and *Pseudomonas*, but are generally considered to be ineffective, in contrast to gold, in *M. tuberculosis*.

The action of mercury salts is bacteriostatic, rather than bacteriocidal, and early clues to their mechanism of action came from the observations that the effects are easily reversed by thiols, such as cysteine, glutathione and sodium thioglycolate [10, 11]. The thiol binding *in vivo* is not specific, and it is generally considered that for bacteria the site of action can range from cell wall components to cytoplasmic constituents such as proteins and enzymes [8]. One effect of binding to cell wall components is that the changes in permeability induced can cause 'leakage' of ions, this resulting in lysis of the cell [12]. Conformational changes in cell membranes can cause loss of K^+ and resultant permeability changes [8]. The effects of any one mercurial will be a result of a number of factors, including solubility, permeability and dissociation constants, and these factors will also affect its rate of reaction with thiols. This aspect has been summarized [8, 13]. Distribution studies with phenylmercuric borate showed relatively fast uptake with high concentrations resulting at the cytoplasmic membrane while protein, RNA and DNA synthesis were all inhibited, suggesting a nonspecific mode of action [14].

The physical alteration of nucleic acids by mercury binding has also been implicated in the mode of action [8, 15]. Certainly, since bacterial DNA is not encased in any nucleus and protected by nuclear membrane, the frequency of binding, upon entry to the cell, can be expected to be favorable.

9.1.2. SILVER

The introduction of silver nitrate for the treatment of infantile blindness (*opthalmia neonatorum*) in 1884 was considered a great medical breakthrough at the time; the development of structural analogues for silver has not been as diversified as that of mercury complexes, and its medieval use not as well documented. However, the microbiological activity of silver has been recognized for many centuries; the absolute minimum inhibitory concentration (MIC) of ionic Ag^+ is remarkably low at less than 0.1 μg/ml. Silver salts are still used to purify water and use of silver vessels in ancient times may also have served this purpose. Continuing research led to the highly useful silver salt of a sulfonamide, silver sulfadiazene, for burn treatment in 1968, a treatment which won remarkably rapid acceptance from researchers and drug agencies.

The use of silver nitrate and colloidal silver in wounds dates from the mid-nineteenth century and the term 'oligodynamic' was applied to the antimicrobial action of heavy metals diluted in water [9]. Over the years, attempts to put silver in contact with wounds produced a number of electrolytic methods [9] as well as use of solutions of silver nitrate, in this case as late as 1965 [16]. However, the imbalance of sodium, potassium and chloride caused by silver nitrate in these cases was unsatisfactory due to the number of resulting side effects. These include in some cases methemoglobinemia from reduction of nitrate to nitrite. In 1968, a complex (AgSD) formed from a sulfonamide and silver was introduced in attempts to combine the 'oligodynamic' action of the heavy metal with the antibacterial effect of the sulfonamide [17]. The insoluble product, which is used as a 1% cream, has suitable properties and excellent wide spectrum antibacterial activity, and is used worldwide for burn prophylaxis and other infectious skin conditions [9]. Side effects are few, the major one being leukopenia. The properties and mode of action of AgSD have been reviewed [9, 18].

The structure of the sulfadiazene and its polymeric complex is shown in Figure 9.2 [19]. Each silver atom is bound to three sulfadiazene molecules, one molecule forming a chelate from O(1) and N(1) of the RSO_2NR'

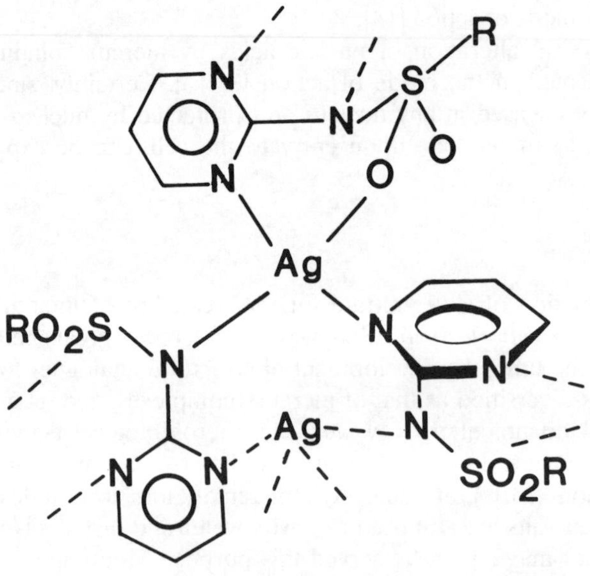

Fig. 9.2. A schematic structure of silver sulfadiazene (AgSD). The R group is p-$C_6H_4NH_2$. The coordination around Ag is approximately tetrahedral. From Reference 19.

portion of a ligand, the other molecules providing one pyrimidine nitrogen each to complete the nearly tetrahedral environment around each metal ion. Each sulfadiazene molecule likewise chelates one Ag, and bridges two others through the pyrimidine nitrogens. The Ag—N (pyrimidine) bonds are shortest, at 2.24 Å and the Ag—Ag distance is 2.916 Å, which is shorter than the 3.64 Å found in the dimer of $[Ag(1\text{-}MeCytosine)]NO_3$ [20].

The polymeric nature of the AgSD complex renders it almost insoluble in water and organic solvents and this property is in fact critical to its activity. The MIC of the Ag^+ ion in culture media is of the order of 1—20 μg/mL, due to binding, and thus inactivation, to serum proteins, etc. The MIC of free sulfadiazene as its sodium salt (NaSD) is 500—1000 μg/mL while that of the complex (AgSD) is 5—100 μg/mL, an approximately 5-fold increase over that of the free Ag^+ ion. Strains resistant to free sodium sulfadiazene or silver are sensitive to the silver salt, and the sulfonamide antagonist *p*-aminobenzoic acid (PABA) does not affect efficacy.

Use of a doubly labelled drug (^{110}Ag and ^{35}S) in a distribution study showed that practically no sulfonamide entered the cell, with the silver being bound mainly to 'cell residue' (proteins, carbohydrates) DNA and somewhat to RNA [21, 22]. The relative affinities for DNA were 10 times greater than for cell residue and 40 times greater than for RNA. The inhibition of DNA synthesis correlated with the amount of Ag^+ bound to the polymer. A value of one Ag atom to 80 base pairs was calculated as the minimum concentration at which growth (DNA synthesis) re-occurred. The authors conclude that silver binding to DNA, upon cellular delivery by the sulfadiazene carrier, is the principle mechanism of cell death. Contradictory results were obtained in an ultrastructural study which showed considerable distortion of the cell surface and little effect on cellular components, although the nuclear material was enlarged. In contrast, $AgNO_3$ demonstrated less effect on the surface and more marked clumping of nuclear material, which coincidentally provides an interesting visualization of the Ag—DNA interaction [23]. No ready explanation for this discrepancy is apparent.

The role of the sulfonamide was examined in further studies [22]. Even sodium sulfadiazene showed little penetration into cells and this presumably explains the lack of correlation between *in vitro* and *in vivo* activity in these species and why, at the concentrations of AgSD used, no antagonism by PABA is seen because the amount of sulfonamide produced is ineffective anyway. There is general agreement that the action of AgSD is due to dissociation and the silver is the active species. The advantages over silver nitrate are apparent in that, because of the insoluble nature of the AgSD, a controlled release of silver ions occurs at a moderate but pharmacokinetically fortuitous rate, producing a 'reservoir' of silver ions. The action of $AgNO_3$ is not as selective and it will further bind to inactivating components. This interpretation supports the contention that AgSD is bacteriocidal, whereas $AgNO_3$ is bacteriostatic [24].

The structure—activity relationship for the unique case of silver sulfadiazene may explain why many other Ag and, indeed, metal—sulfonamide complexes are active *in vitro* but not *in vivo*, although no clear correlations with physicochemical parameters such as acid dissociation constants or solubility products are apparent [23]. Substitution on the amine nitrogen by various heterocycles gives the series of sulfonamide ligands, of general formula $\{p\text{-}NH_2C_6H_4SO_2N(H)Ar\}$, but whether the polymeric nature is facilitated by the positions of the nitrogen atoms in the pyrimidine ring of AgSD, and how this relates to structures with pyridine, for instance, where stacking would be different, are unclear; these factors may affect dissociation and also initial attachment. It is interesting to speculate that a bridging arrangement of pyrimidines is an effective way of metal delivery to cells with slow release, and structures based on such bridging and the Pt—pyrimidine blues (Chapter 5) can be designed.

The dissociation of silver may not in some cases be a prerequisite for activity; silver uracil complexes did not penetrate cells and the *in vitro* antibacterial action was attributed to membrane effects [25]. A survey of the *in vitro* antimicrobial effects of a series of metal sulfanilamides also appointed silver and gold salts as the most active, the gold sulfadiazene being as active as the silver complex [26]. The chemistry and microbiology of these species has been summarized [27].

9.2. Antibacterial Activity of Chelating Agents and Metal Chelates

The interplay between metal sequestration or utilization and overload in bacterial systems has been well documented [1, pp. 416 ff]. The two major possibilities of biological action of chelates, the removal of metals from or masking of metals within a cell, and augmentation by chelation of the toxic effect of metals, have been detailed, as well as delivery by metal chelates with subsequent dissociation.

The activity of metal chelates formed from 1,10-phenanthroline and 2,2′-bipyridine has been fairly extensively studied. The studies by Dwyer and co-workers have been summarized [28]. The chelates examined were those of Ni(II), Co(II), Fe(II), Mn(II), Ru(II), Cu(II), Cd(II) and Zn(II), and the order of decreasing sensitivity of various organisms was *M.*

tuberculosis (most sensitive), *Staph. aureus, Strep. pneumoniae, E. coli* and *Pr. vulgaris* [28—30]. Thus, a general activity against both Gram-positive and Gram-negative strains as well as *M. tuberculosis*, an acid-fast strain,[1] is seen. The complexes in general were more active against Gram-positive strains, a common occurrence found also with the penicillins. Limited results showed little activity *in vivo*, however.

For the 1,10-phenanthroline complexes the activity in many cases does not differ dramatically from that of the free ligands. Thus, for the labile Fe(II) complex, the overall activity may be due to reactions following intracellular dissociation. An interesting feature is the consistent activity against *M. tuberculosis*. The iron chelate of 5-NO_2-1,10-phenanthroline is particularly active ($-\log M = 7.5$, where M is the molar concentration) and, since the growth of this microorganism is Fe-dependent, a sequestering mechanism is indicated [29]. The Ni(II) and Cu(II) complexes of this ligand are also particularly active ($-\log M$ of 7.5 and 7.2, respectively) whereas the Ru chelate was two orders of magnitude less efficient. The inert nature of the Ru complex supports the sequestering mode of action. Other substituted phenanthrolines are, however, also approximately two orders of magnitude less active than the nitro ligand. The tubercule microorganism develops resistance very readily (a problem even with the drugs that are in current clinical use) and it is therefore of interest to note that resistance to the metal chelates was not easily developed.

Although as a class the chelates are definitely bacteriostatic, the relative contributions of uptake and dissociation for both inert and labile species need to be more clearly defined before a full picture is obtained. The results suggest that kinetic reactivity may be more important than thermodynamic stability [30]. In the case of copper(II) complexes with 1,10-phenanthroline and 2,9-dimethyl-1,10-phenanthrolines, more recent studies showed that the major mode of action on *P. denitrificans in vitro* was in fact inhibition of respiratory electron transport in the cytoplasmic membrane, no correlation with inhibition of macromolecular synthesis being apparent [31]. These results are relevant in view of the discrepancy in the reported antitumour activity of these species (see Section 6.2).

Chelating agents such as oxine (8-hydroxyquinoline), Figure 9.3, have had their antibacterial activity extensively explained in terms of metal sequestration and activation and the system is something of a classic case [1]. Oxine is bacteriocidal only in the presence of iron(III) and its action

[1] The bacteria may be differentiated into Gram positive or negative depending on their reaction to a stain developed by the microbiologist Gram in 1884. The differentiation may be due to differing makeup of the cell wall in the two types [27].

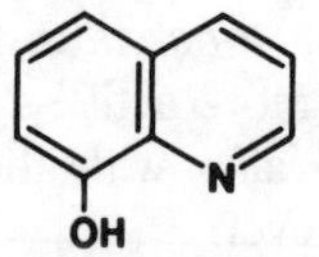

8-Hydroxyquinoline
(98)

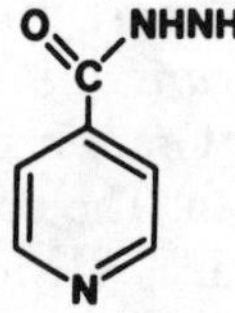

Isoniazid
(99)

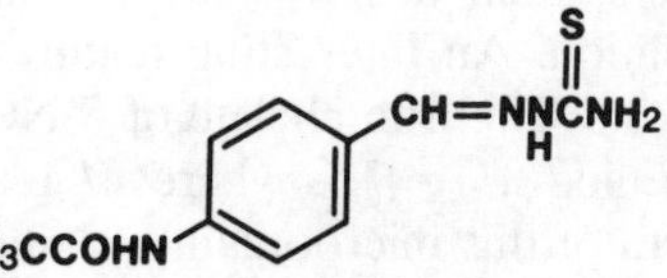

Thiacetazone
(100)

Fig. 9.3. Examples of antibacterial chelating agents whose activity (98) is through metal sequestration and activation or whose metal chelates (99, 100) have greater uptake compared to free ligand, with subsequent intracellular dissociation of metal complex or activation.

is inhibited by cobalt(II). The chelation to give active iron species can result in oxidation of SH groups and, in particular, lipoic acid $\{SH(CH_2)_2CH(SH)(CH_2)_4COOH\}$ an essential coenzyme for the oxidative decarboxylation of pyruvic acid. This mechanism is formally analogous to that of the inhibition of ribonucleotide reductase by thiosemicarbazones discussed in Section 6.2.1.

In fact, some thiosemicarbazones, such as thiacetazone, Figure 9.3, are antitubercular and have their uptake enhanced by copper (II) chelation. A further example of metal binding effecting cell entry, also analogous to the thiosemicarbazones and bisthiosemicarbazones, is found in isoniazid (the hydrazide of isonicotinic acid), also shown in Figure 9.3. This drug is one of the principal antitubercular drugs employed today. Its uptake by *M. tuberculosis* is greatly enhanced by the presence of Cu(II), which increases the lipid solubility [32]. Intracellular dissociation allows for oxidation of the isoniazide anion to an analogue of NAD which subsequently acts as an analogue inhibitor, disturbing lipid metabolism [33].

With repect to metal chelates, the use of naturally occurring iron-containing antibiotics (the siderophores) has been known for some time. The structures of many of these compounds and the mechanisms by which they may act have been reviewed recently [34]. In principle, iron depletion by simple competition among species, introduction into the cell of a toxic substance via the iron uptake pathway, and production of toxic oxygen radicals (see Section 9.4), may all be recognized [34]. The principal compound of interest is albomycin, whose structure has been recently elucidated [35]. The basic structure is a linear pentapeptide with a pyrimidine thionucleoside appended to one of the terminal groups. The octahedral iron binding site is made up of all oxygen atoms from the $CH_3C\underline{O}N\underline{O}R$ side chains of the peptide. The antibiotic has broad-spectrum activity equivalent to the penicillins and is also generally effective against penicillin and tetracycline-resistant strains [34].

Related to these effects of essential metals are the antibacterial effects of some copper compounds [55] and the wound-healing properties of zinc salts [56].

9.3. Antibacterial Activity of Metal—Amine Complexes

A further well defined series of metal complexes with antibacterial activity is that of rhodium(III) complexes of formula *trans*-$[RhX_2(py)_4]Y$ [36, 37]. The differences in activity between free ligand and complex are much more pronounced in this system than for the chelates of either 1,10-phenanthroline or 2,2′-bipyridine. The geometric requirement is also strict; the complexes $[RhCl_2(bipy)_2]^+$ are inactive, perhaps because they possess the *cis*-configuration. Again, Gram-positive microorganisms are more susceptible than Gram-negative ones but where Gram-negative bacteria, such as *E. coli*, are susceptible the inhibitory concentrations are approximately the same. Also, *M. tuberculosis* is inhibited at concentrations of 10^{-6} to 10^{-7} M. The consistent susceptibility of this microorganism to a number of transition metal systems is notable, although no mechanistic explanations have been put forward.

The most lipophilic rhodium complexes are the most bacteriostatic, this perhaps correlating with penetration. The surfactant properties of the rhodium complexes have been studied as a model for membrane perturbation [38]. The studies showed that, although membrane distortion with ion leakage occurred, there was no correlation between this leakage and the bacteriocidal concentrations needed for these complexes, and the mechanism of action was not directly related to the membrane effects. Other possible mechanisms have been mentioned [39].

The original finding of the antibacterial activity of platinum—amine complexes led, of course, to the discovery of their antitumour activity [40]. The initial electrolysis product from the platinum electrode in Rosenberg's initial experiment was the hexachloroplatinate(IV) species, $[PtCl_6]^{2-}$, present in ammoniacal solution as the NH_4 salt. Irradiation of this salt, as the pure compound, gave the 'active' species in filamentous growth induction. The 'active' species was subsequently shown to be the neutral complex $[PtCl_4(NH_3)_2]$. An interesting difference is observed, in that the parent $[PtCl_6]^{2-}$ ion is bacteriocidal at low concentrations without production of filaments, whereas the neutral amine-substituted compound is only bacteriocidal at high concentrations [41]. In general, Gram-negative bacteria are most susceptible to the induction of filamentous growth. The difference in susceptibility is not considered to be platinum uptake, but a study with radiolabelled Pt showed that, in Gram-positive bacteria, the platinum was associated with metabolic intermediates. In *E. coli*, by contrast, *cis*-$[PtCl_4(NH_3)_2]$ was associated further with nucleic acids and cytoplasmic proteins, whereas the inhibitory $PtCl_6^{2-}$ ion was only associated with cytoplasmic proteins [41]. While these results implicate DNA interaction in the filamentous growth process they also, on the other hand, imply that bacterial cell kill can occur without such an interaction. In view of the preponderance and likelihood of metal—thiol interactions, as evidenced by free Hg^{2+} and other metals, and the probable importance of these reactions in modifying the biological response to almost all heavy metal complexes, it would be of interest to know how the Pt(IV) salt is bacteriocidal. The nature of Pt—DNA interactions has been studied elegantly by examining the effects on *E. coli* mutants deficient in repair processes [42, 43].

A random survey of simple salts showed that Rh and Ru complexes, such as $RhCl_3$, $(NH_4)_2[RhCl_6]$, and K_2RuCl_6 were effective in filamentation initiation, but at concentrations an order of magnitude higher than with the platinum salts [44]. The amine complex *mer*-$[RhCl_3(NH_3)_3]$ is significantly less effective than *trans*-$[RhCl_2(py)_4]Cl$, despite reasonable antitumour activity, (see Chapter 6.2). An interesting effect of ligand substitution is seen for the almost ineffectual *mer*-$[RhCl_3(py)_3]$ [45]. At low concentrations, *trans*-$[RhCl_2(py)_4]Cl$ induces filamentous growth in *E. coli* but again, in contrast to data for the platinum—amines, this does not correlate with any antitumour activity. The comparisons between Rh and Pt complexes have been summarized in the 1974 edition of *Recent Results in Cancer Research* and the early bacterial work can be found there [46]. The ruthenium complex, *fac*-$[RuCl_3(NH_3)_3]$, is a very effective inducer of filamentous growth.

From the examples given above, it is clear that the apparent relationship between antitumour activity and filamentous growth is not general; the only strict correlation is that all antitumour active platinum complexes induce filamentous growth in *E. coli.* Certainly, all bacteriocidal or bacteriostatic complexes are not necessarily antitumour active. The interesting relationships between bacteriocidal metal complexes and the apparent greater susceptibility toward inhibition of Gram-positive bacteria by the 1,10-phenanthroline chelates and the rhodium—pyridine complexes — a property shared with penicillin, which luckily is much more active in the positive strains — do not appear to be clear at this time.

9.4. Redox Activity and Antibacterial Action

The production of activated oxygen radicals may, as for mammalian cells, result in cell death, and the antibacterial activity of the antibiotics discussed in Chapter 7 is presumably mediated in this way. The possibility of production of oxygen radicals has also been mentioned in the context of increasing the iron load in the bacterial system and the mechanism of action of oxine. Further, a role for SOD in the bacteriocidal action of phagocytes may also be argued for along the same lines as the antitumour effect (Section 6.2) [47]. The natural defence provided by phagocytes against bacteria is also oxygen-dependent and the microbicidal agent is believed to be an active oxygen radical [48].

Some of these points are illustrated by the fact that the bacteriocidal activity of streptonigrin is enhanced by increased iron [49] and oxygen [50] concentration. Streptonigrin-resistant *E. coli* had increased levels of SOD [50]. Increased susceptibility to H_2O_2, by production of $^{\cdot}OH$, has indeed been shown in *S. aureus* grown with increasing levels of iron [51].

An interesting effect of a redox couple, whereby metals in two oxidation states, e.g. Fe(II)/Fe(III) or mixed-metal systems such as Mn(II)/Fe(III), exert a considerably increased bacteriocidal effect over either component separately, has been observed [52]. Once the redox system is set up, significantly increased killing of *S. aureus* was noted [52]. In some cases, the redox balance can be altered by agents such as ascorbic acid, which potentiates the effect of Cu(II) [53], and deferriferrioxamine B [54].

9.5. Summary

Silver sulfadiazene is a characterized silver complex used as an antibacterial in the treatment of burns. The polymeric aggregation of the material renders it almost insoluble in water but allows for slow release of the active silver ion. Mercury salts, probably acting by non-specific thiol binding, have been known to be bacteriostatic for many years and were of undoubted chemotherapeutic utility in earlier centuries.

Metal chelates, particularly of the 1,10-phenanthrolines, have been studied for their antibacterial activity and in this case a metal-sequestering (iron) mechanism is indicated, analogous to naturally-occurring antibiotics. Other chelating agents such as 8-hydroxyquinoline may chelate iron and activate oxygen as a mode of action.

Discrete metal complexes with bacteriocidal activity are the platinum amines and rhodium complexes of formula *trans*-$[RhX_2(pyridine)_4]Y$ and in the latter case there is greater susceptibility to Gram-positive species. Platinum—amine complexes induce filamentous growth in bacteria and it was this original finding that eventually led to their development as anticancer agents. In contrast to the amines, simple salts such as $[PtCl_6]^{2-}$ are bacteriocidal without induction of filamentous growth.

References

1. A. Albert: *Selective Toxicity* See Chapter 6, 6th Edition. Chapman and Hall (1979).
2. E. F. Gale, E. Cundliffe, P. E. Reynolds, M. H. Richmond, and M. J. Waring: *The Molecular Basis of Antibiotic Action* See Chapter 1, 2nd Edition. Wiley (1981).
3. R. Koch: *Mitteilungen aus dem Kaiserlichen Gesundheitsamt* 234 (1881).
4. P. D'Arcy Hart: *Brit. Med. Journal* **805**, (1946).
5. A. Castiglioni: *A History of Medicine* See pp. 452—460 for an historical account. A. A. Knopf (1941).
6. W. C. Hill: *Syphilis and Local Contagious Disorders* Walton (London) 1868.
7. W. V. Farrar and A. R. Williams: *A History of Mercury* (The Chemistry of Mercury, Ed. C. A. McAuliffe), pp. 3—45. Macmillan (1977).
8. N. Grier: *Mercurials — Inorganic and Organic* (Disinfection, Sterilization and Preservation, Ed. S. S. Block), pp. 346—374. Lea and Febiger (1983).
9. N. Grier: *Silver and Its Compounds* (Disinfection, Sterilization and Preservation, Ed. S. S. Block), pp. 375—389. Lea and Febiger (1983).
10. P. Fildes: *Br. J. Exp. Pathol.* **21**, 67 (1940).
11. A. Cremieux *et al.*: *Ann. Pharm. Franc.* **37**, 101 and 165 (1979).
12. M. Schaecter and K. A. Santomassino: *J. Bacteriol.* **84**, 325 (1962).
13. J. L. Webb: *The Mercurials* Vol. II, pp. 729—1070. Academic Press (1966).
14. M. Cortat: *Zentralb. Bakteriol., Abt. Orig. B*, **166**, 517—539 (1978); *C.A.* 89: 141111m.
15. D. B. Millar: *Biochem. Biophys. Res. Comm.* **28**, 70 (1967).
16. C. A. Moyer *et al.*: *Arch. Surg.* **90**, 812 (1965).
17. C. L. Fox: *Arch. Surg.* **96**, 184 (1968).
18. A. Bult: *Pharm. Int.* **3**, 400 (1982).
19. D. A. Cook and M. F. Turner: *J. Chem. Soc.* (*Perkin Trans.*) **2**, 1021 (1975).
20. L. G. Marzilli, T. J. Kistenmacher, and M. Rossi: *J. Am. Chem. Soc.* **99**, 2797 (1977).
21. S. M. Modak and C. L. Fox: *Biochem. Pharmacol.* **3**, 2391 (1973).
22. C. L. Fox and S. M. Modak: *Antimicr. Ag. Chemother.* **5**, 582 (1974).
23. J. E. Coward, H. S. Carr, and H. S. Rosenkranz: *Antimicr. Ag. Chemother.* **3**, 621 (1973).
24. M. S. Wysor and R. E. Zollinhofer: *Pathol. Microbiol.* **39**, 434 (1973).
25. M. S. Wysor and R. E. Zollinhofer: *Chemotherapy* **17**, 188 (1972).

26. S. Yamashita, Y. Seyama, and N. Ishikawa: *Experientia* **34**, 472 (1978).
27. A. Bult: *Pharm. Weekbl.* **3**, 1 (1981).
28. A. Shulman and F. P. Dwyer: *Metal Chelates in Biological Systems* (Chelating Agents and Metal Chelates, Eds. F. P. Dwyer and D. P. Mellor), p. 383. Academic Press (1964).
29. F. P. Dwyer, I. K. Reid, A. Shulman, G. M. Laycock, and S. Dixson: *Aust. J. Exp. Biol. Med. Sci.* **47**, 203 (1969).
30. H. M. Butler, A. Hurse, E. Thursky, and A. Shulman: *Aust. J. Exp. Biol. Med. Sci.* **47**, 541 (1969).
31. H. Smit, H. van der Goot, W. T. Nauta, P. J. Pijper, S. Balt, M. W. G. de Bolster, A. H. Stouthamer, H. Verheul, and R. D. Vis: *Antimicr. Ag. Chemother.* **18**, 249 (1980).
32. J. Youatt: *Aust. J. Expl. Biol. Med. Sci.* **40**, 201 (1962).
33. J. Seydel, K. Schaper, E. Wempe, and H. Cordes: *J. Med. Chem.* **19**, 483 (1976).
34. J. B. Neilands and J. R. Valenta: *Metal Ions in Biol. Syst.* **19**, 313 (1985).
35. G. Benz, T. Schroder, J. Kurz, C. Wunsche, W. Karl, G. Steffens, J. Pfitzner, and D. Schmidt: *Angew. Chem. Suppl.* 1322 (1982).
36. R. J. Bromfield, R. H. Dainty, R. D. Gillard, and B. T. Heaton: *Nature* **223**, 735 (1969).
37. R. D. Gillard: *Kem. Kozlemenyek* **47**, 107 (1977).
38. T. R. Jack: *J. Inorg. Biochem.* **12**, 187 (1980).
39. R. D. Gillard: *Heavy Metals in Medicine* (New Trends in Bioinorganic Chemistry, Eds. R. J. P. Williams and J. R. R. F. da Silva), p. 355. Academic Press (1978).
40. B. Rosenberg, L. vanCamp, and T. Krigas: *Nature* **205**, 698 (1965).
41. B. Rosenberg, L. vanCamp, E. B. Grimley, and A. J. Thomson: *J. Biol. Chem.* **242**, 1347 (1967).
42. J. Drobnik, M. Urbankova, and A. Krekulova: *Mutation Res.* **17**, 13 (1973).
43. D. J. Beck and R. R. Brubaker: *J. Bacteriol.* **116**, 1247 (1973).
44. B. Rosenberg, E. Renshaw, L. vanCamp, J. Hartwick, and J. Drobnik: *J. Bacteriol.* **93**, 716 (1967).
45. M. J. Cleare: in Reference 46, pp. 29—34.
46. *Recent Results in Cancer Research* (Eds. T. A. Connors and J. J. Roberts, v. 48). Springer-Verlag (1974).
47. I. Fridovich: *New Eng. J. Med.* **290**, 624 (1974).
48. B. M. Babior: *Superoxide and Oxidative Killing by Phagocytes* (The Biology and Chemistry of Active Oxygen, Eds. J. V. Bannister and W. H. Bannister) pp. 190—207. Elsevier, New York (1984).
49. H. N. Yeowell and J. R. White: *Antimic. Ag. Chemother.* **22**, 961 (1982).
50. R. M. Hassan and I. Fridovich: *J. Bacteriol.* **129**, 1574 (1977).
51. J. E. Repine, R. B. Fox, and E. M. Berger: *J. Biol. Chem.* **256**, 7094 (1981).
52. A. J. Salle: *Heavy Metals other than Mercury and Silver* (Disinfection, Sterilization, and Preservation, Ed. S. S. Block) pp. 390—398. Lea and Febiger (1983).
53. D. B. Drath and M. L. Karnovsky: *Infect. Immun.* **10**, 1077 (1974).
54. B. S. van Asbeck, J. H. Marcelis, J. H. van Kats, E. Y. Jaarsma, and J. Verhoef: *Eur. J. Clin. Microbiol.* **2**, 426 (1983).
55. J. R. J. Sorenson: *Prog. Med. Chem.* **15**, 211 (1978).
56. P. A. Simkin: *Lancet* 539 (1976).

CHAPTER 10

ANTIVIRAL AND ANTIPARASITIC EFFECTS OF METAL COMPLEXES

At first glance there may not appear to be much similarity between viruses and what we commonly think of as parasites but in fact the virus may be considered to be the most highly developed parasite. They are further related in the context of this monograph by the paucity of information on the effects of coordination complexes, and indeed chemotherapeutic agents in general, on their reproduction. These studies are of importance from mechanistic viewpoints because both viruses and the pathogenic parasites responsible for tropical diseases use the host cell for reproduction. Thus, selectivity is of increasing importance, because, as with bacteria, the basis of selective toxicity with a clear distinction between invader and host is apparent, unlike the situation with cancer. This is especially true of parasitic diseases, where not only is selectivity necessary but, unlike bacteria without a nuclear membrane, there is a well-defined nucleus to enter, exactly as in human cells, before the desired effect is achieved. Specific approaches to selective toxicity may well be more easily studied in these systems which, apart from potential clinical benefit, must obviously improve our overall understanding on the pharmacological effects of metal complexes.

10.1. Antiviral Therapy

Despite the wealth of knowledge on the reproduction of viruses and their proliferation within host cells, effective antiviral therapy is still considerably less advanced than, for example, antibacterial therapy. The reason for this is, of course, the intimate relationship between virus and host cell. The viruses, as stated, are intracellular parasites and are the smallest biological structures that contain the ability to replicate, consisting of a core of nucleic acid surrounded by a protective protein shell and in some cases an outer envelope of lipid or polysaccharide. Viruses contain no ribosomes and no mitochondria and are totally dependent on the host cell for all energy requirements, and also low-molecular weight precursors. The nucleic acid may be DNA or RNA but not both, and the nucleic acid may be singly or doubly stranded. The viruses are usually classified according to the type of nucleic acid present. Further delineation is obtained by consideration of nucleic acid characteristics such as polarity (whether or not the virus contains RNA which can be directly translated to protein), infectivity and size [1]. The glossary contains details of viruses mentioned in this text.

The viral cycle is represented schematically in Figure 10.1 and involves initial adsorption, penetration, liberation of nucleic acid, followed by an eclipse phase during which the viral components are synthesized. At this point, the viral particles are being synthesized independently and no evidence of a discrete viral particle is apparent. Assembly of viral capsids begins when a critical concentration of viral proteins is obtained, and assembly continues with envelopment and ultimately release of the newly created virus. Up to a thousand new particles may be released by each cell and in a typically acute viral infection the host cell is dead within 10—48 hours after infection. Slow virus infections may remain dormant for many months or may even replicate without killing the host. Descriptions of the viral cycle are to be found in reference [1] and in excellent general articles on synthetic antiviral agnets [2] and chelating agents in antiviral chemotherapy [3, 4].

Systematic strategies for development of antiviral agents tend to emphasize selective inhibition of distinct processes in viral replication [1,2,6,7]. The intimate relationship between virus and host is obviously a

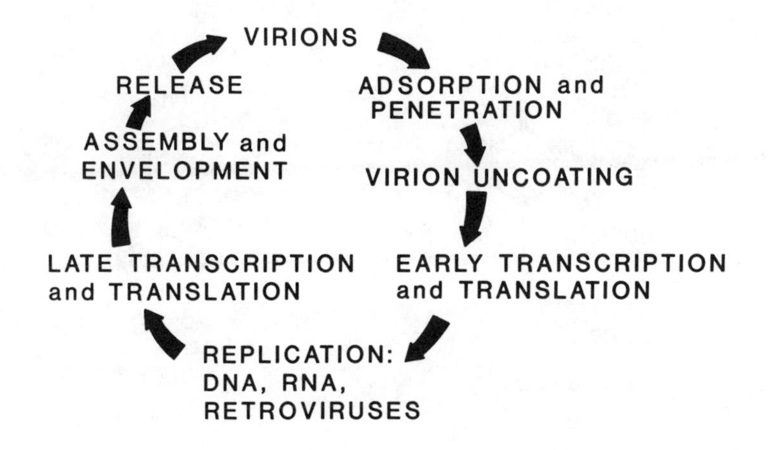

Fig. 10.1. Depiction of the major paths of the viral cycle from entry into the cell to production of new viruses.

severe challenge, since the ideal drug should be not only broad-spectrum with complete inhibition of viral replicative ability but should also be non-toxic to the host. A distinction must be made, therefore, between antiviral agents which act selectively on the virus and those which exert their observed antiviral effect by toxicity to the host cell. The structures of some known antiviral agents are shown in Figure 10.2 A large number of modified nucleosides are present which act as substrate analogues for DNA synthesis, causing inhibition of viral DNA synthesis through incorporation. This is an example of an indirect inhibition of DNA synthesis as outlined in Chapter 1. Drugs of particular interest currently are acyclovir (Compound 101) and ribavairin (Structure 102) [2].

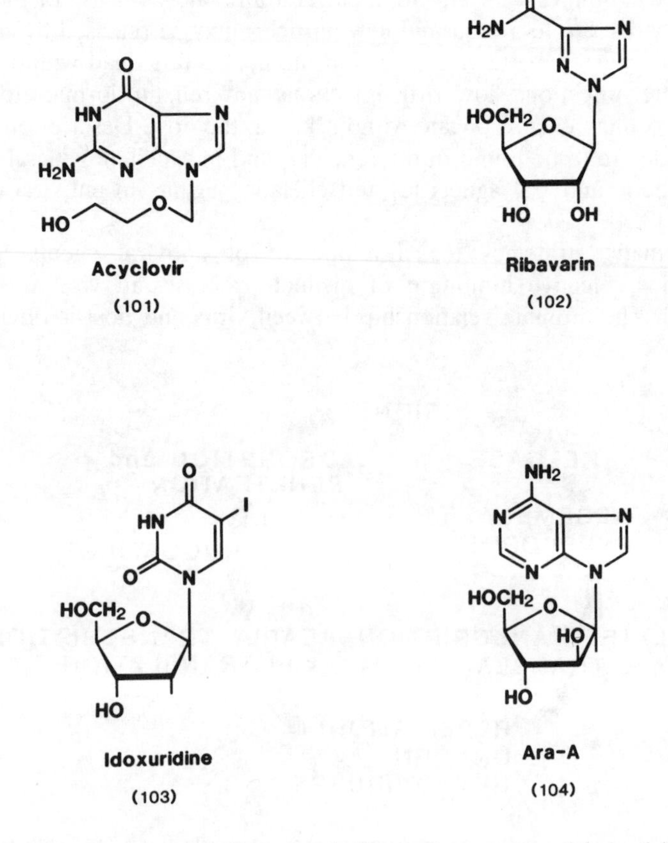

Fig. 10.2. Structure of purine and pyrimidine analogues with antiviral activity.

10.2. Chelating Agents and Metal Complexes as Antiviral Agents

10.2.1. CHELATING AGENTS

While there is a lack of detailed knowledge of the specific actions of robust metal complexes as antiviral agents, a number of chelating agents or potential chelating agents (Figure 10.3), have antiviral activity and this aspect and its relationship to metal binding have been discussed [3, 4]. The antiviral activity of one class of chelators, the thiosemicarbazones containing the clinically used methisazone (Structure 105), has been reviewed [5]. The role of metal sequestration and/or metal activation by excess metal ion is again apparent in the action of some of these agents.

Thus, the antiviral activity of the thiosemicarbazones has been correlated, in one study, with their ability to inhibit ribonucleotide reductase [8]. Methisazone inhibits the RNA-dependent DNA polymerase of Rous sarcoma virus (RSV) and inactivates its ability to transform chick-embryo cells [9], as well as inactivating herpes simplex [10] and some RNA viruses [11]. The activity against RSV is Cu-dependent [9], the Cu complex also being capable of inhibition of the polymerase [12]. Since these enzymes are zinc-dependent, interference with the role of zinc is possible. The inhibition of influenza virus-associated RNA transcriptase by

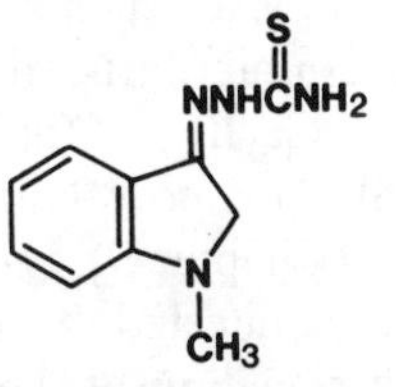

Methisazone
(105)

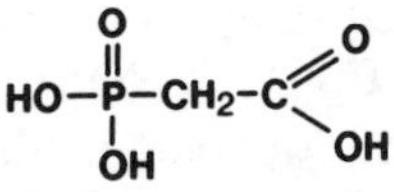

Phosphonoacetic acid
(106)

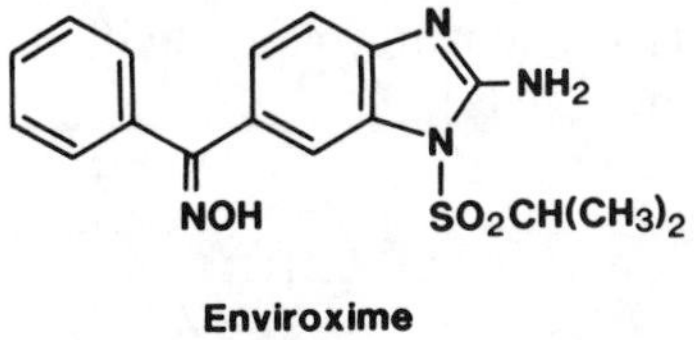

Enviroxime
(107)

Fig. 10.3. Structures of chelating agents whose antiviral activity may be modified by metal binding.

2-acetylpyridine (thiosemicarbazone) has been attributed to zinc chelation [13]. Similarly, the antiviral action of phosphonoacetic acid (Structure 106) is indicated to be inhibition of herpes virus-induced DNA polymerase and the acid, and its formic analogue, selectively inhibit replication of herpes viruses [6, 14, 15]; an analysis of the metal binding of these species indicated that they could bind zinc under simulated *in vivo* conditions and the zinc complex could be the active species [3, 16]. The specific dependence of RNA and DNA polymerases for zinc therefore represents an attractive target for selective toxicity, and in this case the differences in metal requirement between virus and host cell may be exploited, both by zinc sequestration and introduction of excess zinc (see Section 10.2.4).

10.2.2. METAL–(1,10-PHENANTHROLINE) COMPLEXES

The ubiquitous chelates of 1,10-phenanthroline were studied some time ago for their antiviral activity [17]. The inhibitory effect of various chelates on the multiplication of influenza virus (Melbourne strain) in chick chorioallantoic membrane *in vitro* was studied, and the most effective complex was shown to be the $[Ru(acac)(3,5,6,8\text{-}Me_4\text{-}1,10\text{-}phen)_2]^+$ cation, which inhibited multiplication at concentrations of 6×10^{-7} M($-\log M =$ 6.2). Other chelate complexes were active at concentrations of 10^{-5} to 10^{-7} M. A structure–activity study of metal chelates showed, however, that for a given chelate ligand, more labile complexes such as those of Cd or Cu were more active than their inert counterparts, the order for doubly-charged (2+) chelates being Cd(II) > Cu(II) > Zn(II) > Mn(II) > Fe(II) > Co(II) > Ni(II) > Ru(II) [18]. This correlation coincided with that found for some antibacterial effects (Chapter 9). Further studies showed that the virostatic activity may be manifested by either direct inactivation of the virus (possibly through dissociation of the more labile chelates) or by direct action on the host cell (for inert complexes). The latter effect is indicated by the fact that the trend in virostatic activity is similar to that in antitumour activity [19] (see also Chapter 6) and the fact that, of the various 1,10-phenanthrolines studied, the tetramethyl derivatives most easily penetrate cells [20].

10.2.3. PLATINUM– AND PALLADIUM–AMINE COMPLEXES

The demonstrated antitumour activity of cisplatin was naturally extended to studies on antiviral effects. The complex may be used as an activator in the presence of light or oxygen to inactivate Herpes Simplex, Type 1 [21]. Further studies showed that the compound had antiviral action *in vitro* against various enveloped viruses such as Herpes simplex, Type 1, and a range of influenza viruses [22]. The necessity to limit side effects and to obtain localized action for any *in vivo* utility has produced interest in topical applications, and a patent describes approaches to this end [23].

In this context the use of DMSO as carrier has been claimed to be applicable [23]. The clinically used pyrimidine analogue, iodouridine, is only sparingly water-soluble and is used in a DMSO solution for topical application. The reactivity of platinum complexes in DMSO, however, could be a potential problem and the approaches outlined in Section 3.3 may well be applicable for a pharmaceutically suitable formulation of antiviral activity. The toxic side effects of the platinum complexes, if also manifested in a localized application, would be a serious drawback in this present context since the desired selectivity is not present. Again, it is probable that the antiviral effect is manifested by killing of the host cell.

The use of complexes of cisplatin with modified nucleosides such as acyclovir has also been proposed for antiviral applications [23]. This approach is similar to that described for antitumour responses in Chapter 3. For antiviral effects, the fact that many of the antiviral nucleoside analogues are not very specific must be considered a disadvantage in this approach. Thus, although there is a distinct synergy between say, are-C, and cisplatin in antitumour effects, the therapeutic index of ara-C for a virus is almost unity and thus toxic side effects may not be overcome. The complex *cis*-$[Pt(NH_3)_2(Guo)_2]^{2+}$ has also been reported to have antiviral activity [23], and also to have some antitumour activity (Section 3.5) [24]. However, the carrier approach may be more effective in antiviral therapy than in destroying tumour cells.

A further interesting extension of the approach of using metal—nucleoside complexes is the use of oligonucleotides as carriers. This is particularly relevant in antiviral therapy because many oligonucleotides, such as poly(A), have specific antiviral effects related to interference with viral synthesis [6]. In some cases, such as poly(I)-poly(C), the oligonucleotides serve as interferon inducers [25]. Platinum-modified poly(I)-poly(C) has been studied from this aspect and there is a difference in interferon induction by the polynucleotide, depending on the type of platinum species bound, (*cis*- or *trans*-$[PtCl_2(NH_3)_2]$, $[PtCl(dien)]^+$) [26]. This may be related to the different conformational changes of the various platinum complexes on the polynucleotide [27—29]. However, the exact nature of these adducts, as described by the later authors [27—29] has been questioned and re-evaluated [30], and the biological results must be examined in this light. Chemical modification of polynucleotides can be expected to alter template—substrate reaction and, in fact, thiolated polynucleotides significantly inhibit DNA polymerases from RNA viruses [31]. The different conformational changes induced by metal complexes on polynucleotides indicate a role for metal species in probing which conformational changes coincide with interference with the viral cycle.

Some oligonucleotides do not act by interferon induction but rather by inhibition of proviral DNA synthesis, one of the most active being single stranded poly(A) [32, 33]. In view of the demonstrated activity of the poly(A)—Rh acetate complex in tumour systems [34] and its binding specificity for adenine nucleotides [35] (Section 6.2) the effect of highly specific metal complexes on this type of interaction is of considerable interest. The synthesis of complexes specific for singly stranded DNA, especially those with little antitumour effect, should be significant.

Clearly, the use of cisplatin analogues as antiviral agents will require their formulation in some manner that is noninjurious to the host and the biochemical effects of nonantitumour active complexes may present possible differentiation between viral and host cell DNA synthesis. In this respect, the reports on antiviral activity of some palladium complexes containing substituted aminopyridines, of formula *trans*-$[PdX_2(py)_2]$, is interesting because these complexes may be expected to have little antitumour activity [36]. The complex $[PdCl_2$(2,6-diaminopyridine)] has also been reported to be antiviral [37]. Of a series of palladium—amine complexes, *cis*-$[PdCl_2(NH_3)_2]$ and $(NH_4)_2PdCl_4$ showed antiviral activity not related to their antitumour activity [38]. While *in vitro* activity (antiviral or antitumour) is not unexpected, in the context of antiviral activity it would be instructive to compare strictly the platinum and palladium systems.

10.2.4. ZINC COMPLEXES

In view of the requirement for metals in all cells, the optimum concentration for a virus may be different from that of the host cell and specific inhibition of the virus may occur by excess or deficiency of metal ion [3]. Certainly, regulation of inherent biological processes in this manner could produce selective antiviral effects. In a survey with various metal ions, zinc was the only one capable of inhibition of viral replication [39]. The mechanism of action is believed to be interference with the processing of viral polypeptides by inhibition of proteolytic cleavage [6]. The zinc may bind to the polypeptide and prevent the approach of the proteolytic enzyme [40]. Copper and nickel salts, interestingly, also inhibit peptide hydrolysis [41]. Zinc ions may also act on viral DNA polymerases [42,

43], and in one case this was shown to have no effect on cellular DNA synthesis [43].

Chelating agents with high affinity for zinc can also influence viral replication as indicated in Section 10.2.1 [3, 44]. Adjustment of the zinc balance, then, may result in virucidal effects. Zinc salts such as the sulfate have been used in treatment of herpes infections, although the clinical efficacy is still in doubt [45—47].

10.2.5. OTHER METAL COMPLEXES

The demonstrated biological activity of any compound in one system may indicate the possession of general suitable pharmaceutical properties for other systems, and it is not surprising that this 'crossover' is also found in metal complexes. Besides the activity of platinum—amine complexes discussed above, antiviral activity has been demonstrated *in vitro* for $[TiCp_2Cl_2]$ in various enveloped DNA and RNA viruses [48, 49]. A further aspect of the biological effects of these metallocenes has been mentioned in Section 6.4 and this is the intriguing appearance of type A-virus particles in cells treated with $[TiCl_2Cp_2]$, the initiation presumably being activated by the metal complex [50].

Antibacterial complexes such as silver sulfadiazene and some mercurials also manifest antiviral activity [51, 52]. Mersalyl, a diuretic (Chapter 12), has some *in vivo* action when mice treated with lethal doses of coxsackie virus are then administered the mercury complex [53]. The levels needed for 100% inactivation *in vitro* by mercurials is dependent on the virus, and thiols reverse the antiviral effect [54]. Conformational changes and breakdown into subunits have also been observed after mercury treatment [51].

Silver sulfadiazene inhibited the infectivity of both types of HSV (*Herpesvirus hominis*) rapidly *in vitro* and much more effectively than did $AgNO_3$ [55]. Application *in vivo* was effective against development of herpetic keratoconjunctivitis, which leads to encephalitis [56]. A silver metachloridine complex has been claimed to have similar properties to that of the sulfadiazene [57]. Miscellaneous species whose antiviral effects have been reported include the Rh(I) organometallic complex, [Rh(acac)(COD)] [58] and Mn(II) Schiff base complexes [59].

Finally, the unusual mineral condensed ions (MCIs) certainly qualify as inorganic salts, and these polyanions formed from tungstate or molybdate (as MO_4^{2-}) and central ions such as silicon and antimony are being studied for their antiviral activity, although no distinct mode of action is apparent [60, 61].

10.3. Metal Complexes as Antiprotozoal Agents

Parasitic diseases are responsible for much of the endemic suffering in developing countries or what is sometimes called the Third World. The major diseases are malaria, leishmaniasis, schistosomiasis and trypanosomiasis. Their negative consequences in human and economic terms cannot be underestimated. Suffice it to say that malaria is responsible for over a million deaths a year, ranking it as one of the major *world* killers, and that the presence of trypanosomiasis has been instrumental in preventing the establishment of stable agricultural communities in a wide belt of sub-Saharan Africa. In view of the continued statements of the necessity for improved chemotherapeutic agents, especially for the treatment of chloroquine-resistant malaria, the application of transition metals, especially with knowledge gained in the design of antitumour agents, would be timely. The development of resistance to any drug is a major obstacle to its continued use, and the patterns of cross-resistance between established drugs and metal complexes should be of interest. In principle, selective uptake to the parasitic cell or selective interference with the parasitic life cycle is possible. Unlike the virus, these parasites are prokaryotes and thus have well defined cells, nuclei and nuclear membranes, and in this sense are the nearest models to mammalian cells.

Many of the foundations of chemotherapy were laid by Ehrlich's work on trypanosomiasis and, indeed, Salvarsan was being studied for its trypanocidal effects when the antisyphilitic properties were discovered [62]. In a similar vein, the observation of the antimalarial activity of methylene blue in 1891 can be argued to have led the quest for chemotherapeutic agents towards dyes and planar molecules. The German chemical industry in the early 20th century was therefore particularly fertile ground for new chemotherapeutic substances and perhaps from these trends came the azo-dyes and thus the sulfonamides. As an aside, the composition of such drugs as Suramin (discovered in 1917) was considered to be highly secret because of the potential advantages to the holder in tropical wars. Despite this background, the present status of chemotherapy for parasitic diseases reflects a dearth of newer and more effective reagents, partly because of the quite correct emphasis on eradication of the vectors, improved hygiene, and vaccine development, but also because of perceived lack of reward for industrial development. With respect to eradication, continued political upheavals are unlikely to generate the conditions necessary for long term success of these programs and elimination of these diseases does not seem achievable in the near future.

10.3.1. METAL COMPLEXES IN TRYPANOSOMIASIS

Perhaps the best studied parasite in terms of interaction with metal complexes is the trypanosome, responsible for sleeping sickness in man. This system is of considerable interest because there exists a remarkable coincidence in drugs which are active trypanocides and active antitumour agents [63]. This is probably attributable to a similarity between the metabolism of tumour cells and pathogenic trypanosomes caused by an increased rate of respiration which results from an inefficient or nonfunctional mitochondrial system [64]. A further parallel is the lack of catalases and peroxidases in both systems (see Section 10.3.1.3). In the case of drugs, nucleoside analogues are remarkably effective in both tumours and trypanosomes [63]. In a survey of the NCI bank of chemicals, a 'hit rate' of 22.7% of compounds with trypanocidal activity was obtained from a group of 66 antitumour compounds [64, 65]. This coincidence was higher than that found by other selection methods [65], and also extends to both cisplatin and bleomycin.

The observation that cisplatin was active *in vitro* led to an examination of various metal complexes with antitumour activity and again a good correlation was found [67], Table 10.I. The platinum complexes were most

TABLE 10.I
Activity of platinum and ruthenium complexes on *T. rhodesiense in vitro*.[a]

Complex	Loss of motility[b]	Loss of Infectivity	
		Partial	Total
cis- $[PtCl_2(NH_3)_2]$	3	5	<3
$[PtCl_2(dach)]$	3	5	3
$[Pt(DMSO)_2(dach)]^{2+}$	3	5	<3
$[RuCl(NH_3)_5]^{2+}$	<3	4	<3
$[Ru(NH_3)_6]^{3+}$	<3	3	<3
Ruthenium Red	<3	4	<3

[a] Figures are in −log M producing the effect. Thus <3 signifies $<10^{-3}$ M, which is considered a cutoff point because of host toxicity. See Ref. 67 for details.

[b] Motility refers to the motion produced by the flagellate parasites and its inhibition indicates an interference with respiratory processes.

active, with some activity being shown by ruthenium complexes. The difference in trypanocidal activity interestingly parallels that of the antitumour activity of the two metals. The activity of 'ruthenium red' (see Figure 6.1) is of interest because the activity has been discovered independently in a systematic study of inhibitors of calcium transport as trypanocides [68], and in fact had been mentioned in 1932 as active against *T. brucei*, in a study of metal complexes undertaken by Fischl and Schlossberger [69]. Indeed, this mention confirmed an even earlier observation of the activity of 'ruthenrot' by Mesnil and Nicolle in 1906, perhaps attracted by the intense color of the trinuclear cation [70].

Complexes with some antitumour but not trypanocidal activity were *cis*-$[RuCl_2(DMSO)_4]$ and rhodium acetate. Of the relatively small number of complexes studied, the observation of activity in two series indicates that the antitumour/trypanocide analogy holds for metal complexes. In this respect, it is of interest to note that *trans*-$[PtCl_2(NH_3)_2]$ is inhibitory *in vitro* to *T. brucei gambiense* at doses 10 times higher than the *cis*-isomer (40 μM vs 4 μM, respectively) [71].

10.3.2. PLATINUM—AMINE COMPLEXES AS TRYPANOCIDES

Although the platinum complexes are active *in vitro*, they are not curative as single agents *in vivo*. Concurrent administration of disulfiram or physiologic saline allows for administration of high doses, which by themselves would be prohibitively toxic, and *in vivo* curative activity can then be observed [72]. In view of the fact that other amine complexes besides those of NH_3 have activity *in vitro*, the possibility exists that less toxic platinum complexes could have useful therapeutic indices for trypanosomiasis. In a study of Pt(II)-1,2-diaminocyclohexane complexes containing carboxylates more solubilizing and less toxic than the parent chlorides a number of species, such as those containing methotrexate (Structure 34, Figure 3.3), 2,5-dihydroxy-benzenesulfonate and hexahydroxyheptanoate, showed activity *in vivo* at doses ranging from 50—200 mg/kg [73, 76]. Parallel studies on the antitumour activity of these complexes indicated that there was no strict correspondence between activity in the parasite.

The use of macromolecules as carriers to reduce host toxicity was demonstrated by Williamson and co-workers to be very effective in improving the trypanocidal activity of daunorubicin [74] and this approach has been attempted also with cisplatin [75]. The complex of polyglutamic acid with *cis*-$[Pt(NH_3)_2(H_2O)_2]^{2+}$ has been reported to have marked activity *in vivo* in *T. congolense* [75]. The activity has been attributed to lower toxicity of this complex relative to that of cisplatin, although factors such as selective uptake may also be important. Trypanosomes are more cationic on the cell surface than mammalian cells and so polyanionic species may be absorbed, and potentially transported, more readily. The polyglutamic acid complex has anionic residues and this may be a contributing factor. In this respect, it is noteworthy that the clinically used Suramin is one of the few anionic drugs which exist [63]. Since a basis for selectivity therefore exists, a larger correlative study between antitumour active platinum complexes, platinum uptake in trypanosomes and structure seems warranted.

Ultrastructural studies on platinum-treated trypanosomes showed some lesions similar to those of tumour cells [67]. Abnormal division with the presence of anucleate, multinucleate, giant, and incompletely divided forms were observed, indicating effects on nuclear DNA. Similarly, central clumping of nuclear chromatin is also indicative of DNA interaction, and it seems that the cellular effects of platinum complexes on trypanosomes may be expected to be similar to those on mammalian cells [67].

10.3.3. OTHER METAL COMPLEXES

Considering the historical role of trypanosomiasis in the development of chemotherapy and the then economic advantages of effective agents against the disease, it is not surprising that many metal complexes were tested in the pre-antibiotic era of the first thirty years of this century. These results have been summarized in the *Handbuch der Chemotherapie* of 1932 [69] and in the compilation by Steck [77].

In a survey covering both spirochetocidal and trypanocidal action of metal halides, $RhCl_6^{3-}$ and $RuCl_3$ were active trypanocides, while platinum salts ($PtCl_4^{2-}$ and $PtCl_6^{2-}$), had weak activity in both surveys [69]. Palladium and iridium salts, on the other hand, had little activity. Contradictory results on rhodium trichloride and other rhodium salts, however, were also reported [78, 79]. Gold salts, being 'fashionable', were also tested and gold 'sulfo-urea', which was reported to be the disodium salt of 4,4′-carbamidobis(2-auromercaptobenzenesulfonic acid) [80], was reported as active *in vivo* in *T. nagana*. Interestingly, a 'gold phenomenon' was reported, whereby gold salts such as sodium gold thiosulfate, by themselves inactive, were noted to enhance activity of trypanocides, especially arsenicals, when administered jointly:

> Most preparations of Au have no curative effect on trypanosome infections of mice, but when the infected mice are given the Au compounds either before or during the injection of arsenicals the activity of the latter is markedly increased. That the seat of action of the Au compounds is probably the trypanosomes is indicated by the fact that normal mice

infected with trypanosomes taken from mice treated with Au compounds also show an enhancement of arsenical effect when treated with arsenicals [81].

So, as with bacterial infections, many metal salts were indicated to have some activity on this parasitic infection. More recently, rhodium complexes of benzothiazoles, such as [Rh(2-amino-4-chlorobenzothiazole)$_4$Cl$_2$]Cl, were reported to be active *in vivo*, and generally as active as or more so than the corresponding platinum complexes [82]. The rhodium complexes also show little antitumour activity and, in view of the studies on rhodium mentioned earlier, it is tempting to speculate on whether this metal might be more preferred in the case of trypanosomiasis. It is of further interest to note that the pyridine complexes of rhodium, also of general formula $[RhL_4Cl_2]^+$, appear to be quite selective in their antibacterial action. However, the exact nature of the complexes needs to be clarified because linkage isomerism can occur in aminothiazoles of Pd [83], and also presumably with Rh. Silver sulfadiazene was found to be active *in vivo* at tolerable doses against *T. rhodesiense* [84].

10.3.4. PORPHYRINS AND METALLOPORPHYRINS AS TRYPANOCIDES. METAL SEQUESTRATION

An interesting phenomenon, and potentially exploitable, is the trypanocidal activity of certain macrocycles based on the heme moiety (Figure 10.4). There is a lack of both catalase and peroxidase in trypanosomes and this results in a susceptibility of *T. brucei* in their bloodstream forms to lysis

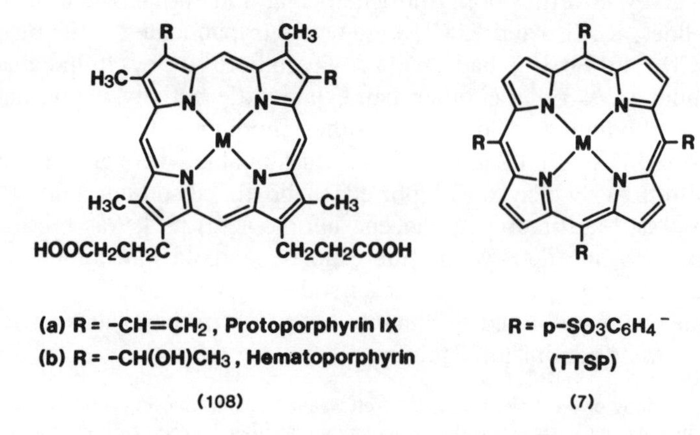

Fig. 10.4. Zinc porphyrins with trypanocidal activity.

by agents which catalyse the production of free radicals from accumulated H_2O_2 [85]. This effect was observed for heme but this was not active *in vivo* [85], although hematoporphyrin D was [86]. A study of various 2,4-disubstituted and meso-substituted porphyrins was undertaken, and the activity is summarized in Table 10.II.

TABLE 10.II
Trypanocidal activity of metalloporphyrins.[a]

M	Porphyrin[b]	LD_{50} (mg/kg)	Dose (mg/kg)	Mean Survival (days)
—	hemato	275	50	25
Cu	"	>800	200	3
Zn	"	175	35	4
—	TTSP	350	75	24
Fe	"	>600	200	10
Zn	"	250	50	28

[a] Adapted from reference 87.
[b] For structures see Figure 10.4.

Of the metalloporphyrins studied only zinc was active *in vitro* and one compound active *in vivo* [87]. There is evidence that free porphyrin is converted to the zinc derivative in the plasma of treated rats and that the zinc complex is the active agent. Similar to heme, the zinc hematoporphyrin complex is considered to induce the lysis of trypanosomes by formation of radicals from H_2O_2. The anionic nature of the sulfonated *meso*-substituted porphyrins, which are active *in vivo* but not *in vitro* and do not lyse trypanosomes, might indicate a different mechanism, similar to that with the anionic Suramin.

Because parasites are dependent on the host cell for nutrients and essential components, the arguments mentioned previously for alteration of the optimal metal concentration and its application to antibacterial and antiviral therapy are equally valid in the case of parasites. The sequestration of zinc and subsequent biochemical effects may present an example of this and the behaviour is somewhat reminiscent of the early and classic demonstration of the mode of action of 8-hydroxyquinoline (oxine). This reagent is also trypanocidal and, as with bacterial systems, cobalt protects against the lethal action [88]. The mode of action of oxine is presumably the same as for bacteria (see Chapter 9). Other chelating agents studied in

T. rhodesiense have been the thiosemicarbazones, which are active *in vivo* [89]. Other diverse chelators studied are caffeic acid and neocupreine [90].

The scope for interference with the bioinorganic chemistry of parasites is underlined by recent findings that many trypanosome species contain an iron-containing superoxide dismutase [91], and this has prompted the study of specific iron chelators. The agent N^1,N^6-bis(2,3-dihydroxybenzoyl)-1,6-diaminohexane is a potent inhibitor of FeSOD, and this and other iron chelators again could serve as very specific antiparasitic agents [92]. As more biochemical differences become apparent between parasites and host, the more readily can these differences be exploited by both organic and inorganic chemists.

10.3.5. METAL COMPLEXES WITH TRYPANOCIDAL DRUGS. DUAL-FUNCTION AGENTS

Chelating agents may serve to effect an imbalance by depletion or activation of an essential metal ion in the target cell, rather than host cell. Cytotoxic metal ions or complexes may also be directed specifically toward the target cell, limiting their toxic effect to the host, and this is clearly of importance in design of antitumour agents.

Many trypanocidal and antimalarial drugs, such as Berenil [93], ethidium [94] (now used widely as a probe of DNA binding but first developed as a trypanocide) and chloroquine [95] are believed to exert their antiparasitic action to some degree by DNA inhibition. The presence of metal binding sites in many of these molecules suggested the possibility of metal complexation and design of metal transport drugs, which would hopefully retain the intrinsic selectivity for the target (parasitic) cell and be capable of delivering the metal ion to the purported cellular target, DNA. The use of the known drug could, theoretically, circumvent the problems of host toxicity, and increase efficiency of action. Furthermore, while this principle is desirable to impose for tumour systems, the validity and factors affecting this approach might be more readily observed where the differences between target cell and host cell are more pronounced.

Initially, berenil was chosen for complexation because:

(1) The drug binds to DNA in an outer sphere manner by polar interaction of the amidine groups and the phosphate oxygens of the nucleic acid, rather than intercalation.
(2) The metal binding site is the triazene bridge, which is distinct from the DNA interacting part of the molecule.

The structures of some rhodium(II) acetate derivatives are given in Figure 10.5. The modes of binding of triazenes to metals are well delineated and the structures of the Rh—berenil complexes were assigned by analogy [96]. The *in vivo* data for one of these complexes, $[Rh_2(OAc)_4(BerH)_2]$ (BerH = neutral berenil), show that metal substitution results in an approximate two-fold increase in therapeutic index, defined as LD_{50}/CD_{50} (compare LD_{50} = 118 mg/kg and CD_{50} = 1.91 mg/kg for the berenil compound *versus* LD_{50} = 140 mg/kg, CD_{50} = 1.90 mg/kg for the free Berenil) [67]. Ultrastructural studies showed extensive damage, not attributable to the free ligand, and indicates, therefore, contribution from the metal part. Although no quantitative data are available for uptake of free rhodium acetate *vs.* the berenil complex, the metal transport is apparent.

An interesting aspect of these complexes is the relative activity of the free complexes and their berenil analogues in tumours and trypanosomes. Thus, while rhodium butyrate is very active *in vitro* in L1210 and not active *vs. T. rhodesiense*, the opposite is true for its bridged berenil analogue, $[Rh_2(OBu)_3(Ber)_a]$ (Ber_a = anion of berenil), which has only slight antitumour activity. Thus, a situation is emerging where the relative activities of two closely related species are dictated by the activity of the carrier group.

The tetrachloroplatinate(II) salt of ethidium, $(Eth)_2PtCl_4$, formed from EthBr and K_2PtCl_4 (see Section 3.1), is considerably less toxic than the simple ethidium bromide derivative, resulting in a five-fold increase in therapeutic index [67]. The curative dose for trypanosomiasis is approxi-

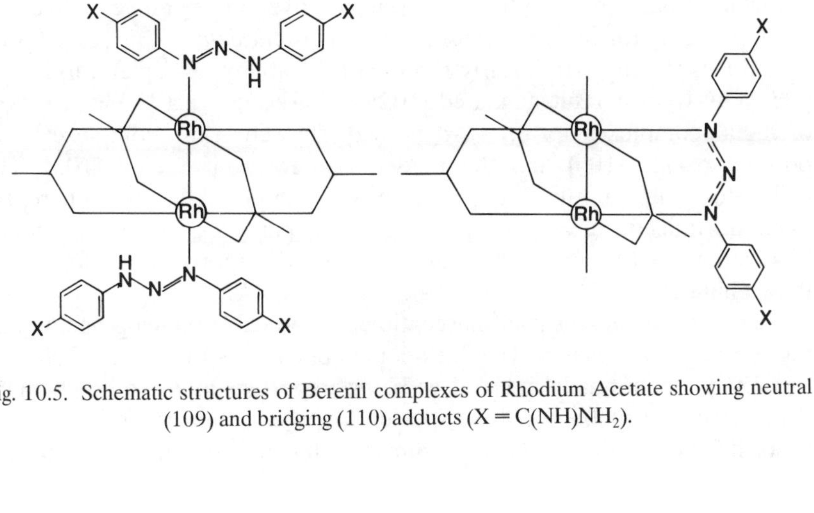

Fig. 10.5. Schematic structures of Berenil complexes of Rhodium Acetate showing neutral (109) and bridging (110) adducts ($X = C(NH)NH_2$).

mately the same, and undoubtedly due to the ethidium portion of the salt, since K_2PtCl_4 has no activity whatsoever. The explanation for this result may be that, like the $PtCl_4^{2-}$ salt of fluoresceinamine [97], the sandwich type structure may allow absorption without dissociation and alteration of the toxic profile. This unexpected finding allowed use of larger doses of this salt (in comparison to the bromide), such that in one experiment in tumour-bearing mice a *T/C* of 156% was obtained, much larger than is possible with the free intercalator [67]. These systems, first observed in the parasites, therefore warrant further study for both tumours and trypanosomiasis.

The interplay of antiparasitic drugs and metal complexes may give a 'synergistic' effect and the way activity is affected is shown by the unusual result that a chloroquine adduct of rhodium acetate, $[Rh_2(acetate)_4(chloroquine)]$, is more active *in vitro* in *T. rhodesiense* than either component [98]. This complex is also as active in malaria *in vivo* as the free chloroquine, although no enhancement over free drug is seen [98]. An interesting historical note to the metallation of trypanocidal drugs is that the platinum salt of Salvarsan was reported by Ehrlich in 1915 in one of his last papers [99].

10.3.6. METAL COMPLEXES IN AMERICAN TRYPANOSOMIASIS

The South American form of trypanosomiasis is due to the parasite *T. cruzi* and is responsible for Chagas' disease, which afflicts up to twenty million people and for which there is no known effective therapy. The status of chemotherapeutic agents has been reviewed [100]. The ethidium tetrachloroplatinate complex mentioned above has curative activity in mice, probably for the same reasons as in *T. rhodesiense* – increased dose due to less toxicity [101]. Early reports of the ability of cisplatin to inhibit infectivity were not substantiated [102] and are incorrect [103]. A survey of metal complexes for *in vitro* activity showed some complexes with positive results [104] and the mode of interaction studied [105]. The ability to inhibit motility does not, however, imply ability to stop reproduction. Indeed, a general problem with parasites is that, although bloodstream forms are more easily studied, the reproductive forms are intracellular.

T. cruzi has shown itself susceptible to radical-producing agents and these have been reviewed [106]. Rational approaches to protozoal chemotherapy are continually stressed as being important as more biochemical data become available for these species. Hypoferremic responses occur in hosts infected with *T. cruzi*, presumably due to iron utilization by the parasite, and amastigote replication was reduced by desferrioxamine, and the explanation is that depletion of Fe is more injurious to the invader, since it requires host Fe for growth [107].

10.4. Summary

Chelating agents may manifest their antiviral activity by metal sequestration and, as with bacteria and parasites, the natural differences between the host and invading cells may be exploited. Chelating agents with antiviral activity include the thiosemicarbazones, 1,10-phenanthrolines and phosphonoacetates. The latter compounds may act by zinc chelation. Metal complexes with antiviral action are platinum and palladium amines, as well as the antibacterial silver and mercury species.

Metal complexes as antiprotozoal agents are also known, including rhodium, platinum and ruthenium amines. The antiparasitic action of porphyrins is also related to metal sequestration and conversion to the zinc chelate may be the initial step in activation.

References

1. J. Vilcek and T. Sreevalsan: In *Antiviral Agents and Viral Diseases of Man* (Eds. G. J. Galasso, T. C. Merigan, and R. A. Buchanan, 2nd Ed.) p. 1. Raven Press (1984).
2. R. K. Robins: *Chem. and Eng. News* Jan. 27, p. 28 (1986).
3. D. D. Perrin and H. Stunzi: *Metal Ions in Biol. Syst.* **14**, 207 (1982); *Int. Encycl. Pharmacol. Ther.* **111**, 255 (1984).
4. W. G. Levinson: *Antibiot. Chem.* **27**, 288 (1980); D. W. Hutchinson: *Antiviral Res.* **5**, 193 (1985).
5. C. J. Pfau: *Handbook Exp. Pharmacol.* **61**, 147 (1982).
6. W. M. Shannon: In *Antiviral Agents and Viral Diseases of Man* (Eds. G. J. Galasso, T. C. Merigan, and R. A. Buchanan, 2nd Ed.) p. 55. Raven Press (1984).
7. E. De Clercq: *Biochem. J.* **205**, 1 (1982).
8. R. W. Brockman, R. W. Sidell, G. Arnett, and S. Shaddix: *Proc. Soc. Expl. Biol. Med.* **133**, 609 (1970).
9. W. Levinson, A. Faras, B. Woodson, J. Jackson, and J. M. Bishop: *Proc. Natl. Acad. Sci. USA* **70**, 164 (1973).
10. W. Levinson, V. Coleman, B. Woodson, A. Rabson, J. Lanier, J. Witcher, and C. Dawson: *Antimicr. Ag. Chemother.* **5**, 398 (1974).
11. A. Haase and W. Levinson: *Biochem. Biophys. Res. Comm.* **51**, 875 (1973).
12. L.-H. Wang and W. Levinson: *Bioinorg. Chem.* **8**, 535 (1978).
13. J. S. Oxford and D. D. Perrin: *Ann. N.Y. Acad. Sci.* **284**, 613 (1977).
14. J. Boezi: *Pharmacol. Ther.* **4**, 231 (1979).
15. B. Oberg: *Pharmacol. Ther.* **19**, 387 (1983).
16. H. Stunzi and D. D. Perrin: *J. Inorg. Biochem.* **10**, 309 (1979).
17. A. Shulman and F. P. Dwyer: In *Chelating Agents and Metal Chelates* (Eds. F. P. Dwyer and D. P. Mellor) p. 383. Academic Press (1964).

18. A. Shulman and D. O. White: *Chem.-Biol. Interactions* **6**, 407 (1973).
19. F. P. Dwyer, E. Mayhew, E. M. F. Roe, and A. Shulman: *Brit. J. Cancer* **19**, 195 (1965).
20. E. Mayhew, E. M. F. Roe, and A. Shulman: *J. Roy. Microsc. Soc.* **84**, 475 (1965).
21. J. P. Davidson, M. J. Cleare, and B. Rosenberg: *Platinum Coordination Complexes in Cancer Chemotherapy* (Eds. T. A. Connors and J. J. Roberts), p. 98. Springer-Verlag (1974).
22. E. Tonew, M. Tonew, B. Heyn, and H. P. Schroer: *Zentralbl. Bakteriol., Mikrobiol. Hyg., Abt. 1, Orig. A* **250**, 417 (1981); *C.A.* **96**: 82565g.
23. R. C. Taylor, S. G. Ward, and P. P. Schmidt: *U.S. Patent* 4, 571, 335, Feb. 18, 1986; *C.A.*
24. R. E. Cramer, P. L. Dahlstrom, M. J. Seu, T. Norton, and M. Kashiwagi: *Inorg. Chem.* **19**, 148 (1980).
25. P. F. Torrence and E. De Clercq: *Antibiot. Chemother.* **27**, 251 (1980).
26. E. De Clercq, D. Hermann, and W. Guschlbauer: *Biochim. Biophys. Acta* **741**, 358 (1983).
27. D. Hermann, G. V. Fazakerley, and W. Guschlbauer: *Biopolymers* **23**, 973 (1984).
28. G. V. Fazakerley, D. Hermann, G. E. Hawkes, and W. Guschlbauer: *Biopolymers* **23**, 961 (1984).
29. D. Hermann, G. V. Fazakerley, C. Houssier, and W. Guschlbauer: *Biopolymers* **23**, 945 (1984).
30. M. D. Reily and L. G. Marzilli: *J. Am. Chem. Soc.* **107**, 4916 (1985).
31. P. Chandra, U. Ebener, L. K. Steel, H. Laube, D. Gericke, B. Mildner, T. J. Bardos, Y. K. Ho, and A. Gotz: *Pharmacol. Ther.* **1**, 231 (1977).
32. N. Stebbing: *Pharmacol. Ther.* **6**, 291 (1979).
33. P. M. Pitha and J. Pitha: *Pharmacol. Ther.* **2**, 247 (1978).
34. J. L. Bear, R. A. Howard, and A. M. Dennis: *Curr. Chemother.* 1321 (1978).
35. N. Farrell: *J. Inorg. Biochem.* **14**, 261 (1981).
36. M. Tonew, E. Tonew, H. P. Schroer, and H. Bodo: *Zentralbl. Bakteriol., Mikrobiol., Hyg., Abt.l, Orig.A* **249**, 421 (1981); *C.A.* **96**: 352y.
37. A. B. Bikhazi, S. M. Aghazarian, and A. H. Tayim: *J. Pharm. Sci.* **66**, 1515 (1977).
38. R. D. Graham and D. R. Williams: *J. Inorg. Nuc. Chem.* **41**, 1245 (1979).
39. B. D. Korant, J. C. Kauer, and B. E. Butterworth: *Nature* **248**, 588 (1974).
40. B. E. Butterworth, R. R. Grunert, B. D. Korant, K. Lonberg-Holm, and F. H. Yin: *Arch. Virol.* **51**, 164 (1976).
41. R. H. Andreatta, H. C. Freeman, A. V. Robertson, and R. L. Sinclair: *Chem. Comm.* 203, (1967).
42. Y. J. Gordon, Y. Asher, and Y. Becker: *Antimicrob. Agents Chemother.* **8**, 377 (1975).
43. J. Shlomai, Y. Asher, Y. J. Gordon, U. Obshevsky, and Y. Becker: *Virology* **66**, 330 (1975).
44. J. S. Oxford and D. D. Perrin: *J. Gen. Virol.* **223**, 59 (1974).
45. A. de Roeth: *Amer. J. Opthalmol.* **56**, 729 (1963).
46. M. Fahim, T. Brawner, L. Millikan, M. Nickell, and D. Hall: *J. Med.* **9**, 245 (1978).
47. A. Wahba: *Acta Dermatovener.* **60**, 175 (1980).
48. P. Kopf-Maier: *J. Cancer Res. Clin. Oncol.* **97**, 31 (1980).
49. E. Tonew, M. Tonew, B. Heyn, and H.-P. Schroer: *Zentralbl. Bakteriol., Mikrobiol. Hyg., Abt.l, Orig. A* **250**, 425 (1980); *C.A.* 96: 82566h.
50. H. Kopf and P. Kopf-Maier: *Platinum, Gold and Other Metal Chemotherapeutic Agents*, Ed. S. J. Lippard. ACS Symposium Series 209, p. 315 (1983).

51. N. Grier: *Mercurials-Inorganic and Organics* (Disinfection, Sterilization, and Preservation, Ed. S. S. Block) pp. 346—374. Lea and Febiger (1983).
52. N. Grier: *Silver and its Compounds* (*ibid.*) pp. 375—389.
53. M. J. Kramer, R. Cleeland, and E. Grunberg: *Antimicrob. Agents Chemother.* **10**, 503 (1976).
54. P. Pfeiffer and B. Dorne: *Biochim. Biophys. Acta* **228**, 456 (1971).
55. T. Chang and L. Weinstein: *J. Infect. Dis.* **132**, 79 (1975).
56. T. Chang and L. Weinstein: *Antimicrob. Agents Chemother.* **8**, 677 (1975).
57. M. S. Wysor: *US Patent* 4, 384, 117, May 17, 1983 (C.A. **99**: 67465t)
58. C. Monti-Bragadin, B. Pani, M. Cantini, T. Giraldi, and G. Mestroni: *G. Ital. Chemiother.* **21**, 109 (1974).
59. N. A. Lagutkin, N. I. Mitin, M. M. Zubairov, V. V. Zelentsov, T. B. Nikolaeva: *Khim.-Farm. Zh.* **18**, 178 (1984); *C.A.* **100**: 209287b.
60. G. H. Werner, C. Jasmin, and J. C. Chermann: *J. Gen. Virol.* **31**, 59 (1976).
61. N. Larnicol, Y. Augery, C. Le Bousse-Kerdiles, V. Degiorgis, J. C. Chermann, A. Teze, and C. Jasmin: *J. Gen. Virol.* **55**, 17 (1981).
62. P. Ehrlich: *Coll. Papers of Paul Ehrlich* (Ed. F. Himmelweit v. 3) p. 240. Pergamon Press, London (1956—60).
63. J. Williamson: *Review of Chemotherapeutic and Chemoprophylactic Agents* (The African Trypanosomiases, Ed. H. W. Mulligan), p. 125. Allen and Unwin (1970).
64. H. van den Bossche: *Nature* (*Lond.*) **273**, 626 (1978).
65. K. E. Kinnamon, E. A. Steck, and D. S. Rane: *Antimicr. Agents Chemother.* **15**, 157 (1979).
66. K. E. Kinnamon, E. A. Steck, and D. S. Rane: *J. Natl. Cancer Inst.* **64**, 735 (1978).
67. N. P. Farrell, J. Williamson, and D. J. McLaren: *Biochem. Pharmacol.* **33**, 961 (1984).
68. J. Williamson and T. J. Scott-Finnigan: *Antimicr. Agents Chemother.* **13**, 735 (1978).
69. V. Fischl and H. Schlossberger: *Handbuch der Chemotherapie* v. II, pp. 836 ff. Fischers Medizinische Buchhandlung (Leipzig) (1932).
70. Nicolle and Mesnil: *Ann. Inst. Pasteur* **20**, 417 (1906).
71. A. E. Balber, S. L. Gonias, and S. V. Pizzo: *Exptl. Parasitol.* **59**, 74 (1985).
72. M. S. Wysor, L. A. Zwelling, J. E. Sanders, and M. E. Grenan: *Science* **217**, 454 (1982).
73. D. Craciunescu, C. Molina, R. M. Carvajal, A. Doadrio Lopez, M. Berger, and C. Ghirvu: *An. Real. Acad. Farm.* **52**, 45 (1986).
74. J. Williamson, T. J. Scott-Finnigan, M. A. Hardman, and J. R. Brown: *Nature* **292**, 466 (1981).
75. S. R. Meshnick, D. Brown, and G. Smith: *Antimicr. Agents Chemother.* **25**, 286 (1984).
76. A. Doadrio, D. Cracienescu, and G. Chirvu: *Inorg. Perspect. Biol. Med.* **1**, 223 (1978).
77. E. A. Steck: *Chemotherapy of Protozoan Diseases* Walter Reed (1960).
78. V. Fischl: *Z. Hyg. Infektions.* **114**, 284 (1932); *C.A.* **27**, 337 (1933).
79. H. Lagodsky: *Bull. Soc. Path. Exot.* **31**, 44 (1938).
80. V. Fischl: *Zentr. Bakt. Parasitenk. I Abt.*, **115**, 383 (1930); *C.A.* **24**, 2506 (1930).
81. W. A. Collier and M. J. Verhoog: *Z. Immunitäts* **88**, 509 (1936); *C.A.* **30**, 7691 (1936).
82. D. Craciunescu, A. Doadrio Lopez, E. Gaston de Iriarte, G. Tena, A. Gomez, R. Tena, and G. Chirvu: *An. Real. Acad. Farm.* **51**, 33 (1985).

83. N. Farrell and T. M. Gomes Carneiro: *Inorg. Chim. Acta* **126**, 137 (1987).
84. M. S. Wysor and J. P. Scovill: *Drugs Exptl. Clin. Res.* **VIII**(2), 155 (1982).
85. S. R. Meshnick, S. H. Blobstein, R. W. Grady, and A. Cerami: *J. Exp. Med.* **148**, 569 (1978).
86. S. R. Meshnick, K. P. Chang, and A. Cerami: *Biochem. Pharmacol.* **26**, 1923 (1977).
87. S. R. Meshnick, R. W. Grady, S. H. Blobstein, and A. Cerami: *J. Pharmacol. Exptl. Ther.* **207**, 1041 (1978).
88. J. Williamson: *Brit. J. Pharm. Chemother.* **14**, 443 (1959).
89. R. A. Casero Jr., D. L. Klayman, G. E. Childs, J. P. Scovill, and R. E. Desjardins: *Antimicr. Ag. Chemother.* **18**, 317 (1980).
90. A. Shapiro, H. C. Nathan, S. H. Hutner, J. Garofalo, S. D. McLaughlin, D. Rescigno, and C. J. Bacchi: *J. Protozool.* **29**, 85 (1982).
91. N. LeTrant, S. R. Meshnick, K. Kitchener, J. W. Eaton, and A. Cerami: *J. Biol. Chem.* **288**, 125 (1983).
92. S. R. Meshnick, K. R. Kitchener, and N. LeTrang: *Biochem. Pharmacol.* **34**, 3147 (1985).
93. B. A. Newton: *Antibiotics III* (Eds. D. Gottlieb and P. D. Shaw) 34, (1975).
94. M. J. Waring: *ibid.* p. 141, (1975).
95. F. E. Hahn: *ibid.* p. 58 (1975).
96. N. Farrell, M. D. Vargas, Y. A. Mascarenhas, and M. T. do P. Gambardella: *Inorg. Chem.* **26**, 1426 (1987).
97. M. J. Abrams, D. H. Picker, P. H. Fackler, C. J. L. Lock, H. E. Howard-Lock, R. Faggiani, B. A. Teicher, and R. C. Richmond: *Inorg. Chem.* **25**, 3980 (1986).
98. N. Farrell: *Chem.-Biol. Inter.* unpublished results.
99. P. Ehrlich and P. Karrer: *Arsenic-Metal Complexes* (Collected Papers of Paul Ehrlich, Ed. F. Himmelweit, v. 3) p. 587. Pergamon Press, London (1956—60).
100. Z. Brener: *Adv. Pharmacol. Chemother.* **13**, 2 (1975).
101. Z. Brener and N. Farrell: personal communication.
102. L. Filardi, W. Leon, F. S. Cruz, and J. Frausto: *Rev. Inst. Med. Trop. Sao Paulo* **20**, 248 (1978).
103. W. Leon: personal communication.
104. L. M. Ruiz-Perez, A. Osuna, S. Castanys, F. Gamarro, D. Craciunescu, and A. Doadrio: *Drug Res.* **36**, 13 (1986).
105. L. M. Ruiz-Perez, A. Osuna, M. C. Lopez, S. Castanys, F. Gamarro, D. Craciunescu, and C. Alonso: *Trop. Med. Parasit.* **38**, 45 (1987).
106. R. DoCampo and S. N. J. Moreno: *Rev. Infect. Dis.* **6**, 223 (1984).
107. V. G. Loo and R. G. Lalonde: *Infect. Immun.* **45**, 726 (1984).

CHAPTER 11

METAL COMPLEXES IN ARTHRITIS

Rheumatoid arthritis is generally described as a chronic inflammatory disease of unknown causation, manifested particularly in the peripheral joints. Approximately 1% of the population of industrialized countries suffer from this disease, although perhaps ten times that number suffer from some sort of rheumatic disorder. The pathogenesis of the disease is linked to a breakdown of the immune system, and disordered immune responses are most likely to be involved in initiation of the disease; the resultant inflammation produces tissue injury and physical debilitation [1].

Bioinorganic aspects of this disease can be considered to include: the process whereby the immunoglobulin becomes deregulated, mediation of inflammation through oxygen metabolites, altered serum levels of Cu, Zn and Fe in patients with the disease, and use of metal compounds, in particular those of gold, in treatment and alleviation of the disease. A supplement to *Agents and Actions* has contributions on many of these topics, although in a somewhat uneven manner [2].

11.1. Gold Complexes in the Treatment of Rheumatoid Arthritis

The original use of gold salts in arthritis therapy (chrysotherapy) stemmed from the opinion that arthritis might be a manifestation of tuberculosis or some related disease. The use of gold salts in tuberculosis (see Section 9.1) in the late 1920s prompted their application in arthritis by Landé [3] and Forestiér [4], and gave impetus to the area. The subsequent travails of gold treatment were only finally eased by the controlled trial of the Empire Rheumatism Council in 1960 which confirmed the utility and established the acceptability of gold treatment [5].

The major gold drugs of utility are shown in Figure 11.1. Until the recent introduction of auranofin (AF, Compound 111) as an oral drug, injection was the accepted route for all other complexes. A number of comprehensive reviews cover the clinical and pharmacological [6—8], as well as the chemical and biochemical aspects of these complexes [9—12]. The ACS Symposium Series v. 209 contains much information from some

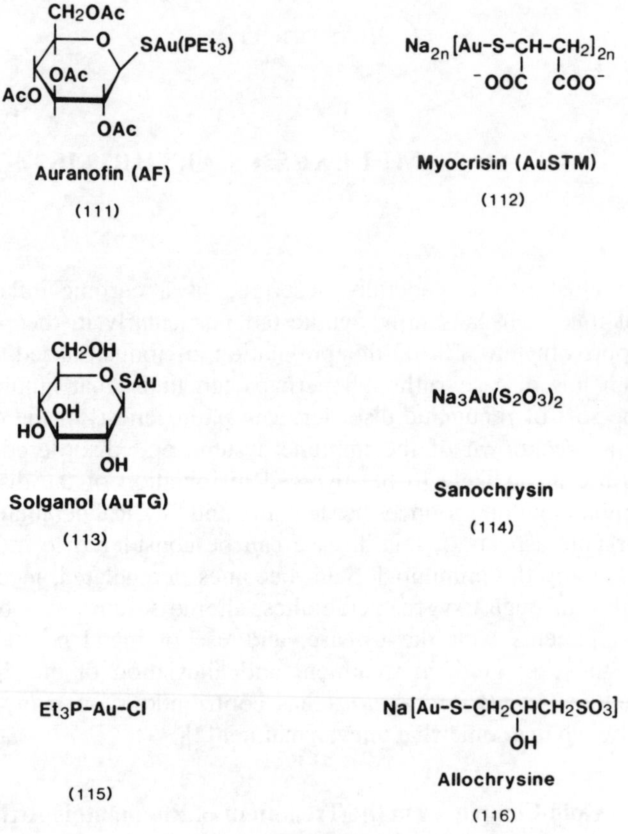

CH_2OAc … $SAu(PEt_3)$ … OAc … AcO … OAc

Auranofin (AF)

(111)

$Na_{2n}[Au\text{-}S\text{-}CH\text{-}CH_2]_{2n}$ (^-OOC, COO^-)

Myocrisin (AuSTM)

(112)

CH_2OH … SAu … OH … HO … OH

Solganol (AuTG)

(113)

$Na_3Au(S_2O_3)_2$

Sanochrysin

(114)

$Et_3P\text{-}Au\text{-}Cl$

(115)

$Na[Au\text{-}S\text{-}CH_2CHCH_2SO_3]$ (OH)

Allochrysine

(116)

Fig. 11.1. Structures of gold antiarthritic compounds.

of the early reviewers [13—16]. This summary will review briefly the clinical, structural and biochemical aspects.

11.1.1. ACTIVITY AND TOXICITY OF GOLD DRUGS

With gold treatment more than 60% of patients will obtain a beneficial effect with some remission and perhaps 25% will obtain substantial remission for over 1 year [8, 17]. Generally, treatment is continuous weekly injections of up to 50 mg (based on Au content) followed by 'maintenance therapy' at less regular intervals upon favorable response. The two major injected drugs are 112 and 113, administered as an aqueous solution and a suspension in oil, respectively. Initial clinical results on auranofin indicate similar activity to injectable gold but with fewer side-effects [18, 19].

Approximately 25% of orally-administered AF is absorbed and serum gold levels of 90—120 g/dL are obtained. Gold from all sources is widely distributed in plasma and serum, perhaps reflecting the lack of natural systems for transport and storage. Peak serum levels are obtained very quickly (within hours). The pharmacokinetics of some gold complexes have been summarized [20]. The serum half-life of complex 112 is 5.5 d [21]. A steady serum level of approximately 300 g/dL is considered desirable, although there is no correlation between serum level and efficacy [13]. Gold may persist in the body for months after treatment and a principal depository is considered to be lysosomes which, upon accumulating the metal, become electron-dense and are called aurosomes [22, 23]. The presence of gold in synovial fluid and the slow formation of synovial aurosomes has received considerable attention, because the synovium is the locus of initial arthritic damage [10, 20]. Synovial levels are approximately half those of serum [21]; equilibration between the two is considered probable. The molecular nature of aurosomes is of some interest. They have been studied by EXAFS [15] and while this indicated an oxidation state of Au(I), the exact description is not clear; metallothionen deposits have been suggested [24].

Gold is mainly excreted by renal and fecal routes: for AF less than 20% of the clearance is renal compared to over 60% for injected gold. Perhaps up to a third of patients experience lateral toxic effects, although some of these are mild. The chief toxic side effect of AF is gastrointestinal, and other reactions noted for all gold complexes are mucocutaneous, nitritoid, hematological and renal. Cessation of treatment is, however, not common.

11.1.2. STRUCTURES OF GOLD COMPLEXES

With the exception of the mixed P and S binding atoms of AF, the majority of gold complexes contain S-bound gold. The simple phosphine derivative (Compound 115, Figure 11.1) is an exception. The crystal structure of AF has confirmed its monomeric structure [13, 25] but the situation is not at all simple for the older drugs. Indeed, the renewed interest of bioinorganic chemists in these species has shown that myocrisin (Compound 112, Figure 11.1) is a mixture of compounds, which may undergo transformation under the influence of light [26, 27], and it is salutary to reflect the difficulties such a preparation would encounter these days in getting passed for clinical use. Mössbauer and EXAFS [14, 15, 28] and NMR [29] studies indicated a polymeric structure in solution for both compounds 112 and 113. Detailed analysis of the EXAFS spectra suggest two major possibilities: open hexamers of alternating Au and S atoms or a closed, ring structure, again with alternating Au and S atoms forming the nucleus [30]. The Au—S bond lengths are 2.3 Å and Au—Au distances are assigned to peaks corresponding to 3.35 Å and a further Au—Au distance of 5.8 Å. The closed structure is favored [30]:

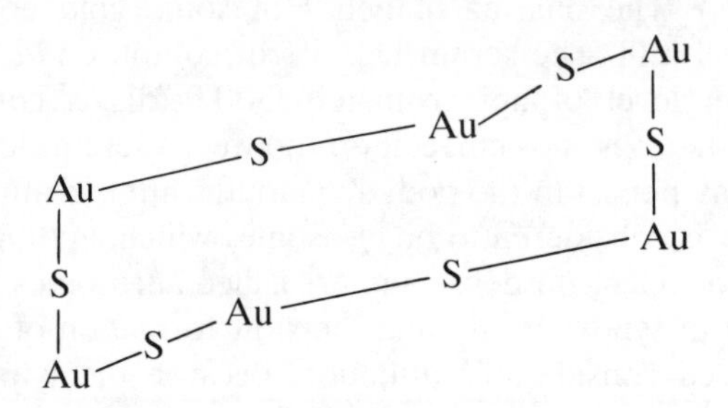

11.1.3. CHEMISTRY AND BIOCHEMISTRY OF GOLD DRUGS

The vast majority of intravascular gold is protein bound, and most of this (> 85%) is bound to albumin, with the remainder linked to globulin moieties. Gold complexes bind to many proteins with inhibition of activity, and the reactions have been extensively tabulated [10]. Studies on bovine serum albumin (BSA) confirmed, both for AuSTM and AF, that the binding site was the unique cysteine residue, Cys34 [31, 32, 33]. The PEt_3 of AF is not lost [32, 33] and the reaction is best considered as a thiol displacement:

$$Et_3PAu(TATG) + BSA \rightarrow Et_3PAu(BSA) + TATG$$

(TATG = tetraacetatothioglucopyranose)

With AuSTM, the major 'strong binding' product is indicated to have a AuS_2 binding core:

$$AuSTM \quad + \quad BSA \quad \rightarrow \quad (BSA)AuSTM$$

Weaker binding occurs, either via bridging thiomalates or noncovalent interactions and models for these have been proposed [31].

Mechanisms whereby the polymeric AuSTM, and presumably AuTG, are broken down to react with albumin in an essentially monomeric form come from studies on the reactions of these complexes with low molecular weight thiols. NMR studies showed slow exchange between AuSTM and thiomalate, *N*-acetylcysteine, mercaptoacetate and glutathione [34, 35].

Similarly, a five-fold excess of cysteine can displace thiomalate at 100 μM concentrations [36]. The bis(ligand) complexes, $Au(Cys)_2^-$ and (Cys) AuSTM have been indicated as the products of these reactions [36—38], although small clusters of the type $[Au_4(SR)_7]^{3-}$ or of stoichiometry $Au(SR)_{1.75}$ have also been proposed [34, 35]. Gold thiolate chemistry is covered in reviews [9—11], and more details on Au(I) and Au compounds are presented in ref. [39].

Model reactions on $[AuCl(PEt_3)]$ with GSH show chloride displacement but no phosphine displacement [40], and intracellular GSH binding to auranofin was also observed by NMR studies on red cells [14]. The presence of the phosphine, necessary for oral administration, may also dictate some of the biochemical differences between auranofin and the injectable gold complexes. The breakdown of auranofin eventually produces Et_3PO, a reaction interestingly mimicked by 2,3-dimercaptopropanol (or British Anti-Lewisite) in its reactions with gold—phosphine complexes [14]. The phosphine oxide has recently been detected in human serum [41], supporting the theory that, upon cellular distribution, loss of phosphine results in behavior similar to that of non-phosphine containing complexes.

There is general agreement on the immediate fate of administered gold in biological systems, although exact binding modes are still needed. The propensity for Au—thiol interactions has been shown by displacement of CN^- from $[Au(CN)_2]^-$ using thiols (indicative of its antibacterial mechanism?) [42]. The interaction of AuSTM with CN^- has also been studied [43]. These interactions are of interest because of the clear demonstration that tobacco smokers have gold distribution different to that of non-smokers [44], and that cyanide from inhaled smoke alters gold metabolism [45].

A further relevant point in Au(I)—SH chemistry is the interaction of Au(I) with D-penicillamine (D-pen, Compound 19, Figure 2.2), which is also of use in arthritis treatment (see Section 11.2.3) as well as for treatment of Wilson's disease [46]. Interestingly, the general pattern of actions of both drugs is remarkably similar and may imply a similar mode of action.

11.1.4. MECHANISM OF ACTION OF GOLD DRUGS

There is no unique mechanism of action for antiarthritic drugs, which reflects the diffuse nature of the initiation of the disease. The correction of the imbalance in the immune system, as evidenced by remission, and thus the implied modulation of the immune response, may be linked to the regulation of the RSH/RSSR balance in the cell. According to one idea, the thiol/disulfide interchange may trigger groups on immunoglobulin which are similar to those which are immunologically active [47]. In support of this, it has been shown that IgG undergoes RSH—RSSR interchange most efficiently at its isoelectric point, $\varepsilon = 7.4$ [48].

The role of gold, and other agents such as D-pen and chloroquine may be interference with the 'natural' RSH/RSSR balance by binding to and thus blocking free thiol sites. The thiol sites will not likely bind to gold equally, because of steric and electronic factors, but the widespread inhibition of protein activity and thiol interchange reactions described above point to an attainable equilibrium, with the possibility of an *in vivo* scrambling amongst available sites [9]. This could affect, for instance, membrane function responsible for transport [16] as well as protein conformation [49]. The occurrence of such interchange reactions would also strengthen the analogy between D-penicillamine (with thiol naturally affecting RSH/RSSR equilibrium and Au complexes affecting the equilibrium by free thiol removal).

Recent results on auranofin [41] expand on these general principles in what is described as a 'shuttle' mechanism. Here, entry of gold into the cell is facilitated by exchange between membrane, thiol and complex. The membrane thiol complex then exchanges with cytosolic thiol to produce intracellular gold, which may be taken out of the cell by the reverse action. The cells most damaged are those of the membrane and lysosomes. Their damage may initiate the sequence of events leading to correction of the thiol imbalance and thus immunological response.

No specific enzyme inhibition is likely to account for the mode of action of gold, and the differing drugs also display quite different abilities of inhibition [10, 50]. Some of the specific biochemical mechanisms possible have been reviewed [51—55]. These include suggestions for direct action on the immune system, action on hydrolytic enzymes, inhibition or alteration of prostaglandin synthesis [56, 57] and inhibition of phagocyte activity [58]. The deposition of gold in synovia and the inhibition of lysosomal enzymes *in vitro* [59] make an attractive proposition.

11.2. Copper in Arthritis

Of the trace elements, the role of copper has been most extensively studied. A number of reviews [60—64] and symposium proceedings [2, 65] are available.

Comparison of normal copper physiology and its physiology in rheumatoid arthritis give consistent agreement that levels of the metal are raised in the disease state with a concomitant lowering of iron and zinc levels; the findings have been extensively catalogued [60, 66]. The principal form of serum copper is ceruloplasmin, which accounts for up to 90% Cu with albumin and amino acid binding accounting for the rest. The increase in Cu is due mainly to an increase in ceruloplasmin content. This protein is responsible for Cu transfer to apoenzymes and mobilizes iron from liver to blood. Since concentrations of Cu in the liver actually fall in the disease state, with a rise in serum levels it has been suggested that the rise in Cu is a physiological response to the disease.

11.2.1. COPPER COMPLEXES IN ARTHRITIS

Early uses of copper complexes have been well reviewed [60, 63]. Interestingly, the early complexes with ligands such as those of Figure 11.2 developed in the same way as the gold complexes, having been tested for antitubercular activity [67, 68]. Forestiér compared copper and gold complexes based on studies done in the the late 1940s and early 1950s [69, 70]. Analysis of the data suggested equal efficacy for the two sets of complexes although the exact parameters for evaluation and comparison are hard to define [60, 63]. Sorenson [71] has pointed out, however, that this era coincided with the introduction of corticosteroids [72] and, in fact, until the Empire Rheumatism Council gold study [5], an exhaustive survey of metal complexes was not thought to be propitious.

An interesting survey of copper salicylate, marketed as Permalon, is to be found in the reviews [60, 63, 71]. The structures of both the salicylate and aspirinate (acetylsalicylato) analogues [72, 73] confirm the expected dinuclear cage structure as in copper acetate (Compound 65, Chapter 6.2.3.) [74]. This study was initiated following the observation that active Finnish miners of copper were unaffected by rheumatism, despite the widespread incidence of the disease in the country. Purulent infections

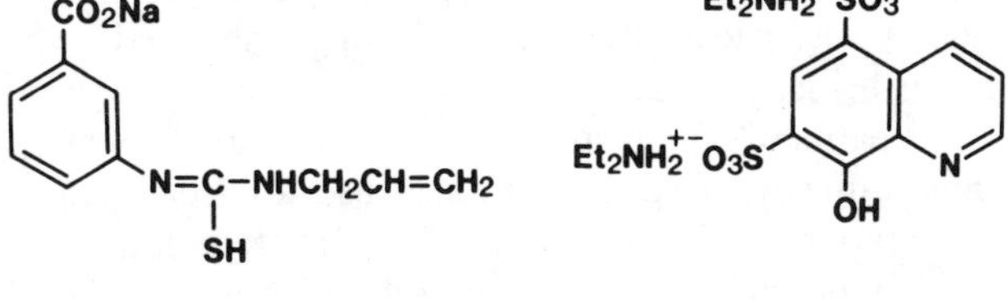

CO_2Na / $N{=}C{-}NHCH_2CH{=}CH_2$ / SH

$Et_2\overset{+}{N}H_2\,\overset{-}{S}O_3$ / $Et_2\overset{+}{N}H_2\,\overset{-}{O}_3S$ / N / OH

(117) (118)

Fig. 11.2. Ligands whose copper complexes have demonstrated antiarthritic activity.

were also absent. Both effects were attributed to the presence of copper, after the initial observations of the rise and fall of serum copper with disease and remission. The results of the use of Permalon, which terminated in 1971, indicated better effects than with aspirin alone [71]. Simple copper salts were re-evaluated by Bonta *et al.* [61] and an extensive series has been tabulated by Sorenson [64], of which the $[Cu_2(O_2CR)_4]$ species seems to be structurally the best defined.

Thirty to ninety per cent of patients (out of approximately 1500) treated with copper are considered to have shown some clinical improvement. Adverse effects include alterations of taste and smell, nausea and anemia. The continued interest in application of copper is shown by intramuscular use of SOD (Orgotein, from bovine liver) [75, 76] and also by ongoing studies on ceruloplasmin [77, 78]. The use of copper has been considered 'potential but unproven' [79].

11.2.2. COPPER BRACELETS

The interest in copper stems, to some degree, from the traditional use of copper bracelets to alleviate muscular pain. An investigation of the therapeutic value has been undertaken and copper uptake demonstrated [80, 81]. Up to 50 mg a month of the copper bracelet may be lost, although the average weight loss is 13 mg. The ^{64}Cu-labelled $[Cu(glycinato)_2]$ complex can perfuse intact cat skin [82], and this complex may be formed under aerobic conditions [83]:

$$Cu^0 + 2\,H_2NCH_2COOH + \tfrac{1}{2}\,O_2 \rightarrow [Cu(H_2NCH_2COO)_2] + H_2O$$

The feasibility of Cu oxidation by amino acids present in sweat is therefore apparent. The references should be consulted for further details and discussion.

11.2.3. COPPER AND D-PENICILLAMINE

The somewhat paradoxical situation of having enhanced copper levels in advanced arthritis, and the moderate advantages of using copper complexes in treatment, is mirrored by the apparent contradiction of D-penicillamine efficacy, despite its clinical use for treatment and excretion of excess copper in Wilson's disease [46]. There is little evidence that penicillamine copper complexes have any relevance *in vivo*. Ceruloplasmin copper is unavailable for chelation and, indeed, D-penicillamine is not particularly effective at scavenging Cu(II) from BSA [84].

A number of Cu-pen complexes are known, including both Cu(I) and

Cu(II) complexes [85], disulfide complexes [86], and an interesting mixed-valence species of stoichiometry $[Cu(I)_8Cu(II)_6(D\text{-pen})_{12}Cl]^{5-}$ [87]. A Cu-pen complex can modulate lymphocyte response [88], a mixture of the two agents can inhibit denaturation of human gamma globulin [89], and SOD-like activity has also been demonstrated [90]. The mechanism of action of D-penicillamine itself is considered to be through a direct anti-inflammatory effect [91]. The fact that O_2^- is considered to be a mediator of inflammation [92] could point to a role for SOD-like activity in reversing the autoimmune response, but much more work will be needed before a definite role for these interactions is accepted. The arguments for metal-binding of antiarthritic agents have been discussed [93].

11.2.4. MECHANISM OF ACTION OF COPPER COMPLEXES

As with all antiarthritic drugs, the situation is not clear. Biochemical effects of copper are general, and no one target, such as a particular protein, is recognizable. The copper complexes are presumably a means of further increasing the copper content, because the species are expected to be rather labile. The introduction of exogenous copper will also affect thiol content and redox state of the cell, and some biochemical responses listed above may be a consequence of this altered state. Besides ceruloplasmin and albumin, major binding sites of Cu(II) are histidine and cysteine [94, 95] and some possibilities for the mechanism of action have been summarized [64].

Since lysyl oxidase, the enzyme responsible for tissue repair, is copper-dependent, the possibility exists that lysyl oxidase activity could be induced by copper. The SOD-like activity of many active copper complexes, and the antiinflammatory effects of SOD itself, offer another possibility. Modulation of prostaglandin synthesis by Cu salts [96, 97]; stabilization of lysosomal membrane by membrane thiol oxidation [98], stabilization of globulin [89], and alteration of histaminic activity and T-lymphocyte responses have been invoked [64].

An interesting suggestion which would incorporate aspects of mechanisms for several of the disease-modifying agents such as gold, penicillamine, and copper, has been put forward by Gerber [99]. The low concentrations of serum histidine observed in arthritis [100, 101] could result in inability to synthesize a naturally occurring sulfhydryl-blocking agent, postulated to contain histidine, cystine and copper [102]. Thiol-containing proteins such as the globulins could then be more susceptible to denaturation. The amount of denatured IgG required to initiate the onset of arthritis is probably very small, and sulfhydryl reacting agents may redress this imbalance. Gold salts and the Cu—histidine—cystine complex inhibit the denaturation of human serum [99], and indeed the Cu/D-penicillamine mixture inhibits denaturation of gamma globulin [89]. Intracellular thiol binding will not be random because of steric and electronic effects, but those thiols which engage in thiol/disulfide interchange may be exposed and, thus, more susceptible to complexation. Inhibition of this exchange would then be reflected in protein protection.

11.3. Other Metals

The immunosuppressant activity of cisplatin, as with most antitumour agents which inhibit nucleic acid synthesis, suggests application in alteration of the immune system associated with arthritis, and platinum complexes have been shown to improve some signs of inflammation [103, 104]. Osmium tetroxide, as an intraarticular injection, is essentially of historical interest [105, 106].

The rationale for the use of zinc is that serum zinc levels are lowered in arthritis sufferers, zinc promotes wound healing, and that zinc is necessary for maintenance of some immune responses [107—109]. Some benefit has been observed but with no dramatic improvements, and efficacy is unconfirmed. Despite the logic of the arguments used for zinc, it has been classed as 'of unlikely benefit', along with other attempts such as acupuncture and prayer [110]. In this respect, taking heart from the traditional use of copper bracelets, we can say that the use of metals probably outdates that of prayer.

11.4. Summary

Auranofin, a gold complex containing phosphine and thioglucose as ligand, is clinically used in the treatment of rheumatoid arthritis. Other complexes with activity are also sulfur-based. The activity of gold complexes was first noted in the 1930s. There is no unique mechanism for the activity of gold complexes but their role may be in interference with the natural thiol/sulfide balance with an eventual return to normal of the malfunctioning autoimmune response. Gold—thiol binding has been demonstrated in biological systems with most gold bound to albumin.

Copper complexes also have demonstrated potential utility in treatment of arthritis. Copper salicylate (permalon) has been marketed but presently no other metal salts besides those of gold are in clinical use. The immunosuppressant activity of cisplatin is implicated in its improvement of some signs of inflammation.

References

1. *Rheumatoid Arthritis. Etiology, Diagnosis, Management* (Eds. P. D. Utsinger, N. J. Zvaifler, and G. E. Ehrlich), Lippincott (1985).
2. *Trace Elements in the Pathogenesis and Treatment of Inflammation* (Eds. K. D. Rainsford, K. Brune, and M. W. Whitehouse). Agents and Actions Supps. v. 8 Birkhauser Verlag (1981).
3. K. Landé: *Munch. Med. Wochenschr.* **74**, 1132 (1927).
4. J. Forestier: *Bull. Soc. Med. Hop. Paris* **44**, 323 (1929).
5. Empire Rheumatism Council: *Ann. Rheum. Dis.* **20**, 315 (1961).
6. N. L. Gottlieb: *Bull. Rheum. Diss.* **27**, 912 (1977).
7. N. L. Gottlieb: *J. Rheumatol.* **6**, 51 (1979).
8. P. E. Lipsky: *Disease-Modifying Drugs* (Ref. 1) pp. 601—634.
9. P. J. Sadler: *Struct. and Bonding* **29**, 171 (1976).
10. C. F. Shaw, III: *Inorg. Perspect. Biol. Med.* **2**, 287 (1979).
11. D. H. Brown and W. E. Smith: *Chem. Soc. Rev.* **9**, 217 (1980).
12. K. C. Dash and H. Schmidbaur: *Metal Ions in Biol. Syst.* **14**, 179 (1982).
13. B. M. Sutton: *Overview and Current Status of Gold-Containing Antiarthritic Drugs* (Platinum, Gold, and Other Metal Chemotherapeutic Agents, ACS Symposium Series v. 209, Ed. S. J. Lippard), p. 355. ACS, Washington (1983).
14. M. T. Razi, G. Otiko, and P. J. Sadler: *Ligand Exchange Reactions of Gold Drugs in Model Systems and in Red Cells* (*ibid.*) pp. 371—384.
15. R. C. Elder, M. K. Eidsness, M. J. Heeg, K. G. Tepperman, C. F. Shaw III, and N. Schaeffer: *Gold-Based Antiarthritic Drugs and Metabolites* (*ibid.*) pp. 385—400.
16. D. H. Brown and W. E. Smith: *Gold Thiolate Complexes In Vitro and In Vivo* (*ibid.*) pp. 401—418.
17. N. O. Rothermich and R. L. Whisler: *Rheumatoid Arthritis* pp. 190—208. Grune and Stratton (1985).
18. J. L. Abruzzo: *Ann. Int. Med.* **105**, 274 (1986).
19. R. C. Blodgett: *Am. J. Med.* **75**, 86 (1983).
20. N. L. Gottlieb and R. G. Gray: Ref. 2, pp. 529—538.
21 R. C. Gerber, H. E. Paulus, and R. Bluestone: *Arthritis Rheum.* **15**, 625 (1972).
22. F. N. Ghadially: *J. Rheumatol.* **6**, 25 (1979).
23. F. N. Ghadially: *J. Rheumatol.* **6**, 45 (1979).
24. H. Nakamura and M. I. Garashi: *Ann. Rheum. Dis.* **36**, 209 (1977).
25. D. T. Hill and B. M. Sutton: *Cryst. Struct. Comm.* **9**, 279 (1980).
26. D. A. Harvey, W. F. Kean, C. J. Lock, and D. Signal: *Lancet* **470**, (1983).
27. M. C. Grootveld and P. J. Sadler: *J. Inorg. Biochem.* **19**, 51 (1983).
28. D. T. Hill, B. M. Sutton, A. A. Isab, M. T. Razi, P. J. Sadler, J. M. Trooster, and G. H. M. Calis: *Inorg. Chem.* **22**, 2936 (1983).
29. A. A. Isab and P. J. Sadler: *J. Chem. Soc.* (*Dalton*) 1657 (1981).
30. R. C. Elder, K. Ludwig, J. N. Cooper, and M. K. Eidsness: *J. Am. Chem. Soc.* **107**, 5024 (1985).

31. C. F. Shaw III, N. A. Schaeffer, R. C. Elder, M. K. Eidsness, J. M. Trooster, and G. H. M. Calis: *J. Am. Chem. Soc.* **106**, 3511 (1984).
32. M. T. Coffer, C. F. Shaw III, M. K. Eidsness, J. W. Watkins II, and R. C. Elder: *Inorg. Chem.* **25**, 333 (1986).
33. D. J. Ecker, J. C. Hempel, B. M. Sutton, R. Kirsch, and S. T. Crooke: *Inorg. Chem.* **25**, 3139 (1986).
34. A. A. Isab and P. J. Sadler: *J. Chem. Soc. Chem. Comm.* 1051 (1976).
35. A. A. Isab and P. J. Sadler: *J. Chem. Soc. (Dalton)* 135 (1982).
36. C. F. Shaw III, G. Schmitz, H. O. Thompson, and P. J. Witkiewicz: *J. Inorg. Biochem.* **10**, 317 (1979).
37. C. J. Danpure: *Biochem. Pharmacol.* **25**, 2343 (1976).
38. C. F. Shaw, III, J. Eldridge, and M. P. Cancro: *J. Inorg. Biochem.* **14**, 267 (1981).
39. R. J. Puddephatt: *The Chemistry of Gold*, Elsevier (1978).
40. N. A. Malik, G. Otiko, and P. J. Sadler: *J. Inorg. Biochem.* **12**, 317 (1980).
41. R. M. Snyder, C. K. Mirabelli, and S. T. Crooke: *Biochem. Pharmacol.* **35**, 923 (1986).
42. G. Lewis and C. F. Shaw, III: *Inorg. Chem.* **25**, 58 (1986).
43. G. G. Graham, J. R. Bales, M. C. Grootveld, and P. J. Sadler: *J. Inorg. Biochem.* **25**, 163 (1985).
44. D. Lewis, H. A. Capell, C. J. McNeil, M. S. Iqbal, D. H. Brown, and W. E. Smith: *Ann. Rheum. Dis.* **42**, 566 (1983).
45. G. G. Graham, T. M. Haavisto, H. M. Jones, and G. D. Champion: *Biochem. Pharmacol.* **33**, 1257 (1984).
46. J. M. Walshe: *Wilson's Disease, A Review* (The Biochemistry of Copper, Eds. J. Peisach, P. Aisen, and W. E. Blumberg) Academic Press (1966).
47. E. V. Jensen: *Science* **130**, 1319 (1959).
48. D. A. Gerber: *J. Immunol.* **92**, 885 (1964).
49. A. Lorber, C. C. Chang, D. Masuoka, and I. Meacham: *Biochem. Pharmacol.* **19**, 1551 (1970).
50. P. K. Fox, A. J. Lewis, P. McKeown, and D. D. White: *Br. J. Pharmacol.* **66**, 141P (1979).
51. A. Lorber: in Ref. 2, pp. 539—571.
52. B. Vernon-Roberts: *J. Rheumatol.* **6** (Supp 5), 120 (1979).
53. D. T. Walz and D. E. Griswold: *Inflammation* **3**, 117 (1978).
54. P. Bresloff: *Adv. Drug Res.* **11**, 1 (1977).
55. J. H. Leibfarth and R. H. Persellin: *Agents and Actions* **11**, 458 (1981).
56. M. A. Bray and D. Gordon: *Br. J. Pharmacol.* **63**, 635 (1978).
57. D. S. Newcombe and Y. Ishikawa: *Prostaglandins* **12**, 849 (1976).
58. J. D. Jessup, B. Vernon-Roberts and J. Harris: *Ann. Rheum. Dis.* **32**, 294 (1973).
59. R. H. Persellin and M. Ziff: *Arthritis Rheum.* **9**, 57 (1966).
60. J. R. J. Sorenson: *Prog. Med. Chem.* **15**, 211 (1978).
61. I. L. Bonta, M. J. Parnham, J. E. Vincent, and P. C. Bragt: *Prog. Med. Chem.* **17**, 185 (1980).
62. A. J. Lewis: *Agents and Actions* **15**, 513 (1984).
63. J. R. J. Sorenson and W. Hangarter: *Inflammation* **2**, 217 (1977).
64. J. R. J. Sorenson: *Metal Ions in Biol. Syst.* **14**, 78 (1982).
65. *Inflammatory Disease and Copper* (Ed. J. R. J. Sorenson), Humana Press (NJ) (1982).
66. R. Milanino, E. Passarella, and G. P. Velo: *Copper and the Inflammation Process*

(Advances in Inflammation Research, Eds. G. Weissman, B. Samuelsson, and R. Paoletti, v. 2) p. 281. Raven Press (1979).
67. E. Fenz: *Munch. Med. Wochenschr.* **18**, 398 (1941).
68. E. Fenz: *Munch. Med. Wochenschr.* **41**, 1101 (1951).
69. J. Forestier, A. Certonicy, and F. Jacqueline: *Stanford Med. Bull.* **8**, 12 (1950).
70. J. Forestier: *Ann. Rheum. Dis.* **8**, 27 (1949); *ibid.* 132.
71. W. Hangarter: *Copper-Salicylate in Rheumatoid Arthritis and Similar Diseases* in Ref. 2, pp. 439—452. See Discussion.
72. M. Kato, H. B. Jonassen, and J. C. Fanning: *Chem. Rev.* **64**, 99 (1964).
73. L. Manjilovic-Muir: *J. Chem. Soc. Chem. Comm.* 1057 (1967).
74. J. N. van Niekerk and F. R. L. Schoening: *Acta Cryst.* **6**, 227 (1953).
75. K.-M. Goebel, U. Storck, and F. Neruath: *Lancet* **1015** (1981).
76. B. Wolf: *Therapy of Inflammatory Diseases with Superoxide Dismutase* in Ref. 2, pp. 453—467.
77. C. W. Denko: *Agents and Actions* **9**, 333 (1979).
78. M. J. Laroche, P. Chappuis, Y. Henry, and F. Rousselet: *Ceruloplasmin: Experimental Antiinflammatory Activity and Physicochemical Properties* in Ref. 2 pp. 61—74.
79. R. S. Panush and S. Longley: *Therapies of Potential but Unproven Benefit* in Ref. 1 pp. 695—709.
80. W. R. Walker and D. M. Keats: *Agents and Actions* **6**, 454 (1976).
81. W. R. Walker: *The Results of a Copper Bracelet Clinical Trial and Subsequent Studies* in Ref. 65 pp. 469—482.
82. W. R. Walker, R. R. Reeves, M. Brosnan, and G. D. Coleman: *Bioinorg. Chem.* **7**, 271 (1977).
83. S. J. Beveridge and W. R. Walker: *Aust. J. Chem.* **33**, 2331 (1981).
84. B. Sarkar, A. Sass-Kortsak, R. Clarke, S. H. Laurie, and P. Wei: *A Comparative Study of In Vitro and In Vivo Interaction of D-Penicillamine and Triethylenetetramine with Copper* (Penicillamine at 21; Its Place in Therapeutics Now, Eds. W. H. Lyle and R. L. Kleinman) Proc. Royal Soc. Med., Suppl. 3, 70, pp. 13—18 (1977).
85. A. Gergely and I. Sovago: *Metal Ions in Biol. Syst.* **9**, 77 (1979).
86. J. A. Thich, D. Mastropaolo, J. A. Potenza, and H. J. Schugar: *J. Am. Chem. Soc.* **96**, 726 (1974).
87. P. Birker and H. C. Freeman: *J. Am. Chem. Soc.* **99**, 6890 (1977).
88. P. E. Lipsky: in Ref. 2 pp. 85—102.
89. D. A. Gerber: *Biochem. Pharmacol.* **27**, 469 (1978).
90. M. Younes and U. Weser: *Biochem. Biophys. Res. Comm.* **78**, 1247 (1977).
91. P. E. Lipsky: in Ref. 1, pp. 601—634.
92. R. Zurier: in Ref. 1 pp. 106—132.
93. J. R. J. Sorenson: *J. Med. Chem.* **19**, 135 (1976).
94. P. M. May, P. W. Linder, and D. R. Williams: *J. Chem. Soc.* (*Dalton*) 588 (1977).
95. C. Furnival, P. M. May, and D. R. Williams: in Ref. 2 pp. 241—257.
96. E. Boyle, P. C. Freeman, A. C. Goudie, F. R. Magan, and M. Thomson: *J. Pharm. Pharmacol.* **28**, 865 (1976).
97. B. B. Vargaftig, Y. Tranier, and M. Chignard: *Eur. J. Pharmacol.* **33**, 19 (1975).
98. J. Chayen, L. Bitensky, R. G. Butcher, and L. W. Poulter: *Nature* **222**, 281 (1969).
99. D. A. Gerber: in Ref. 2, pp. 165—184.
100. D. A. Gerber: *J. Clin. Invest.* **55**, 1164 (1975).
101. D. A. Gerber, R. S. Pinals, and E. D. Harris Jr.: *The Relationship of a Low Serum*

Histidine Concentration to Rheumatoid Arthritis (Histidine. Metabolism, Clinical Aspects, Therapeutic Use, Eds. R. Kluthe and N. R. Katz) p. 90. Georg Thieme (1979).
102. D. A. Gerber: *Arthr. and Rheum.* **19**, 593 (1976).
103. J. R. Bowen, G. R. Gale, W. A. Gardner, and W. A. Bonner: *Agents and Actions* **4**, 108 (1974).
104. D. P. Fairlie and M. W. Whitehouse: in Ref. 2, pp. 399—434.
105. N. Mitchell, L. Carroll, and N. Shepard: *J. Bone Joint Surgery* **55B**, 814 (1973).
106. C. C. Hinckley, J. N. BeMiller, L. E. Strack, and L. D. Russell: *Osmium Carbohydrate Polymers as Potential Antiarthritic Drugs* (Platinum, Gold and Other Metal Chemotherapeutic Agents, ACS Symposium Series v. 209, Ed. S. J. Lippard), p. 421. ACS, Washington (1983).
107. C. Job, C. J. Menkes, and F. Delbarre: *Arthr. Rheum.* **23**, 1408 (1980).
108. P. A. Simkin: *Lancet* **539** (1976).
109. P. A. Simkin: in Ref. 2, pp. 587—596.
110. R. S. Panush and P. Katz: in Ref. 1, pp. 819—823.

CHAPTER 12

MISCELLANEOUS USES OF METAL COMPLEXES

The variety of clinically used inorganic agents may be gauged by the number of miscellaneous reagents included in this chapter, ranging from diuretics to vasodilators. Although mercurial diuretics are nowadays not as frequently used as previously, they are worth remembering for the specific effects that can be induced by toxic substances when selective reactivity occurs, as well as for the contributions their study have made to the techniques and principles of renal pharmacology. The vasodilation properties of sodium nitroprusside are an important aid in heart surgery and treatment. The curare-like effect of charged metal complexes, while certainly not of clinical advantage, is an effect which can be predicted from the nature of cellular membrane receptors and, in the case of ruthenium red, does have useful attributes as a probe of biochemical reactions. As with the gold salts discussed in the previous chapter, the biochemical effects of these agents may be traced back to the enzymatic level and involving disruption of the balance of essential ions such as Na^+, K^+, and Ca^{2+}.

12.1. Mercurial Diuretics

The diuretic effect of organomercurial complexes was discovered in 1919 as a side-reaction during treatment of a syphilitic patient with the then new mercurial, Novasurol. An account of this discovery has been given [1]. The observation of the same effect of mercurous chloride can be traced back to the 16th century, the effect being rediscovered in 1849 and again in 1885 [2]. A number of older reviews and symposium proceedings are available, and these still basically reflect the current understanding [3, 4, 5]. The 1968 review by Cafruny concentrates on the historical development of the now accepted theories of action, while the 1969 review of Heidenreich contains an extensive bibliography [4].

A general structure for the organomercurial diuretics is $R'—CH_2—CH(OR)—CH_2—HgX$ and some of the most representative complexes are shown below ($R = CH_3$):

	R′	X
Mersalyl ⟶	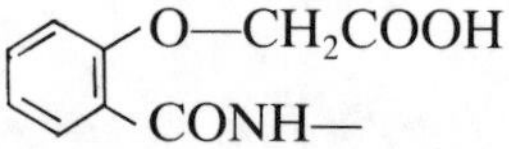 C_6H_4(—O—CH_2COOH)(—CONH—)	Theophylline
Meralluride	COOH—$(CH_2)_2$—(CONH—$)_2$	Theophylline
Chlormerodrin	H_2N—CO—NH—	Cl
Merbiurelidin	H_2N—(CONH—$)_2$	OH

Other organomercurials, and inorganic mercury salts, exert only very mild diuresis compared to that covered by the substituted propan-2-ol structure. The peptide linkage presumably aids water solubility and the use of theophylline as counteranionic group stems from observations of diuretic potentiation by the combination but, although still used, this interpretation is in doubt [6]. The R′ group is the principal determinant of activity and encompasses a wide variety of structural types from aryl to heterocyclic and amide [6]. Few studies on isomeric effects, either geometric or optical, appear to have been made despite the wide range of R′ groups.

The site of action of mercurial diuretics is predominantly intrarenal. An increased release of sodium ion is found along with water, and the conclusion is that the reabsorption of Na^+ is inhibited by binding at transport sites. The interference with sodium transport from the urine to the renal tubules, which cleanse the blood and effect recycling of the glomerular filtrate produced by the kidney, thus decreases the reabsorption of both the salt and water and results in increased efflux.

Early studies on mechanism of action of the mercury compounds had searched for reactions with thiol-containing enzymes, or in general protein-bound thiol, because of the high affinity of Hg for thiol and the demonstrated mode of antibacterial action as involving such an interaction [7]. Thiols such as dimercaprol inhibit the action. The activating process for Na^+ and K^+ transport through membranes and thus ion regulation are mediated by the sulfhydryl enzyme ATPase [8], which is located in the kidney in the peritubular region where sodium transport takes place. The structure and function of the ($Na^+ + K^+$)—ATPase has been reviewed [8a]. The *in vitro* inhibition of this enzyme by both diuretic and non-diuretic mercurials was demonstrated [9, 10] and later results correlated *in vitro* and *in vivo* data [11]. In the *in vivo* experiments subcellular fractions from rats pretreated with mercurials and subsequently sacrificed showed a diminution of ATPase activity and a concomitant inability to stimulate cytoplasmic glycolysis [11]. The reduction in the rate of glycolysis, a process which supplies energy for transport, subsequently decreases the capacity for sodium reabsorption [12].

Non-diuretic mercurials were not active in the above *in vivo* experiments, leading to the conclusion that they were not bound to the same receptor sites as the active diuretics. This raises the question of the nature of the active moiety involved in the inhibition of ATPase and the diuretic action. The fact that the nondiuretic *p*-chloromercuribenzoate (PCMB) blocks the *in vivo* action of diuretic mercurials contradicts the idea of different binding sites. The fact that organic mercurials are excreted mainly as R—Hg—cysteinate, whereas $HgCl_2$ and some organics with labile Hg—C bonds are excreted as the bis (cysteinate) complexes might suggest that the active moiety is the RHg group, rather than the secreted Hg^{2+} ion [13]. However, the demonstration that all active diuretics are acid labile, and that compounds such as PCMB have very stable Hg—C bonds, argue for mercuric ion release [14]. Further, the hypothesis of toxic ion release explains the potentiation of diuresis during acidosis, when administration of NH_4Cl or NH_4NO_3 or other effects lower urinary pH. In a similar manner, the decrease in activity at high pH is consistent with toxic ion release. The observation of free Hg^{2+} ion upon administration of chlormerodrin reinforces the hypothesis [15]. The intact organic mercurials need not be totally ineffective but the much more reactive Hg^{2+} ion, produced in a region of low pH such as the kidney, can exert a much more efficient action on thiol binding, this leading to the observed effects. Excess of mercuric ion may be diuretic because of gross renal damage but controlled liberation can reduce the nephrotoxic side effects. A final interesting point is that, even if mercuric ion or organomercuric ion is cysteine bound by endogenous amino acids, competitive binding to ATPase receptor sites must take place, either by liberation or direct replacement. The nephroprotective action of thiols for other heavy metals such as platinum, (See Chapter 2), may also, therefore, not necessarily be due to competitive binding and other factors may be involved.

12.2. Neuromuscular Effects of Metal Complexes

The neuromuscular effects of metal chelates may also be attributed to interference, in many cases, of ion transport across membranes.

The highly charged complex, ruthenium red (Structure 59, Chapter 6) is a powerful and specific inhibitor of calcium transport across mitochondrial membranes [16, 17], similar to the inhibition induced by lanthanides [18]. Significant amounts of the complex may also penetrate into nerve fibers *in vivo* and *in vitro*, and induce neurotoxicity [19, 20]. Interestingly, the toxicity is dependent on the route of administration; intracranial administration produces convulsions while intraperitoneal injection induces paralysis [21].

Calcium ions are critical in the release of neurotransmitter substances at chemical synapses, and intracellular buffering mechanisms, possibly through a Na/K—ATPase mediated exchange [22], are important in controlling the release of Ca^{2+} ions and thus release of the neurotransmitter. The blocking of calcium transport by ruthenium red increases the spontaneous release of neurotransmitter [23]. The utility of these effects as probes is shown by the fact that acetylcholine release induced by tityustoxin, a scorpion neurotoxin, was shown to be enhanced by ruthenium red; the stimulation was Ca-dependent and thus the conclusion was reached that free calcium, caused by competitive binding of ruthenium red to receptors, increased the release of acetylcholine [24].

The effects of ruthenium red are of interest because they are consistent with some earlier work by Dwyer and co-workers where they examined the neuromuscular blocking of a number of metal complexes [25]. Large complex cations, for example $[Ru(terpy)_2]$ $(ClO_4)_2$, and divalent cationic complexes derived from 1,10-phenanthroline and 2,2′-bipyridine such as $[Ru(phen)_3]I_2$ and $[Ru(bipy)_3]I_2$, are very effective in inducing paralysis, which is reversible when the complex is washed out. The complexes are approximately an order of magnitude less active than *d*-tubocurarine, the active ingredient of the South American arrow poison, curare. Synthetic organic cations such as decamethonium iodide have equivalent activity. An increase in acetylcholine, attributed to the inability of acetylcholinesterase to hydrolyze its substrate, accompanies this effect [26]. A similar mechanism to that considered for ruthenium red would also explain this result. Complexes with amines, such as the cobalt(III) ion $[Co(NH_3)_5(NO_2)]^{2+}$, are less effective than the chelates, and increasing peripheral charge on the complex increases curariform-like activity. As an aside, the preparation of the nitro complex is commonly employed in laboratory inorganic courses and the potential side-effects of such complexes should be mentioned. A further interesting observation is that optically active forms are stereospecific — the *d* forms of $[M(phen)_3]^{2+}$ (M = Ni, Ru, Os) are 1.5—2 times more effective than their corresponding *l* forms. This result is yet another example of the importance of conformation in dictating biological properties.

In view of these results, the strong neuromuscular blocking effect of cations such as $[Pt(en)_3]^{4+}$, discovered during the synthesis and testing of platinum complexes, is not really surprising. The unusual difference between the chemically very similar [Pt(en)(oxalato)] (strong neuromuscular inhibition) and [Pt(en)(malonato)] (good antitumour activity, no neuromuscular toxicity at active doses) is however unexplained. A further complex with marked neuromuscular effects is the metallointercalator

$[Pt(terpy)X]^+$ [27]. The ability of charged metal complexes to block membrane sites and to act as neurotoxins is also consistent with the findings of Hoeschele *et al.* who demonstrated high brain uptake of $[Pt(NH_3)_4]^{2+}$ (Chapter 3) [28].

12.3. Nitroprusside as Vasodilator

The hypotensive effect of the simple inorganic anion, nitroprusside, $Na_2[Fe(CN)_5NO]$ was described almost a hundred years ago but modern use begins with the report by Page *et al.* on its use to treat hypertensive patients [29]. A number of reviews treat its application and pharmacology [30—35] and a very useful summary of relevant chemistry is also available [61].

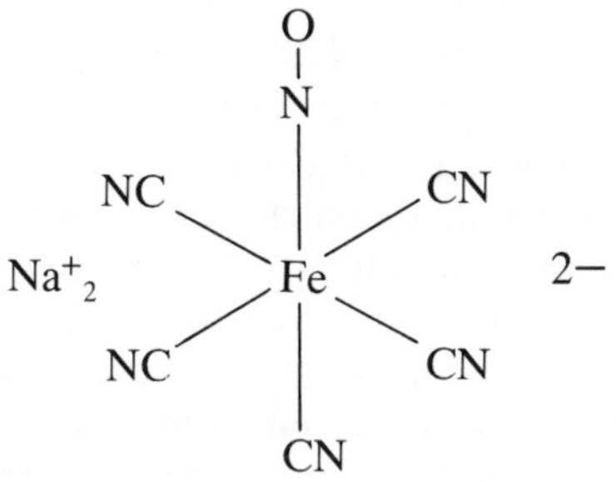

O
N
NC CN
Na^+_2 Fe 2−
NC CN
CN

(119)

12.3.1. ACTIVITY AND TOXICITY OF NITROPRUSSIDE

The primary action of sodium nitroprusside is smooth-muscle relaxation, and intravenous administration produces a very rapid but easily reversible hypotension. Thus, the blood pressure can be regulated by altering infusion rate. Upon cessation of infusion the blood pressure returns very quickly to its original value, implying a rapid breakdown of the complex. The complex is 30—1000 times more potent than simple nitrites [36], with which it is most often compared. Nitroprusside is light sensitive [37] and needs to be used quickly after dilution.

There is no known contraindication, and an upper limit of 3 mg/kg for the total dose is recommended [38]. The major source of toxicity comes from the rapid breakdown and release of cyanide ion. Studies *in vitro* show that this occurs when nitroprusside is incubated with plasma, serum, hemoglobin and whole blood [34]. There is evidence that the principal source of CN^- liberation is via reduction of the nitroprusside by hemoglobin, resulting in methemoglobin and reduced nitroprusside [39, 40]. It was thought early on that reaction with endogenous thiol was responsible for reduction. Thiols do react readily but the reaction is considered to be too slow to account for the *in vivo* cyanide release [34]. However, other reports suggested that CN^- is liberated from the initial *in vitro* photoproduct of nitroprusside, $[Fe(CN)_5(H_2O)]^{2-}$, which could imply safer use of higher doses [41, 42]. Further analysis confirmed release of CN^- both *in vivo* and *in vitro*, and CN^- toxicity was confirmed as a real complication [35, 42a]. Indeed, there have been three recorded cases of suicide by nitroprusside up to 1975 [34] and these unfortunate facts underline yet again the potential toxicity of common laboratory inorganic chemicals, facts that are becoming increasingly well disseminated through various agencies and laboratory safety programs.

The released CN^- is transformed into SCN^- by a hepatic and renal enzyme, rhodanese [43—45], this sulfuryl transferase being discovered in 1933 [46]. The enzymatic reaction proceeds slowly unless sulfur is supplied and is stimulated by thiosulfate, which is therefore a powerful antidote for CN^- poisoning. Another antidote is vitamin B_{12}, and results indicate that as plasma cyanide increases the vitamin B_{12} level decreases suggesting that the vitamin may be a cofactor of rhodanese. Vitamin B_{12} will be in the aqua (not hydroxo) form at physiological pH [47], and cyanocobalamin formation is believed to be responsible for the antidotal properties [48—50]. Side effects have also been noted, however, in this connection [43, 51], and low plasma B_{12} levels may complicate treatment. The direct interaction between nitroprusside and vitamin B_{12} has been examined by NMR and 1 : 1 and 1 : 2 adducts have been observed [47].

12.3.2. MECHANISM OF ACTION OF NITROPRUSSIDE

In bioinorganic terms, the mechanistic interest lies in the involvement, and necessity of the NO group. The biochemical explanations have evolved along with the more detailed understanding of the factors involved in muscle function. The presently held opinion on nitroprusside, and nitrites, is that activation of guanylate cyclase (cyclic GMP) [52] occurs which results in relaxation, perhaps by a chain of events eventually reflected as a lowering of Ca^{2+} concentration and resulting in muscle contraction [53].

Structure—activity relationships have received scant attention. There is general agreement on the rapid breakdown of nitroprusside *in vivo*, yet all biological data indicate a 'nitrite' (RNO_2)-type action. Nitroprusside must pass into the cell intact [54], but then the question is whether liberated

NO^+ or bound NO is responsible for the biological effects. Experiments *in vitro* showed that $[Fe(CN)_5L]^{3-}$, L = H_2O or NH_3, and the Ru and Mo analogues of nitroprusside, relaxed smooth vascular muscle, albeit to a lesser extent than the parent iron complex [31, 32]. Further substitution of NO by NO_2^- and NOS^- also gave active iron-containing complexes [55]. In this latter study, $[Mn(CN)_5NO]^{3-}$ was found to be inactive. The variation of charge, and perhaps also ν(NO), in these complexes make any evaluation difficult.

A recent fascinating mechanism has been proposed whereby compounds containing NO, released perhaps by thiols [56] and in the active form of nitrosothiols [57, 58], react with heme to give nitrosylheme, which is a potent activator of guanylate cyclase [59]. Guanylate cyclase is considered to be heme-dependent, and heme-enriched enzyme is markedly more activated than heme-deficient enzyme [59]. Activation may be similar to that shown by Fe(protoporphyrin IX) and there are similarities with respect to kinetic parameters [60]. Reference 59 details the argument.

12.4. Summary

Miscellaneous biological effects of well-defined metal complexes include nitroprusside as vasodilator, neuromuscular blocking by large metal cations, and diuretic effects of mercury complexes. Their mechanisms of action all involve complicated inhibition of enzymes.

In the case of mercury-induced diuresis interference with ATPase and subsequent imbalance of the Na/K transport system may be involved. The large cations may also affect ATPases, in this case that responsible for Ca transport. The calcium blocking could increase the release of neurotransmitters such as acetylcholine. The mechanism of action of nitroprusside may involve a chain of events eventually leading to a lowering of calcium ion concentration and muscle contraction.

References

1. A. Vogl: *American Heart Journal* **39**, 881 (1950).
2. E. Jendrassik: *Deut. Arch. Klin. Med.* **38**, 449 (1885).
3. E. Cafruny: *Pharmacol. Rev.* **20**, 89 (1968).
4. O. Heidenreich: *Quecksilberhaltige Diuretica* (Handbuch der Experimentellen Pharmakologie Band XXIV, Diuretica, Ed. H. Herken), pp. 62—194. Springer-Verlag (1969).
5. *Ann. N. Y. Acad. Sci.* (Mercury and its Compounds v. 65, Consulting Ed. C. V. King), pp. 357—652 (1957).
6. H. L. Friedman: Reference 5, pp. 461—470.

7. P. Fildes: *Brit. J. Exp. Pathol.* **21**, 67 (1940).
8. C. Walsh: *Enzyme Reaction Mechanisms* pp. 194–198. Freeman, San Francisco (1979).
8a. C. M. Grisham: *Adv. Inorg. Biochem.* **1**, 193 (1979).
9. C. B. Taylor: *Biochem. Pharmacol.* **12**, 539 (1963).
10. E. J. Landon and R. L. Norris: *Biochim. Biophys. Acta* **71**, 266 (1963).
11. V. D. Jones, G. Lockett, and E. J. Landon: *J. Pharmacol. Exp. Ther.* **147**, 23 (1965).
12. V. D. Jones, J. L. Norris, and E. J. Landon: *Biochim. Biophys. Acta* **71**, 277 (1963).
13. R. H. Kessler, R. Lozano, and R. F. Pitts: *J. Clin. Invest.* **36**, 656 (1957).
14. I. M. Weiner, R. I. Levy, and G. H. Mudge: *J. Pharmacol. Exp. Ther.* **138**, 96 (1962).
15. T. W. Clarkson, A. Rothstein, and R. Sutherland: *Brit. J. Pharmacol. Chemother.* **24**, 1 (1965).
16. C. L. Moore: *Biochem. Biophys. Res. Comm.* **42**, 298 (1971).
17. C. S. Rossi, F. D. Vasington, and E. Carafoli: *Biochem. Biophys. Res. Comm.* **50**, 846 (1973).
18. K. C. Reed and F. L. Bygrave: *Biochem. J.* **140**, 143 (1974).
19. M. Singer, N. Krishman, and D. A. Fyfe: *Anat. Rec.* **173**, 375 (1972).
20. W. Bondareff: *J. Neurosurg.* **32**, 145 (1970).
21. R. Tapia, G. Meza-Ruiz, L. Duran, and R. R. Druckercolin: *Brain Res.* **116**, 101 (1976).
22. E. M. Stephens and C. M. Grisham: *Adv. Inorg. Biochem.* **4**, 263 (1982).
23. R. Rahamimoff and E. Alnaes: *Proc. Natl. Acad. Sci. U.S.A.* **70**, 3613 (1973).
24. M. V. Gomez and N. Farrell: *Neuropharmacology* **24**, 1103 (1985).
25. F. P. Dwyer, E. C. Gyarfas, R. D. Wright, and A. Shulman: *Nature* **179**, 425 (1957).
26. A. Shulman and F. P. Dwyer: *Metal Chelates in Biological Systems* (Chelating Agents and Metal Chelates Eds. F. P. Dwyer and D. P. Mellor), pp. 400–404. Academic Press, New York (1964).
27. L. S. Hollis: personal communication.
28. J. D. Hoeschele, T. A. Butler, and J. A. Roberts: *Correlations of Physico-Chemical and Biological Properties with In Vivo Biodistribution Data for Platinum-195m-Labelled Chloroammineplatinum(II) Complexes* (Inorganic Chemistry in Biology and Medicine, ACS Symposium Series v. 140, Ed. A. E. Martell) p. 181. ACS, Washington (1980).
29. I. H. Page, A. C. Corcoran, H. P. Dustan, and T. Koppanyi: *Circulation* **11**, 188 (1955).
30. O. R. Leeuwenkamp, W. P. Bennekom, E. J. Van der Mark, and A. Bult: *Pharm. Weekbl.* **24**, 129 (1984).
31. V. A. W. Kreye and F. Gross: *Antihypertensive Drugs* (Ed. F. Gross) *Handbook of Pharmacology* **39**, 418 (1977).
32. V. A. W. Kreye: *TIPS Reviews* **384**, (1980).
33. I. R. Verner: *Postgraduate Medical Journal* **50**, 576 (1974).
34. J. H. Tinker and J. D. Michenfelder: *Anaesthesiology* **45**, 340 (1976).
35. V. Schulz: *Clin. Pharmacokinetics* **9**, 239 (1984).
36. C. C. Johnson: *Arch. Intern. Pharmacodyn. Ther.* **35**, 489 (1929).
37. J. H. Swinehart: *Coord. Chem. Rev.* **2**, 385 (1967).
38. D. W. Davies, L. Greiss, and D. J. Seward: *Can. Anaesth. Soc. J.* **22**, 553 (1975).
39. R. P. Smith and H. Kruszyna: *J. Pharmacol. Exp. Ther.* **191**, 557 (1974).
40. R. P. Smith and H. Kruszyna: *N. Engl. J. Med.* **292**, 1081 (1975).
41. W. I. K. Bisset, A. R. Butler, C. Glidewell, and J. R. Reglinski: *Br. J. Anaesth.* **53**, 1015 (1981).

42. W. I. K. Bisset, M. G. Burdon, A. R. Butler, C. Glidewell, and J. R. Reglinski: *J. Chem. Res. Miniprint* 3501 (1981).
42a. W. P. Arnold, D. E. Longnecker, and R. M. Epstein: *Anaesthesiology* **61**, 254 (1984).
43. C. J. Vesey, P. V. Cole, J. C. Linnell *et al. Brit. Med. J.* **2**, 140 (1974).
44. R. Mintel and J. Westley: *J. Biol. Chem.* **241**, 3381 (1966).
45. R. P. Smith: *Proc. Soc. Exp. Biol. Med.* **142**, 1041 (1973).
46. K. Lang: *Biochemische Z.* **259**, 243 (1933).
47. A. R. Butler, C. Glidewell, A. S. McIntosh, D. Reed, and I. H. Sadler: *Inorg. Chem.* **25**, 970 (1986).
48. M. A. Posner, R. E. Tobey, and H. McElroy: *Anaesthesiology* **44**, 157 (1976).
49. J. Wilson and D. M. Matthews: *Clin. Sci.* **31**, 1 (1966).
50. F. Lutier, P. Dusoleil, J. DeMontgros: *Arch. Mal. Prof.* **32**, 683 (1972).
51. J. R. Krapez, C. J. Veey, L. Adams, and P. Cole: *Br. J. Anaesth.* **53**, 793 (1981).
52. N. D. Goldberg and M. K. Haddox: *Ann. Rev. Biochem.* **46**, 823 (1977).
53. L. J. Ignarro, H. Lippton, J. C. Edwards, W. H. Baricos, A. L. Hyman, P. J. Kadowitz, and C. A. Gruetter: *J. Pharmacol. Exp. Ther.* **218**, 739 (1981).
54. C. A. Greutter, B. K. Barry, D. B. McNamara, D. Y. Greutter, P. J. Kadowitz, and L. J. Ignarro: *Cyclic Nucleotide Res.* **5**, 211 (1979).
55. F. Marckwardt, E. Glusa, J. Sturzebecher, W. Jehn, and B. Kaiser: *Acta Biol. Med. Germ.* **37**, 469 (1978).
56. L. J. Ignarro and C. A. Greutter: *Biochim. Biophys. Acta* **631**, 221 (1980).
57. L. J. Ignarro, J. C. Edwards, D. Y. Greutter, B. K. Barry, and C. A. Greutter: *FEBS Lett.* **110**, 275 (1980).
58. L. J. Ignarro, B. K. Barry, D. Y. Greutter, J. C. Edwards, E. H. Ohlstein, C. A. Greutter, and W. H. Barricos: *Biochem. Biophys. Res. Comm.* **94**, 93 (1980).
59. L. J. Ignarro, C. A. Greutter, A. L. Hyman, and P. J. Kadowitz: *Molecular Mechanisms of Vasodilation* (Vasodilator Mechanisms Eds. P. M. Vanhoutte and S. F. Vatner) pp. 259—288. Karger (1983).
60. M. S. Wolin, K. S. Wood, and L. J. Ignarro: *J. Biol. Chem.* **257**, 13312 (1982).
61. A. R. Butler and C. Glidewell: *Chem. Soc. Rev.*: **16**, 361 (1987).

NAMES AND ABBREVIATIONS

AA	atomic absorption
acac	acetylacetonate
acyclovir	9-(2-hydroxyethoxymethyl)guanine
ADR	adriamycin
AF	auranofin
AgSD	silver sulfadiazene
allochrysine	gold sodium thiopropanolsulfonate
am	any neutral amine
amch	2-aminomethylcyclohexylamine
AMP	adenosine-5′-monophosphate
ANT	2-amino-5-nitrothiazole
Ara-A	9-β-D-arabinofuranosyladenine
Ara-C	9-β-D-arabinofuranosylcytosine
auranofin	1-thio-β-D-glucopyranose-2,3,4,5-tetraacetato-*S*-(triethylphosphine)gold
azomycin	2-nitroimidazole
benznidazole	*N*-benzyl-(2-nitro-1-imidazolyl)acetamide
berenil	4,4′-diazoaminodibenzamidinebis (*N*-acetylglycinate)
bipy	2,2′-bipyridine
BLM	bleomycin
BSA	bovine serum albumin
CBDCA	diammine (1,1-cyclobutanedicarboxylato)platinum(II)
CHIP	*cis*-dichloro-*trans*-dihydroxy-*cis*-bis(isopropylamine)platinum(IV)
Cisplatin	*cis*-diamminedichloroplatinum(II)
COD	1,5-cyclooctadiene
Cp	cyclopentadienyl
cys	cysteine
dach	1,2-diaminocyclohexane
damch	1,1-diaminomethylcyclohexane

DDTC	diethyldithiocarbamate
dien	diethylenetriamine
DIP	4,7-diphenyl(1,10-phenanthroline)
diphos	1,2-bis(diphenylphosphino)ethane
DMSO	dimethylsulfoxide
DTT	dithothreitol
EDTA	ethylenediamine tetraacetic acid
en	ethylenediamine
enviroxime	2-amino-1-(isopropylsulfonyl)-6-benzimidazolephenylketoneoxime
ethidium	homidium bromide (2,7-diamino-10-ethyl-9-phenylphenanthridium bromide)
EtNC	ethylisocyanide
EXAFS	Extended X-ray absorption fine structure spectroscopy
Furosemide	5-(Aminosulfonyl)-4-chloro-2[2-(furanylmethyl)amino]benzoic acid
GSH	glutathione
Gy	Grays (a unit of radiation)
HET	hydroxyethanethiol
Hermophenyl	mercuric sodium *p*-phenolsulfonate
H_2KTS	3-ethoxy-2-oxobutyraldehydebis(thiosemicarbazone)
HMB	*p*-hydroxymercuribenzoate
hth	head-to-head
htt	head-to-tail
HU	monoanion of uracil
idoxuridine	5-iodo-2′-deoxyuridine
i.p.	intraperitoneal
isoniazid	hydrazide of isonicotinic acid (pyridine-4-carboxylic acid)
lapachol	2-hydroxy-3-isoprenyl-1,4-naphthoquinone
lawsone	2-hydroxy-1,4-naphthoquinone
LET	linear energy transfer
lipoic acid	thioctic acid (DL-1,2-dithiolane-3-pentanoic acid)
mal	malonato

mercurichrome	disodium 2,7-dibromo-4-hydroxymercurifluorescein
merthiolate	Sodium ethylmercurithiosalicylate
methisazone	1-methylisatin-3-thiosemicarbazone
methotrexate	4-amino-10-methylfolic acid
metronidazole	1-(2-hydroxyethyl)-2-methyl-5-nitroimidazole
1-Me-Cyt	1-Methylcytosine
1-MeHyd	1-Methylhydantoin
1-MeT	anion of 1-MethylThymine
1-MeU	anion of 1-MethylUracil
MIC	minimum inhibitory concentration
misonidazole	1-(2-nitro-1-imidazolyl)-3-methoxypropanol
myochrisin	gold sodium thiomalate
NCI	National Cancer Institute
N,N-Me_2-en	*N,N*-dimethylethylenediamine
N,N-Me_4-pn	*N,N*-tetramethyl-1,3-propanediamine
4-*N*-Mepy	4-(*N*-methylpyridyl)$^+$
neocupreine	(2,9-dimethyl-1,10-phenanthroline)copper(II)
NO_2Im	any substituted nitroimidazole
orgotein	Commercial form of superoxide dismutase in treatment of arthritis
ox	oxalato
oxine	8-hydroxyquinoline
PALA	*N*-(phosphonoacetyl)-L-aspartic acid
PCMB	*p*-chloromercuribenzoate
permalon	bis(salicylato)copper(II)
D-pen	D-penicillamine (D-2-amino-3-mercapto-3-methylbutanoic acid)
phen	1,10-phenanthroline
phthiocol	2-hydroxy-3-methyl-1,4-naphthoquinone
PMA	phenylmercuriacetate
PR_3	a tertiary phosphine
proflavine	3,6-diaminoacridinium hydrochloride
py	pyridine
α-pyrH	neutral α-pyridone
α-pyr	anionic α-pyridone
α-pyrl	α-pyrollidone
Ribavarin	1-β-D-ribofuranosyl-1,2,4-triazole-3-carboxamide

R_2SO	any substituted sulfoxide
RSU 1069	1-(2-nitro-1-imidazolyl)-3-(1-aziridinyl)-2-propanol
sanochrysin	gold sodium thiosulfate
SOD	superoxide dismutase
solganol	gold thioglucose
SR 2508	1-{*N*-(2-hydroxyethyl)-acetamido}-2-nitroimidazole
STM	thiomalate, sodium salt
sulfiram	Bis(diethylthiocarbamoyl)sulfide
TATG	2,3,4,5-tetraacetato-thio-β-D-glucopyranose
terpy	2,2:6,2″-terpyridine
thiacetazone	4′-formylacetanilidethiosemicarbazone
TM	thiomalate
TMP	3,4,7,8-tetramethylphenanthroline
TU	thiourea
WR2721	*S*-2-(3-aminopropylamino)ethylphosphorothioic acid

GLOSSARY OF TERMS AND DEFINITIONS

Acidosis A process which results in an increase in hydrogen ion concentration in body fluids. Also called acidemia.
Acute Refers to a disease or symptoms which last a short period of time. Opposite of chronic.
Adenoviruses A family of doubly stranded DNA viruses causing keratoconjunctivitis and pneumonia.
Anorexia Loss of appetite
Arthritis Inflammation of the joints
Ascites Accumulation of free serous fluid in the abdominal cavity.
Aurosome Gold containing bodies accumulating in synovial membrane during chrysotherapy.

Bacteriocide A substance which kills bacteria.
Bacteriostat A substance which arrests bacterial growth without necessarily killing them.
Brachytherapy Treatment with radiation where the source of radiation is near to or implanted in the tissue being irradiated.

Carcinoma Malignant tumour spreading to surrounding tissue and also to other parts of the body by metastasis.
Chorioallantoic Refers to an extraembryonic membrane structure of vertebrates responsible for respiratory exchange and which in many mammals gives rise to the placenta.
Chronic Refers to a disease of slow progress which persists over a long period of time. Opposite of acute.
Chrysotherapy Therapeutic treatment with gold salts.
Coxsackie virus One of a group of Picornviridae viruses which exert pathological effects on the brain, heart, muscle and epithelial tract.

Diuretic A substance which promotes the excretion of urine.

Escherichia coli (*E. coli*) A motile species of Escherichia, a genus of Gram-negative bacteria, normally present in human intestines.

Emesis Vomiting
Encephalitis Inflammation of the brain.

Glycolysis The process of producing energy by breakdown of sugar into lactic acid, especially in muscles.
Gray (Gy) The SI unit of absorbed radiation dose equal to one joule per kilogram. Equal to 100 rads.

Herpes simplex An acute localized eruption of the skin and mucous membrane caused by viruses of Herpesviridae.
Herpesviridae The family of viruses of which the herpes simplex group is one subfamily. They contain doubly-stranded DNA.
Herpesvirus hominis (HSV 2) Genital herpes.
Herpetic keratoconjuctivitis Inflammation of the cornea caused by herpes virus. The epidemic form of keratoconjunctivitis is usually caused by adenoviruses.
Hyperpyrexia Extremely high fever.
Hypotension Low blood pressure.

Intracranial In the skull.
Intraperitoneal In the peritoneal cavity.

Leukemia Disease reflected by abnormal white blood cells in bone marrow.
Leukopenia An abnormal decrease in the number of white blood corpuscles.
Lysogeny The process whereby viral DNA is incorporated into the bacterial genome, resulting in transmission of the virus upon reproduction.

Melanin Dark pigment of skin and hair.
Melanoma A tumour of melanin-containing cells.
Methemeoglobinemia An increase in blood concentration of methemeogbolin.
Mucocutaneous Referring to mucous membrane and skin, and especially their line of meeting such as in nasal, oral, etc. orifices.
Metastasis The transfer of a disease from its primary point of origin to a distant location.
Mucositis An inflammation of the mucous membrane.
Murine Refers to animals of family Muridae, especially rats and mice.
Mutagenesis The formation of a genetic mutation. In a frame shift mutation nucleotides are inserted or deleted resulting in alteration of

sequence. In a base-pair mutation the base is substituted by another base.

Mycobacterium tuberculosis (*M. tuberculosis*) The infective agent of tuberculosis, causing formation of tubercles in affected tissues, especially lungs.

Myelosuppression Inhibition of production of bone marrow.

Nephrotoxicity Toxicity to kidney cells.

Neuropathy Disease of the nervous system.

Opthalmia neonatorum Infantile blindness caused by acute conjunctivitis of the newborn acquired from a mother with gonorrhea.

Ototoxicity Toxicity to the ear, especially its neural components.

Pathogenesis Origin and development of disease.

Phagocyte A cell that ingests foreign particles.

Proteus vulgaris (*Pr. vulgaris*) One of the family of enterobacteria characterized by urease production, common in soil and sewage.

Paracoccus denitrificans (*P. denitrificans*) A fungus of the family Paracoccoides.

Psuedomonas aeruginosa (*Ps. aeruginosa*) Bacteria common in soil of the genus Psuedomonas, and dominant in opportunistic infections.

Pulmonary fibrosis Formation of fibrous tissue in the lung.

Rad A unit of radiation which represents absorption of 100 ergs of energy per gram and equal to 0.01 gray.

Rheumatism General term for diseases which cause pain in the muscles, joints and fibrous tissues.

Rheumatoid arthritis Chronic disease of connective tissue of the body.

Rous sarcoma virus (*RSV*) A sarcoma-like disease in fowl caused by viruses.

Salmonella A genus of Gram-negative bacteria causing acute intestinal inflammation.

Sarcoma Malignant tumour of connective tissue.

Staphylococcus aureus (*S. aureus*) Bacteria of the Staphylococcus genus (Gram positive) causing boils and abscesses.

Streptococcus pneumoniae (*Strep. pneumoniae*) A member of the Gram-positive genus Streptococcus causing pneumonia and meningitis.

Trepanoma pallidum Causative agent of syphilis.

Trypanosoma brucei(*T. brucei*), *T. congolense* and *T. nagana* are all trypanosomal diseases affecting cattle.

Trypanosoma cruzi(*T. cruzi*) The species of Trypanosoma responsible for Chagas' disease or American trypanosomiasis.

Trypanosoma gambiense(*T. gambiense*) A species of Trypanosoma causing sleeping sickness. Generally in mid- or West Africa.

Trypanosoma rhodesiene(*T. rhodesiense*) A species of the genus Trypanosoma causing East African sleeping sickness.

Urinalysis Analysis of urine.

Vasoconstrictor A substance which causes narrowing of the inner space of the blood vessels.

Vasodilator A substance which causes widening of the inner spaces of blood vessels, thus leading to increased blood flow.

Adapted from Melloni's *Illustrated Medical Dictionary* 2nd Ed. I. Dox, B. J. Melloni, and G. M. Eisner, Williams and Wilkins, Los Angeles (1985) and *International Dictionary of Biology and Medicine*, Editor-in-Chief S. I. Landau, Wiley, New York (1986).

APPENDIX 1

The structures of the principal DNA and RNA bases are shown, along with the numbering scheme. Other subsitution products, either in the purine N_9 or pyrimidine N_1 position, or substitution of exocyclic amine hydrogen atoms by alkyl groups, are named in standard fashion.

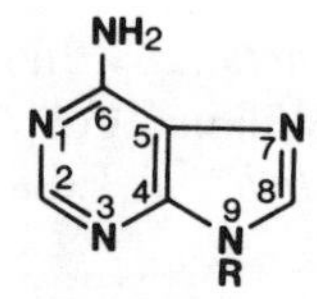

R = H, Adenine
R = CH_3, 9-Me-Adenine

R = H, Guanine
R = CH_3, 9-Me-Guanine

R = H, R′ = H, Uracil
R = H, R′ = CH_3, Thymine
R = CH_3, R′ = H, 1-Me-Uracil
R = CH_3, R′ = CH_3, 1-Me-Thymine

R = H, Cytosine
R = CH_3, 1-Me-Cytosine

The common hydrogen bonding schemes, beside those of the normal Watson—Crick convention are shown below:

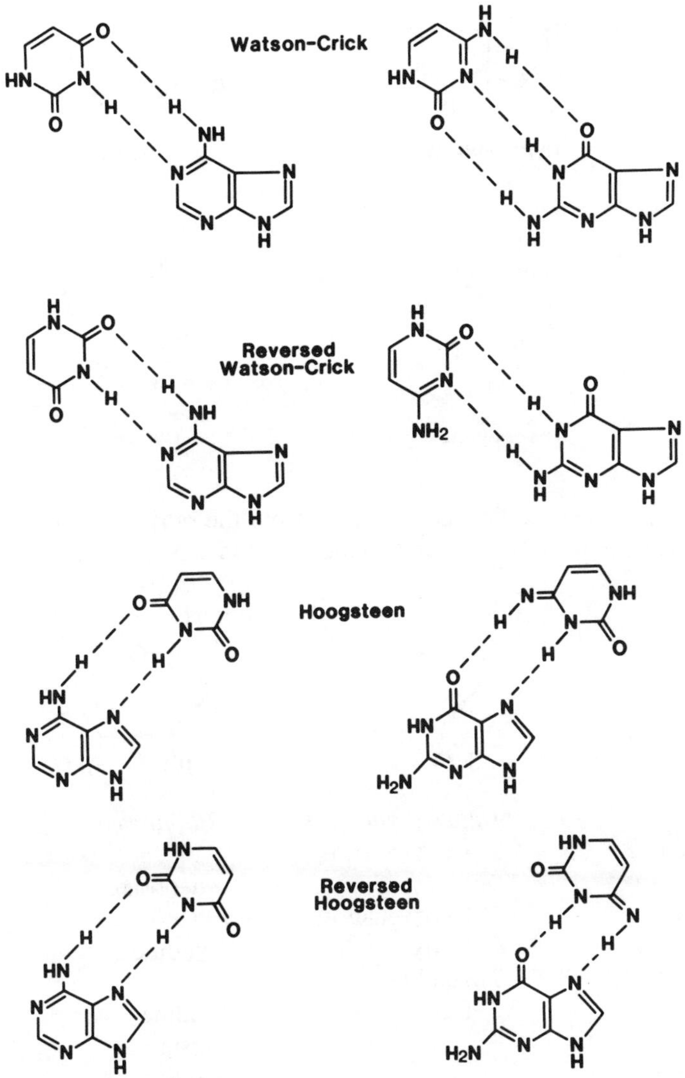

Watson-Crick

Reversed Watson-Crick

Hoogsteen

Reversed Hoogsteen

The structures of some common purine analogues are given below:

Hypoxanthine **Xanthine**

Theophylline **Caffeine**

The nucleoside linkage is formed from the purine N_9 or pyrimidine N_1 to the C_1' of the ribose sugar, and the common names are given below:

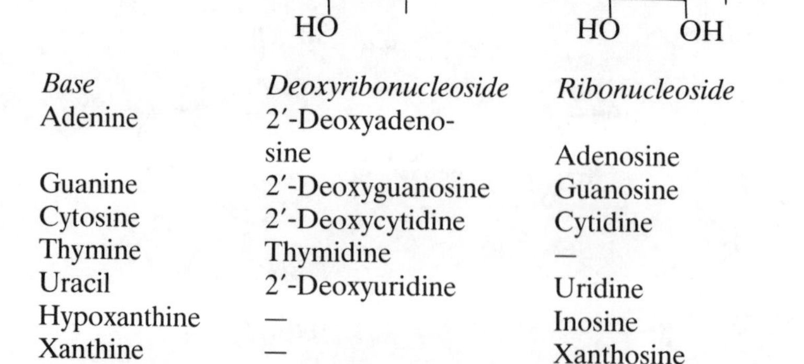

Base	*Deoxyribonucleoside*	*Ribonucleoside*
Adenine	2′-Deoxyadenosine	Adenosine
Guanine	2′-Deoxyguanosine	Guanosine
Cytosine	2′-Deoxycytidine	Cytidine
Thymine	Thymidine	—
Uracil	2′-Deoxyuridine	Uridine
Hypoxanthine	—	Inosine
Xanthine	—	Xanthosine

For DNA the sugar is deoxyribose and for RNA it is ribose. Note that adenosine, guanosine, and cytidine in fact contain the ribose moiety while thymidine contains deoxyribose.

Rotation around the sugar—base bond permits various conformations. In the *anti* conformation the N_1—C_6 bond of pyrimidines and the N_9—C_8 bond of the purine projects on to the sugar ring whereas, in the *syn* conformation, the N_1—C_2 pyrimidine bonds and N_9—C_4 purine bonds project on to the sugar. The *anti* conformation is favored and occurs in double-helical DNA.

The nucleotide unit is formed by attachment to either the 5′ or 3′ carbon oxygens. The *anti* conformation around the sugar—base bond is shown for the representative nucleotides 5′-cytidine monophosphate (5′-CMP) and 5′-guanosine monophosphate (5′-GMP):

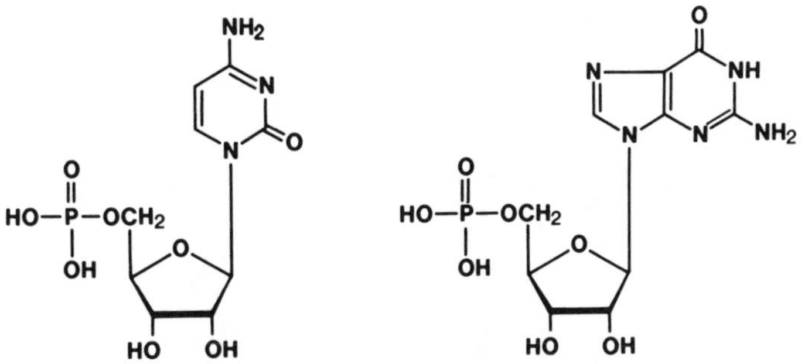

The repeating dinucleotide units are formed by attachment of a nucleoside to a 5′-nucleotide via a phosphoester linkage in the 3′-position of the second sugar. Polymer formation continues in the same way and this can be described [after Kornberg, *DNA Synthesis*, Freeman, San Francisco (1974)]:

sugar-3′ sugar-3′

phosphate-5′ phosphate-5′ phosphate-5′

Since there is now a phosphodiester linkage, cleavage can give the 5′-mononucleotide or its 3′-isomer. The two chains of the double helix are antiparallel and run in opposite directions, i.e. 5′—3′ linkage is opposite to a corresponding 3′—5′ unit. The arrangement for a (GpG) dinucleotide unit in the 3′—5′ direction is shown. The unit with the deoxyribose sugar would be d(GpG). For mixed dinucleotides and oligonucleotides it is conventional to start from the 3′—5′ direction. Thus (CpG) is cytidyl-(3′—5′)-guanosine etc.

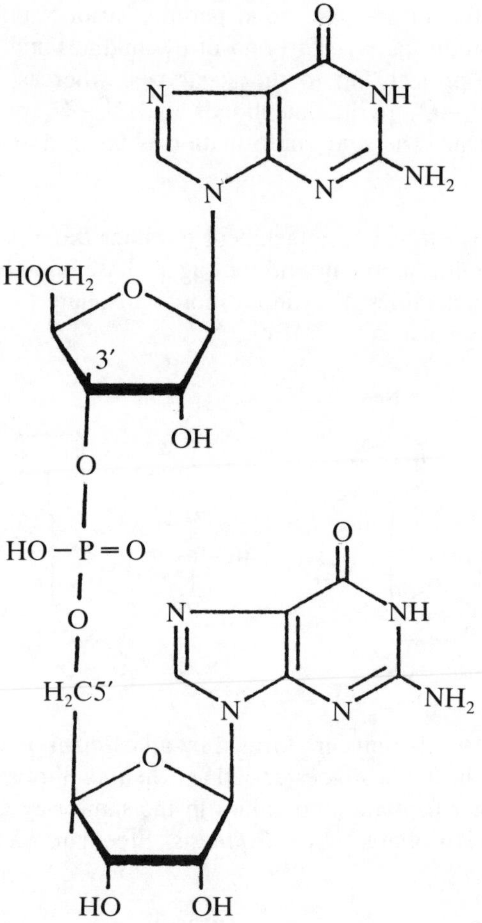

The abbreviations used in this text are Ade, Cyt, Gua, Thy and Ura for the simple bases or substituted bases. The nucleosides are abbreviated Ado, Cytd, Guo, Thyd and Urd.

APPENDIX 2

Of Mice and Men

A number of murine tumours are used as preliminary screens for evaluation of antitumour activity. The principal screens are L1210, P388, S180 and Ehrlich ascites, which are all classified as leukemias and thus are usually disseminated. In some cases (e.g. S180) solid tumour lines are also available. Solid tumours include B16 melanoma, Lewis lung, M5076 (sarcoma) and mammary carcinoma. The ADJ/PC6 screen is a solid plasma cell tumour and the Walker 256 line is a rat solid tumour.

In *in vitro* screening, effective doses (ED's) or inhibitory doses (ID's) are usually calculated as concentrations required to inhibit growth of a certain percentage of colonies — ID_{50} is the dose for 50% inhibition. In the case of *in vivo* ascites and leukemias, the mean survival time is compared to that of controls and the percentage increase is usually given as ILS, increase in life span.

Inhibitory doses and increase in life span parameters are also used in other disease types. In some case (viruses, parasites) inhibition of infectivity can be measured by incubation of disease-causing agent with compound and comparing infectivity with that of controls.

For *in vivo* solid tumours regression or inhibition is judged by comparison of mean weights of treated animals (T) to mean weights of untreated, control animals (C) and is represented as a percentage % T/C. Values above 150% are considered significant.

The variation between these murine screens, their relative susceptibilities and especially their applicability as estimates of effectiveness in humans is a permanent source of discussion [1]. There is considerable interest in development of human solid tumours as primary screens and a recent article surveys known antitumour agents in a human tumour colony-forming assay, and compares with P388 data [2]. Initial results were considered promising and some tumour types may give rise in the near future to evaluable assays.

At present, selection of a drug for clinical trials is based on activity in a number of murine screens, comparison with similar agents or novelty of structure (or proposed mechanism), and suitable formulation properties such as stability and aqueous solubility. Phase 1 clinical trials evaluate human toxicology, fixing tolerated doses and dose-limiting toxicity. Phase 2 studies evaluate initial efficacy in a range of tumours at tolerated doses, usually in poor risk patients. Once a spectrum of activity has been found, Phase 3 studies concentrate on selected efficacy in specific tumours and in combination regimes.

References

1. E. Frei: *Science* **217**, 600 (1982).
2. R. H. Shoemaker *et al.*: *Cancer Res.* **45**, 2145 (1985).

INDEX OF SUBJECTS